# Annonas

## Botany, Production and Uses

# Annonas

## Botany, Production and Uses

*Edited by*

**Sisir Mitra**

*Retired Professor, Bidhan Chandra Krishi Viswavidyalaya (State Agricultural University), Mohanpur, West Bengal, India*

CABI

**CABI is a trading name of CAB International**

CABI
Nosworthy Way
Wallingford
Oxfordshire OX10 8DE
UK

Tel: +44 (0)1491 832111
E-mail: info@cabi.org
Website: www.cabi.org

CABI
200 Portland Street
Boston
MA 02114
USA

Tel: +1 (617)682-9015
E-mail: cabi-nao@cabi.org



A catalogue record for this book is available from the British Library, London, UK.

Library of Congress Cataloging-in-Publication Data

Names: Mitra, S. K. (Horticulturist) editor
Title: Annonas : botany, production and uses / edited by Sisir Mitra.
Description: Boston, MA : CAB International, 2025. | Series: Botany, production and uses | Includes bibliographical references and index. | Summary: "An overview of the genus Annona in the tropical plant group Annonaceae (Soursop family), which includes fruits such as cherimoya, soursop, sweetsop and custard apple. The book reviews the current state of knowledge on the important commercially grown species, and includes production technology, postharvest management and processing"-- Provided by publisher.
Identifiers: LCCN 2025038180 (print) | LCCN 2025038181 (ebook) | ISBN 9781800627604 hardback | ISBN 9781800627611 ebk | ISBN 9781800627628 epub
Subjects: LCSH: Annona
Classification: LCC QK495.A6 A566 2026 (print) | LCC QK495.A6 (ebook)
LC record available at https://lccn.loc.gov/2025038180
LC ebook record available at https://lccn.loc.gov/2025038181

ISBN-13: 9781800627604 (hardback)
9781800627611 (ePDF)
9781800627628 (ePub)

DOI: 10.1079/9781800627628.0000

Commissioning Editor: Rebecca Stubbs
Editorial Assistant: Emma McCann
Production Editor: Shankari Wilford

Typeset by Straive, Pondicherry, India
Printed and bound in the UK by CPI Group (UK) Ltd, Croydon, CR0 4YY

# Contents

# Contributors

**Gabriela Aguilar-Hernández**, Division of Agricultural Sciences and Engineering, Centro Universitario de los Altos (CUALTOS)-University of Guadalajara, Guadalajara, Jalisco, Mexico. Email: gaby.mca2017@gmail.com

**Ivan De-La-Cruz-Chacón**, Laboratorio de Fisiología y Química Vegetal, Instituto de Ciencias Biológicas, Universidad de Ciencias y Artes de Chiapas, Tuxtla Gutiérrez, CHIS, México. E-mail: ivan.cruz@unicach.mx

**Mansheppa Devindrappa**, Division of Crop Protection, ICAR-IIHR, Hessaraghatta Lake, Bengaluru 560089, India. Email: dnayak4505@gmail.com

**Vittorio Farina**, Università degli Studi di Palermo, Department of Agricultural, Food and Forest Sciences, Viale delle Scienze, Palermo, Italy. Email: vittorio.farina@unipa.it

**Gisela Ferreira**, Department of Biodiversity and Biostatistics, Institute of Biosciences, São Paulo State University (UNESP), Botucatu, São Paulo, Brazil. E-mail: gisela.ferreira@unesp.br

**Hirokazu Higuchi**, Graduate School of Agriculture, Kyoto University, Kyoto, Japan. Email: higuchi.hirokazu.2a@kyoto-u.ac.jp

**Siti Izera Ismail**, Department of Plant Protection, Faculty of Agriculture, Universiti Putra Malaysia, 43400 Serdang, Selangor, Malaysia.

**Muhammad Iqbal Hakim Johari**, Department of Plant Protection, Faculty of Agriculture, Universiti Putra Malaysia, 43400 Serdang, Selangor, Malaysia.

**Manikyahalli Chandrashekar Keerthi**, Division of Crop Protection, ICAR-IIHR, Hessaraghatta Lake, Bengaluru 560089, India. Email: keerthimanikya@gmail.com

**Stephane Catarine Rosa Kim**, Department of Biodiversity and Biostatistics, Institute of Biosciences, São Paulo State University (UNESP), Botucatu, São Paulo, Brazil. E-mail: stephane.kim@unesp.br

**Hiroshi Matsuda**, Tropical Agriculture Research Front, Japan International Research Center for Agricultural Sciences, Ishigaki, Okinawa, Japan. Email: matsudah0196@jircas.go.jp

**Libier Meza-Espinoza**, Department of Agricultural Sciences, Technological University of the Coast, Santiago, Ixcuintla, Nayarit, Mexico. Email: lmeza@utdelacosta.edu.mx

**Carolina Ovile Mimi**, Department of Biodiversity and Biostatistics, Institute of Biosciences, São Paulo State University (UNESP), Botucatu, São Paulo, Brazil. E-mail: carolina.mimi@unesp.br

**Samuel Bassani Lins de Miranda**, Department of Biodiversity and Biostatistics, Institute of Biosciences, São Paulo State University (UNESP), Botucatu, São Paulo, Brazil. E-mail: samuel.miranda@unesp.br

**Sisir Mitra**, Former Professor, Faculty of Horticulture, Bidhan Chandra Krishi Viswavidyalaya (State Agricultural University), Mohanpur, West Bengal, India. Email: sisirm55@gmail.com

**Efigenia Montalvo-González**, Integral Laboratory of Food Research, National Technologic of Mexico/Technological Institute of Tepic. Tepic, Nayarit, Mexico. Email: emontalvo@ittepic.edu.mx

**Vijayaram Eyarkai Nambi**, Department of Food Plant Operations Incubation and Entrepreneurship, National Institute of Food Technology Entrepreneurship and Management-Thanjavur, Tamil Nadu, India. Email: eyarkainambi@iifpt.edu.in

**Roberta Passafiume**, Università degli Studi di Palermo, Department of Agricultural, Food and Forest Sciences, Viale delle Scienze, Palermo, Italy. Email: roberta.passafiume@unipa.it

**Nagalapura Ramakrishnappa Prasannakumar**, Division of Crop Protection, ICAR-IIHR, Hessaraghatta Lake, Bengaluru 560089, India. Email: prasannaent@gmail.com

**Norsazilawati Saad**, Department of Plant Protection, Faculty of Agriculture, Universiti Putra Malaysia, 43400 Serdang, Selangor, Malaysia.

**Alka Sharma**, Institute Technology Management Unit, ICAR-Central Institute of Post Harvest Engineering and Technology, Ludhiana, Punjab, India. Email: eralkasharma@gmail.com

**Ilenia Tinebra**, Università degli Studi di Palermo, Department of Agricultural, Food and Forest Sciences, Viale delle Scienze, Palermo, Italy. Email: ilenia.tinebra@unipa.it

**Dzarifah Zulperi**, Department of Plant Protection, Faculty of Agriculture, Universiti Putra Malaysia, 43400 Serdang, Selangor, Malaysia; Laboratory of Sustainable Resources Management, Institute of Tropical Forestry and Forest Products, Universiti Putra Malaysia, 43400, Serdang, Selangor, Malaysia. Email: dzarifah@upm.edu.my

# Preface

The family Annonaceae A.L. de Jussieu is one of the most diverse and primitive angiosperms, reported to contain from 110 to 200 genera and from 2300 to 2500 species. The majority of the *Annona* spp. are considered underutilized, and hence very little scientific information is available. However, the areas under production have increased more rapidly than the contributions from science and technology. The most widely cultivated species of the family are *Annona cherimola* (cherimoya), *A. squamosa* (sugar apple), *A. muricata* (soursop), *A. cherimola* × *A squamosa* (hybrid atemoya) and *A. reticulata* (custard apple). Other species that have possible commercial interest include *A. macrophyllata* (ilama), *A. purpurea* (soncoya), *A. glabra* (pond apple), *A. mucosa* (biriba) and *A. senegalensis* (wild soursop).

With the exception of cherimoya, which is a distinct subtropical species, most of the fruit crops in the genus *Annona* originate from warmer lowland tropical regions with subtropical and tropical climates. In the last 25 years, cherimoya, sugar apple, soursop and atemoya researchers around the world have selected many good varieties for cultivation and have standardized the propagation, production and postharvest technologies. A monograph on *Annona* spp. was compiled in 2005 by the International Centre for Underutilized Crops, University of Southampton, UK, in the series 'Fruits for the Future'. Since then, no other compilation has been produced that includes comprehensive information on all aspects of commercially important *Annona* spp.

This publication presents the current state of knowledge concerning the history, physiology, culture, postharvest management and value addition of *Annona* spp. The book is targeted mainly at researchers, teachers and academics, students, advisors and industry support personnel.

I am confident that this work will be a reference book for a broad spectrum of users, and I look forward to seeing readers' feedback to enhance future editions of this book.

Sisir Mitra
Kalyani, India, 2025

# Acknowledgements

I wish to express my appreciation and gratitude to all the authors who have given their time, without any financial reward, to contribute to this book. I am grateful to the scientists who have provided photographs, information and/or reviewed the chapters: Dr Juan Jose Hueso Martin, Estacion Experimental Cajamar, Almeria, Spain; Dr Peter Achunike Akah, University of Nigeria, Nigeria; Dr Henrique Silvano Aruda, University of Campinas, Brazil; Dr Jose Newthon Pico Mendoza, Universidad Tecnica de Manabi, Ecuador; Dr Isaac Aiyelaagbe, University of Agriculture, Ogun State, Nigeria; Dr R. Sinchana, Bharati College of Farmacy, Karnataka, India; Dr T. Sakthival, Indian Institute of Horticultural Research, India; Dr Arun Kumar Kamble, University of Horticultural Science, Karnataka, India; Dr Iran Alia-Tejcal, Universidad Autonoma, Morelos, Mexico; Dr Debora Souza Mendes, Universidade Estadual de Montes Claros, Brazil; Dr Alma Rosa Gonzalez-Esquinca, Universidad de Ciencias y Artes de Chipas, Mexico; Dr Efigenia Montalova Gonzalez, Instituto Technologico de Tepic, Mexico; Dr Hannia Bridg, Bridg-Networking Consultancy, Berlin, Germany; Dr Prawit Chumchu, Kasetsart University, Sriracha, Thailand; Dr Alexandre Reis Machado, Universidade Federal de Pernambuco, Recife, Brazil; Dr Hernandez Fuentes Luis Martin, INIFAP, Nayarit, Mexico; Dr Nirmal Kumar Meena, Agricultural University, Kota, Rajasthan; Dr Y.S. Balgude, Pune, India; Dr Vu Viet Huang, Fruit and Vegetable Research Institute, Hanoi, Vietnam; Dr Lynton L. Vawdrey, Department of Agriculture and Fisheries, Mareeba, Queensland, Australia; Dr Rodrigo Lisa, Institute de Ecologia, Veracruz, Mexico; Dr Narayan Chawda, VNR Seeds, Raipur, India; Leila Aparecida Salles Pio, Federal University of Lavras, Brazil; Dr Ben Faber, UC Cooperative Extension, California; Dr Binu Mathew, North Eastern Hill University, Tura Campus, Meghalaya, India; Dr Grant Bignell, Department of Primary Industries, Maroochy Research Facility, Queensland, Australia; Dr Ahmad Muslim, Universitas Sriwijaya, South Sumatra, Indonesia; Dr Raffi Stern, MIGA, Galilee Technology Center, Kiryat Shmona, Israel; Dr Gary S. Bender, UCCE San Diego County, USA; Dr V. Yadav, ICAR-CIAH-CHES, Godhara-Vejalpur, Gujarat, India; Dr Mona Arfa, Pharmaceutical Drug Research Institute, NRC, Dokki, Cairo, Egypt; Sri P. Mukhopadhyay, Visva Bharati University, India.

My special thanks to Dr P.K. Pathak, and Sri Niloy Talukder, West Bengal University of Animal and Fisheries Science, India, Dr S. Debnath, Dr Debalina Majhi and Sri Susanta Roy, Bidhan Chandra Krishi Viswavidyalaya, India, for their editorial assistance.

The continued support and encouragements received from various editors and assistants of CABI, particularly from Rebecca Stubbs, Kin Yunn Ho and Emma McCann, are gratefully acknowledged.

On a personal note, I thank my wife, Ipsa. Without her sacrifices, I could not possibly have finished this book.

Sisir Mitra

# 1 Taxonomy, Origin and Distribution

**Sisir Mitra***
*Former Professor, Faculty of Horticulture, Bidhan Chandra Krisi Viswavidyalaya (State Agricultural University), Mohanpur, West Bengal, India*

## 1.1 Introduction

Plants in the family Annonaceae are found in the tropics and subtropics of both the Old and New World. The family Annonaceae contains some of the most diverse and primitive angiosperms, and is included within the early-divergent Magnoliidae clade with around 2400 species in 208 genera of trees, shrubs and lianas (Chatrou *et al.*, 2004, 2012). It is estimated that about 900 species are from the Caribbean and Central America, 450 from tropical Africa and close to 1200 species from tropical Asia and Australia (Chatrou, 1999; Couvreur *et al.*, 2011; Guo *et al.*, 2017). There are several large genera in the family: *Annona* (150 spp.), *Guatteria* (265 spp.), *Duguetia* (100 spp.), *Uvaria* (100 spp.) and *Polyalthia* (100 spp.) (Hormaza *et al.*, 2020).

Annonaceae is one of the most primitive families within the spermatophytic (syn. Phanerogamous) plants and sometimes is called Annoniflorae, because species frequently have an indefinite number of free floral parts and spiral stamens (Leal and Paull, 2023). The genus *Annona* is the second largest genus in the family with about 60 species distributed throughout tropical America and four species in Africa (Mass *et al.*, 2019). The genus has undergone numerous taxonomic changes, and the 44 species of the genus *Rollinia* (Mass *et al.*, 1992) have been included in the genus *Annona* (Rainer, 2007).

The Annonaceae are trees, shrubs and vines (liana) with aromatic foliage, resinous channels in their stems, and a septate medulla or pith. The stems have alternate simple leaves that occur in two rows, with deciduous or persistent stipules (Leal and Paull, 2023). The root system has abundant thin lateral roots and a taproot that goes about 2–3 m deep in the soil. The flowers of *Annona* spp. are super axillary, regular, solitary or in fragrant racemes, being bisexual (rarely unisexual) and protogynous (because the stigmas lose their receptivity before pollen dispersal). There are three green sepals and six petals arranged in two verticiles. The external verticil has three yellow-greenish petals and the internal one has three yellow petals. The flowers have several conglomerated and spirally arranged stamens below and around the upper globose (conical)-shaped dome of numerous fused carpels, which have one ovule each. After fertilization, the fused carpels will form a syncarp or composite fruit (Pinto *et al.*, 2005; Leal and Paull, 2023). The fruit are berries, capsules or a dry fruit, although in some genera the pistils are located in an elongated receptacle that transforms into an aggregated and pulpous fruit.

* Email: sisirm55@gmail.com


DOI: 10.1079/9781800627628.0001

The seeds are 5–35 mm long. They are ellipsoid, oblong or ovate, and fleshy. The testa is present or absent (when sclerotesta present), the surface is smooth or unsmooth; and the embryo is differentiated from the food reserve, and well developed.

The most important species of the genus *Annona* are *A. cherimola* Mill., *A. muricata* L., *A. squamosa* L., *A. reticulata* L. and the interspecific hybrid atemoya (*A. cherimola* × *A. squamosa*) (Sanewski, 1988). Some other species such as *Annona diversifolia* Saff., *Annona montana* Macfad., *A. purpurea* Moç. & Sessé ex Dunal, *A. crassiflora* Mart., *A. glabra* L. and *A. senegalensis* Pers. produce edible fruit and are grown in some countries, but commercial plantings are not available. In addition to their importance for the production of edible fruit, interest in different species of the family Annonaceae has increased in recent years due to the presence in different tissues of polyketide secondary metabolites called acetogenins, which are exclusive to this family and have shown promising cytotoxic, anti-tumour, antimalarial and pesticidal properties (Alali *et al.*, 1999; Liaw *et al.*, 2010; Vidal-Lezama *et al.*, 2019; Leung *et al.*, 2022; Ramos *et al.*, 2022).

## 1.2 Description of Important *Annona* Species

### 1.2.1 *Annona cherimola* Mill.

Common names: cherimoya, cherimola. Other names: chérimolier (French), annona, custard apple, noina ostrelia (Thai), cerimolia (Italy), anona del Peru, catuche (Spanish), lakshamanphal (India).

The origin of *A. cherimola* is still under discussion. One hypothesis is that it originated from the Andes at an altitude of 700–2400 m, while an alternative hypothesis postulates that it originated in Central America due to common presence of wild relatives (Perrone *et al.*, 2022). The linguistic data, botanical remains and ceramics of the Chimu-Inca and Moche cultures in northern Peru indicate the presence of cherimoya in Peru as early as 2500 BC (Bonavia *et al.*, 2004). The existence of natural populations of cherimoya in the mountain valleys in southern Ecuador and northern Peru provide evidence to support the origin of cherimoya in these inter-Andean valleys, (Popenoe, 1921; Chatrou *et al.*, 2004). The inter-Andean origin would make *A. cherimola* the only one of the 12 species of the Atta section of the genus *Annona* not native to Central America and the Caribbean (Larranga *et al.*, 2019).

The early Spanish explorers introduced cherimoya to continental Spain from America between the 16th and 18th centuries, with the primary record in 1957. Most likely, it was then distributed to Italy, Madeira (Portugal), the Canary Islands, Algeria and Egypt, and from Italy it was introduced to Libya, Eritrea and Somalia (Morton, 1987). The Spanish explorers also introduced cherimoya to Asia via Africa (Ochse *et al.*, 1974).

Cherimoya is a small, erect and/or somewhat spreading, deciduous tree, reaching a height of about 7–8 m (Fig. 1.1A). It has simple, alternate, two to four ranked leaves, which are ovate-lanceolate to elliptical in shape, 10–25 cm long, glabrous on the ventral surface and pubescent dorsally, with leaf shading in the spring (Pinto *et al.*, 2005). Cherimoya flowers are fragrant, and dichogamous and protogynous often opposite to the leaves. They occur in most cases on 1-year-old branches, on new shoots and on older wood. The flowers are solitary or in clusters (two to three), nodding on short peduncles. They are long (~2.5 cm) and narrow, with brown or yellowish tomentose outside. The outer petals (six) are three-parted and narrow, and the inner ones are small and scale-like (Bridg, 2000). Fruit crops of *Annona* spp. are fleshy and aggregated with several seeds formed mainly by a reticulate endosperm and a tiny embryo (Perrone *et al.*, 2022). The fruit is normally heart-shaped, conical or somewhat irregular in form due to irregular pollination. The fruit measure 7.5–12.5 cm in length and weigh 200–700 g (Fig. 1.1B). The greenish-yellow to purple skin is thin or thick and may be smooth with bumps or oblong scales, which overlap each other like roof tiles or a fingerprint depression. The fruit is easily broken or cut open. The pulp is creamy while the juicy flesh has a pleasant aroma and taste, with a sour-sweet flavour, and contains numerous hard, brown or black bean-like glossy seeds (1–2 cm long) (Bourke, 1976; Morton, 1987; Bridg, 2000).

**Fig. 1.1.** *Annona cherimola* (cherimoya) tree (A) and fruit (B). Photographs courtesy of Dr Juan Jose.

### 1.2.2 *Annona muricata* L.

Common name: soursop. Other names: guanabana (Spanish), corossolier (French), graviola (Portuguese).

*A. muricata* most likely originated in Central America, the Antilles or northern South America, and is found in the Andean valleys in Peru, presumably as an ancient introduction (Pinto *et al.*, 2005). Morton (1987) mentioned that soursop is a very abundant plant in northern South America and the Antilles and Spanish colonizers distributed it to other tropical regions of the world (Popenoe, 1939; Purseglove, 1968). The existence of several wild types of soursop in the Amazon region (Cavalcante, 1976) suggests that this may be a primary centre of diversity, but the types could be remnants from cultivated introductions. It is also found both wild and cultivated in the Bahamas and Bermuda. It can be located from sea level up to an attitude of 1150 m above sea level (m.a.s.l.) throughout the Antilles and from southern Mexico to Peru and Argentina. Soursop was one of the first fruit trees to be brought from the Americas to the Old World tropics where it was widely distributed from south-western China, Australia, and the lowlands of eastern and western Africa (Fuentes *et al.*, 2022). In India, it grows at a limited scale in southern states such as Tamil Nadu, Andhra Pradesh, Karnataka and Kerala.

Soursop has an erect growth habit and is about 5–10 m tall. The stems are rounded, rough and not pubescent, with a dark brown colour. The leaves have short petioles and are oblong-ovate to cylindrical, 14–16 cm in length and 5–7 cm in width (Pinto *et al.*, 2005). The soursop fruit is an edible collective fruit that is typically ovoid or ellipsoid, usually irregular, oblique or curved, heart-shaped berry, and is dark green in colour. The average fruit weight ranges from 0.4 to 1.0 kg (Fig.1.2A) and each fruit may contain 55–170 black seeds (Awan *et al.*, 1980) when fresh, which become light brown when dry. The pulp is creamy-white (Fig.1.2B), fleshy, juicy and subacid, with soft, cottony strands (Bridg, 2000).

### 1.2.3 *Annona squamosa* L.

Common names: sugar apple, sweetsop. Other names: custard apple (English), pinha (Portuguese), attier (French), sitaphal, ata (India) noi-na (Thailand), atis (the Philippines), fan-li-chi (Taiwan) saramuyo (Mexico), fruta do conde (Brazil).

*A. squamosa* is a small tropical tree originating in the New World tropics, probably Central America. From there, it was distributed to Mexico and throughout tropical America. In the lowlands of Mexico, it is found in a naturalized or wild state. It is grown from Central America southwards to northern South America, extending to north-eastern Brazil. It is believed that it was first introduced into Brazil via Bahia state, in 1626, by Conde de Miranda, which explains the vernacular name in Brazil – fruta do conde (Pinto *et al.*, 2005). In Mexico, it was naturally distributed on the Pacific slope from Michoacán to Oaxaca, throughout the Yucatán peninsula, between 0 and 1000 m.a.s.l. It was later taken to the Philippines and Asia via the West Indies (Antilles) and the Cape of Good Hope (Popenoe,

1974; Leon, 1987). In India, it is now grown commercially in many states including Gujarat, Karnataka, Telangana, Bihar and West Bengal.

The sugar apple tree is smaller than the soursop, reaching a maximum of 6 m in height, with many lateral branches (Fig. 1.3A). Lenticels are present on the stems, and the young shoots are pubescent and the older ones smooth (Pinto *et al.*, 2005). The leaves are brilliant green above and bluish-green below, with petioles 0.7–1.5 cm in length. The leaves are oblong-elliptical in form, measuring 5–17 cm in length and 2–7 cm in width, with an obtuse or acuminate apex. The blade has 15–17 pairs of veins (Ochse *et al.*, 1974).

The flowers measure 2.0–2.5 cm in length. Pollination and fruit-set problems are similar to those of other annonas. Pollen germination is low and may influence final fruit set, which varies from 5.4% to 5.6% (Thakur and Singh, 1965). The fruit is rounded, heart-shaped, ovate or conical, 5–7.5 cm in diameter, 6–10 cm in length and weighing 120–330 g (Fig. 1.3B). The pulp has a pleasant sweet-sour flavour. The fruit contains 35–45 black seeds, each 1.5–2.0 cm in length and 0.6–0.8 cm in width (Pinto *et al.*, 2005).

**Fig. 1.2.** *Annona muricata* (soursop) fruit (A) and pulp (B). Photographs courtesy of Dr Issac Aiyelaagbe.

**Fig. 1.3.** *Annona squamosa* (sugar apple) tree (A) and fruit (B). Photographs courtesy of Dr A.K. Kamble (A) and VNR Seeds, Raipur, India (B).

### 1.2.4 *Annona reticulata* L.

Common name: custard apple. Other names: bullock's heart, corazon (English), condessa, coracao-de-boi (Portuguese), buah nona (Indonesia), ramphal (India).

Custard apple is believed to have originated in the Antilles (Fouqué, 1972) and other parts of the circum-Caribbean region. The Spanish explorers may have distributed this species to Mexico and to Asia and Africa. It is commonly found in home gardens in coastal towns throughout tropical America. It is also found growing in India, Sri Lanka, the Malay Archipelago, Polynesia, the Philippines, Australia and most of the countries of Africa (Pinto *et al.*, 2005). The vernacular name 'custard apple' is usually applied to sugar apple in India.

The tree reaches 6.0–7.5 m in height, with many lateral branches; the stems are cylindrical, with lenticels and very short coffee-coloured hairs. It can be distinguished from cherimoya by its long, narrow, glabrous leaves (Fig. 1.4A), from sugar apple by its solid compact fruit as well as its larger leaves, and from *A. glabra* (see below) by its small, dark brown seeds (Leon, 1987). The leaves are oblong-lanceolate and dark green, measuring 25–30 cm in length and 7 cm in width (Pinto *et al.*, 2005).

The flowers are similar in form to those of sugar apple, except that they are grouped in a short inflorescence with two to ten flowers. The fruit weigh 0.1–1.0 kg and are commonly heart-shaped, but may be conical, ovate or irregular in form, and are 10–12 cm long (Fig. 1.4B). They are coriaceous and have a reddish-yellow surface colour. The flesh is milky-white and sweet, although the flavour is insipid. The seeds are dark coffee-coloured and oblong in shape (Leon, 1987). Various plant parts are used as source of medicine and also for industrial products (Jamkhande and Wattamwar, 2015).

### 1.2.5 *Annona cherimola* × *Annona squamosa* hybrid

Common name: atemoya. Other name: chirimorinon (Venezuela).

Atemoya is grown in the USA, Australia, Israel, South Africa, the Philippines, China, Egypt and in some other countries. The hybrid was developed in Miami, Florida, in 1910, and shows superior quality to sugar apple. It was given the name 'atemoya' as a combination of 'ate', an old Mexican name for sugar apple, and 'moya' from cherimoya.

**Fig. 1.4.** *Annona reticulata* (custard apple) tree (A) and fruit (B). Photographs courtesy of Dr V. Yadav.

The tree closely resembles that of the cherimoya; it is fast growing and reaches up to 9 m in height (Fig. 1.5A), with drooping branches. The leaves are deciduous, alternate, elliptical, leathery and less hairy than those of cherimoya; they are up to 15 cm in length. The flowers are long-stalked, triangular, yellow, and about 6 cm long and 4–5 cm wide. The fruit is conical or heart-shaped (Fig. 1.5B), generally 8–10 cm long and 9–10 cm wide. It weighs from 400 g to 2.25 kg, and is pale bluish-green or pea-green and slightly yellowish between the areoles. The rind is thick (3 mm), and is composed of fused areoles that are more prominent and angular than those of sugar apple, with tips that are rounded or slightly up turned; it is firm, pliable and indehiscent. The fragrant flesh is snowy-white, of fine texture, almost solid and not conspicuously divided into segments, with fewer seeds than the sugar apple. it is both sweet and subacid and resembles the cherimoya in flavour. The seeds are hard and smooth, cylindrical, 2 cm long and 8 mm wide, and are dark brown, almost black, in colour.

The fruit are consumed fresh and also processed for various products such as juice and ice cream.

### 1.2.6 *Annona purpurea* Moc. & Sessé ex Dunal

Common name: soncoya. Other names: axtier (French), cabeza de negro (Portuguese).

The species originated in the highlands of Mexico and is fairly tropical in its requirements. It is a Mesoamerican tree, distributed from the Pacific and Atlantic slopes of Mexico to South America (Topete-Corona *et al.*, 2020).

The tree is tall (Fig. 1.6A), and the fruit is large (Fig. 1.6B) and spiny, similar to soursop but with a bright orange, aromatic flesh. The flesh has a cottony texture and clings to the seed (Sanewski, 1988). The leaves, stems, seeds, bark and fruit have high medicinal value and are a source of secondary metabolites including acetogenins, alkaloids and essential oils with anti-tumour, antibacterial, antifungal, antioxidant, antiparasitic, cardiotonic, anxiolytic and insecticidal effects (Vidal-Lezama *et al.*, 2019).

### 1.2.7 *Annona glabra* L.

Common name: pond apple. Other names: alligator apple, monkey apple, swamp apple, low apple, corkwood, bob wood, mangrove annona.

This species has been reported from the mangrove swamps of coastal America, from Florida to Brazil, and across the Atlantic in West Africa from Senegal to Congo (Sivrajan and Sunil, 1993). It has been reported as introduced and subspontaneous in Asiatic countries including Sri Lanka (Huber, 1985), India (Sivrajan and Sunil, 1993) and Australia (Sanewski, 1988). Sivrajan and Sunil (1993) observed that the

**Fig. 1.5.** *Annona cherimola* × *Annona squamosa* (atemoya) tree (A) and fruit (B). Photographs courtesy of Dr Grant Bignell.

**Fig. 1.6.** *Annona purpurea* (soncoya) tree (A) and fruit (B). Photographs courtesy of Dr Ram Cuves Guzman.

fruit has a leathery pericarp and that they are capable of floating across oceans. The occurrence of seedlings in shallow water along the shoreline suggests that seeds released after decay of the fruit wall are brought to shore by water currents where they germinate and give rise to new plants.

It is a small, glabrous tree, 5–6 m tall. The branches are brownish-green, and the leaves are distichous and petioled. The flowers are on short, extra-axillary tubercles, solitary or paired; the young buds are subglobose, and the open flowers are broadly ovoid, creamy-yellow and fragrant. The fruit are subglobose or ovoid, smooth (Fig. 1.7), fleshy, lemon-yellow when ripe and 7 × 5 cm in size. The fruit has many straw-coloured, lenticular seeds, which are often strongly margined, ovoid, obovoid or oblong, rounded at the ends, smooth and embedded in a yellow pulp with ruminate endosperm (Sivrajan and Sunil, 1993).

The fruit is edible and can be made into jam and fresh fruit drinks. Sinchana *et al.* (2024) isolated 28 phytochemicals from the fresh fruit including 19 kaurane diterpenoids, which have different pharmacological responses.

**Fig. 1.7.** *Annona glabra* (pond apple) fruit. Photograph courtesy of Dr R. Sinchana.

### 1.2.8 *Annona deliciosa* Saff. (previously known as *Rollinia deliciosa* Saff.) and *Annona mucosa* (Jacq.) Baill. (previously known as *Rollinia mucosa* (Jacq.) Baill.)

Common name: biribá. Other names: biribá de Pernambuco, fruta da condessa, araticu, araticum, araticum pitaya (Brazil); mulato (Colombia); rinon or rinon de monte (Venezuela), annona babosa, zambo (Mexico); anon cimarron, corossol, cahiman, cashina, candongo anon de monte, araticum, condessa, graviola brava; wild sweetsop, wild sugar apple (English).

*Rollinia* is thought to have originated in northern Brazil along the banks of the Amazon.

It is found in the natural state in Guiana, southern Mexico, Peru and northern Argentina. Seeds from Brazil were sent to Florida and the Philippines in the early 1900s (Love and Paull, 2011). It also grows in Queensland, Australia (Sanewski, 1988).

Both *R. deliciosa* and *R. mucosa* have long, thin, glossy leaves (Fig. 1.8A). The flowers are round rather than long and thin like those of the atemoya. The fruit is medium-sized with either smooth or pointed protuberances (areoles). As the fruit softens, the skin turns yellow (Fig. 1.8B) and then black within a few days. The flesh is semi-translucent and has a glutinous texture (Sanewski, 1988), and is white, juicy, and sub-acid to sweet. There is a slender, opaque-white core and numerous dark brown elliptic or obovate seeds, 1.6–2.0 cm long.

The fruit is eaten fresh and is also fermented to make wine in Brazil. The fruit is regarded as analeptic and antiscorbutic (Morton, 1987).

Other fruiting *Rollinia* spp. found mostly in Brazil include *R. sericea*, *R. sylvatica*, *R. salicifolia*, *R. rugulosa* and *R. emarginata* (Love and Paull, 2011).

### 1.2.9 *Annona senegalensis* Pers.

Common name: wild soursop. Other names: wild custard apple (English), pomme canella du Senegal (French), gishta gaba (Arabic), nchakwe (Tanzania), nta ntnyerere, mthopa (Zambia), gishita (Ethiopia).

There is no precise information on the origin of wild soursop (Pinto *et al.*, 2005). The species is most widely distributed in Tanzania, Kenya, Mozambique, and in the Zanzibar and Pemba Islands, suggesting that eastern Africa may be the region of origin and diversity (FAO, 1983). In Sudan, this species is found where rainfall is greater than 500 mm, typically in tall-grass savannah areas. It also grows in Angola, Senegal and Mauritania (Vogt, 1995).

It is a shrub or small tree (2–6 m). The leaves are alternate, simple, oblong, ovate or elliptic, green to bluish-green, almost without hairs on top but often with brownish hairs on the underside; the apex is rounded or slightly notched, the base square to slightly lobbed, the margin entire and the petiole short (0.5–2.5 cm). The flowers are solitary or in groups, up to 3 cm in diameter, arising above the leaf axis, and comprise six fleshy cream to yellowish petals in two whorls, greenish outside, creamy or crimson, with the inner whorl of the petals curving over the stamens and ovary. The fruit is formed from many fused carpels and is fleshy, lumpy, egg-shaped (Fig. 1.9), 2.5–5 × 2.5–4 cm, and ovoid or globose. The unripe fruit is green, turning yellow to

**Fig. 1.8.** *Annona deliciosa* (biribá) leaves (A) and fruit (B). Photographs courtesy of Dr T. Sakthivel.

orange on ripening, with numerous seeds that are cylindrical or oblong and orange brown in colour.

The edible white pulp of the ripe fruit has a pleasant pineapple-like taste. In Africa, the leaves and barks are commonly used in the treatment of malaria. Ointments prepared from the fruit and leaves are used for treatment of skin disorders (Akah, 2023).

**Fig. 1.9.** *Annona senegalensis* (wild soursop) fruit. Photograph courtesy of Dr Peter Achunike Akah.

### 1.2.10 *Annona crassiflora* Mart.

Common names: marolo, araticum. Other names: araticum do cerrado, bruto, cabeca de negro, cascudo, pana, pinha do cerrado and pasmada (Arruda *et al.*, 2023).

It is a native and endemic to Brazilian Cerrado. *A. crassiflora* is widely distributed along the Cerrado biome areas, covering the states of Minas Gerais, São Paulo, Goiás, Bahia, Mato Grosso do Sul, Tocantins, Maranhão, Piauí, Pará and Federal District (Cota *et al.*, 2011).

*A. crassiflora* is a deciduous, heliophytic, xerophytic, hermaphrodite tree reaching up to 8.0 m in height (Fig. 1.10A) (Lorenzi, 1998) and about 4.0 m in crown diameter (Arruda *et al.*, 2023). The fruit is subglobose (Fig. 1.10B) and

**Fig. 1.10.** *Annona crassiflora* (marolo) tree (A), fruit (B), pulp (C) and seeds (D). Photographs courtesy of Dr Henrique Arruda.

brown when ripe. The pulp is arranged on a fleshy and tapered receptacle (Fig. 1.10C) and ranges from white to yellowish-pink, with a characteristic strong and slightly sweet flavour and a pleasant smell. In addition, the pulp is composed of cone-shaped buds (90–190 carpels) containing a single seed (Arruda *et al.*, 2023). The seeds have a flattened obovoid shape (Fig. 1.10D) and a boney consistency with a smooth, opaque and light brown surface (Arruda and Pastore, 2019).

The fruit are enjoyed fresh or in delicacies, as the pulp is sweet and very aromatic. The leaves have anti-diarrhoea and antimicrobial properties (Silva Junior, 2005). The pulp can be processed and used in ice creams, ice lollies, jellies and juices (Arruda and Pastore, 2019). The peel of the fruit is treated as waste, despite studies showing that it has a high antioxidant (Roesler, 2011), antibacterial (da Silva *et al.*, 1996) and hepatoprotective (Justino *et al.*, 2017) properties. The pulp, peel and seed of the fruit contain compounds with noticeable biological activities, such as alkaloids and acetogenins, among other secondary metabolites (Ramos *et al.*, 2022).

### 1.2.11 *Annona montana* Macfad.

Common name: mountain soursop. Other names: guanaba, false graviola.

*A. montana* is native to Central America, the Amazon and the Caribbean Islands (Leung *et al.*, 2022). The species grows in South America (Brazil, Bolivia, Peru, Ecuador, Colombia, the Guyanas), Central America (Panama, Costa Rica, the Caribbean), Australia (Sanewski 1988), China and India (Wu *et al.*, 1995).

It is an evergreen tree, resembling soursop (*A. muricata*), but is less susceptible to cold and is able grow at higher elevations. The tree height is about 10 m and the outer bark has a pungent smell. The globose fruit is up to 15 cm in diameter; its rind is green and composed of tiny spicules (Barbalho *et al.*, 2012). The pulp may be white or yellowish, with a soft texture. The seeds are brownish-yellow in colour (Mootoo *et al.*, 2000). Eating quality is reported to be similar (but inferior) to soursop (Sanewski, 1988). The fruit are consumed fresh for desserts when fully ripe or mixed with ice cream or milk to make a drink.

Numerous alkaloids have been isolated from this species, some of which have tranquilizing properties, while others influence the nervous system. Leaf extracts of *A. montana* have been found to strongly inhibit *in vitro* differentiation of human pre-adipocytes, which may indicate useful applications to human health by helping to lower the incidence of chronic degenerative disease (Leung *et al.*, 2022).

### 1.2.12 *Annona diversifolia* Saff.

Common names: ilama (Central America), papausa (southern Mexico).

*A. diversifolia* is a native fruit of south-eastern Mexico, Guatemala and El Salvador and is considered one of the most important subtropical fruits from warm-weather locations (Julián-Loaeza *et al.*, 2011).

The trees grows up to 6–8 m in height (Lefebvre *et al.*, 2018). The trunk is usually about 20–25 cm thick, often branching from the ground to form three to six main stems. The fruit is oval in shape and weighs 500–100 g. Each tree produces an average of 50 fruit. The fruit surface is rough, with carpellary areas indicated by deeply incised lines. A short, thick protuberance is attached to the areoles (Duke and du Cellier, 1993). The flowers vary from cream to red. The pulp is pink, white, red, purple or blue (Lefebvre *et al.*, 2018), with large tan-coloured seeds. The flavour is described as between that of cherimoya and sweetsop (Sanewski, 1988) *A. diversifolia* is considered as an important source of fibre, sugars, potassium, zinc and biologically active components with antioxidant activity such as flavonoids (Julián-Loaeza *et al.*, 2011).

In Australia, severed hybrids of ilama and cherimoya have been developed, as well as the available cultivars 'Imery' and 'Perez' (Sanewski, 1988).

### 1.2.13 *Annona scleroderma* Saff.

Common name: poshte. Other names: annona del monte (Spanish), cawesh, cahuex (Mexico).

*A. scleroderma* is grown mainly in south-western Guatemala (FAO, 1994). In Mexico, it grows in areas with a warm climate, such as

Tabasco, Chiapas, Nayarit, Michoacán, Yucatán and Veracruz (Safford, 1913). The fruit has a very tough skin, allowing it to be easily handled and making it resistant to insects. The fruit may be cut and the flesh removed with a spoon. Its potential value is in its high-quality flesh, hard skin and high yield (FAO, 1994).

It is a tall tree, reaching 15–25 m high, and has tough, lanceolate leaves measuring 10–25 × 5–8 cm. They are shiny on the upper side, slightly pubescent on the underside and have fragile, 3 cm long petioles. The flowers are greenish-yellow, and the outer petals have a longitudinal prominence, which arises in the small branches or in groups in the old part of the thicker branches. The fruit occurs in compact spherical groups; it is 5–10 cm in diameter and generally falls off when ripe, without a noticeable colour change. The cream-coloured flesh has a bitter-sweet flavour and soft texture (FAO, 1994). Seed extracts have great potential as a source of cytotoxic compounds such as squamins C–F, rolliniastatin 2 and bullatacinone, which are believed to have marked activity against several cancer cell lines (Sosa-Rueda *et al.*, 2023).

### 1.2.14 *Annona deceptrix* (Westra) H. Rainer

Common names: chirimoya de monte or anonilla (Pico-Mendoza *et al.*, 2020).

*A. deceptrix* is a tree found distributed in coastal wet and deciduous forests at an elevation range of 0–85 m.a.s.l. (Muriel and Pitman, 2003). The tree produces white to yellowish flowers and conical, elongated, sweet fruit with different diameters and a smooth-to-rough appearance (Fig. 1.11A–D). The fruit can be consumed fresh or in processed products (Pico-Mendoza *et al.*, 2024).

## 1.3 Chromosome Number

The ancestral basic chromosome number in Annonaceae is seven (Walker, 1972). Most *Annona* spp. are diploid ($2n$ = 14 or 16) with the exception of *A. glabra* (Bowden, 1945, 1948; Miege, 1960; Morawetz, 1986a; Datta and De, 1990) and *A. lutescens* (Morawetz, 1986a), which have been reported as tetraploid and the previously named genus *Rollinia* (now included in *Annona*) (Rainer, 2007), with tetraploid (*A. neolaurifolia* and *A. exsucca*) (Morawetz, 1986b) and hexaploid (*A. mucosa* and *A. pulchrinervis*) species (Morawetz, 1986a,b). *A. glabra* is a tetraploid ($2n = 28$) species (Datta and De, 1990). Martin *et al.* (2019) observed a high percentage of triploid hybrid atemoya (35%) from an interspecific cross between *A. cherimola* × *A. squamosa*. However, atemoya has been considered diploid (Bowden, 1948). The triploid hybrids may result from the fusion of reduced ($n$) and unreduced ($2n$) gametes (Martin *et al.*, 2019).

## 1.4 Distribution

Statistics on minor fruits such as *Annona* spp. are unavailable in many countries, and where reported, they often lack reliability, uniformity and continuity. Generally, production data relate only to plantation or orchard crops grown for sale for international markets, such as citrus, banana, grape and mango. Data on production of minor fruits from scattered trees used mainly for home consumption is not collected (Pinto *et al.*, 2005). Table 1.1 shows some areas and production of *Annona* spp. in different countries.

The Australian atemoya (*A. squamosa* × *A. cherimola*) industry produced around 1422 t of fruit in 2020 with an estimated farm gate value of A$6million. There are about 150 active growers. Most of the production is consumed domestically, but around 5% (71 t) is exported, mainly to Asian markets. Industry plantings are currently made up of the following cultivars: 20,000 'African Pride', 8900 'Pink's Mammoth', 43,000 'KJ Pinks', 2000 'Pinks Blush' and 1000 'Maroochy Gold'. The major growing regions are northern New South Wales (47%), the Sunshine Coast (26%), Tablelands of North Queensland (11%), Wide Bay (9%), Central Queensland (3%) and Western Australia (4%) (Bignell, 2022).

Spain is considered as one of the most important cherimoya producers in the world. It produced 44,081 t in 2020 (Anon., 2021a), of which 99.6% was produced in the Andalusia region (in the province of Granada, Malaga and Cadiz). The majority of production (80–90%) is consumed domestically, and the remaining

**Fig. 1.11.** *Annona deceptrix* (chirimoya de monte) tree (A), leaves (B), fruit (C) and seeds (D). Photographs courtesy of Dr Jose Pico-Mendoza.

10–20% is exported to Portugal, France and Germany.

In Portugal, cherimoya production is concentrated primarily in the Archipelagos of Madeira and Azores. Production in 2021 was reported to be 584 t in Madeira and about 245 t in Azores. A small amount (~14 t) was exported to the European continent in 2019 (Anon., 2021b).

The main areas where cherimoya is produced in Ecuador are the provinces of Carchi, Imbabura, Tungurahua, Bolívar, Azuay and Loja. According to statistics, the production was 1057 t from an area of 372 ha, with an average yield of 2841 kg ha$^{-1}$ (MAGAP, 2019). Ecuador has a large area (2400 ha) under soursop cultivation.

Soursop is widely consumed locally in Mexico and particularly in the Soconusco region (Chiapas, Mexico). According to the Agro-Food and Fisheries Information Service (SIAP), in 2019 in Mexico, the established soursop area was 3612.54 ha with a told yield of 30,790.70 t.

**Table 1.1.** Areas and production of *Annona* spp. in different countries.

| Country | Year | Crop | Area (ha) | Production (t) | Reference |
|---|---|---|---|---|---|
| Australia | 2020 | Atemoya | – | 1,422 | Bignell (2022) |
| Bolivia | 2019 | Cherimoya | 1,000 | 7,000 | Anon. (2019) |
| Brazil | 2017 | Sugar apple | – | 8,700 | IBGE (2022) |
| | 2014 | Atemoya | 1,500 | – | Lemos (2014) |
| Ecuador | 2022 | Soursop | 2,400 | – | R.G.M. Macias (2023, personal communication) |
| India | 2018–2019 | Sugar apple | 45,000 | 449,000 | NHB (2021) |
| Mexico | 2022 | Soursop | 3,269 | – | Anon. (2023) |
| | 2022 | Sugar apple | 91 | – | |
| | 2022 | Cherimoya | 60 | – | |
| Peru | 2021 | Cherimoya | – | 20,000 | Anon. (2021c) |
| Portugal | 2021 | Cherimoya | 117 | 584 | Anon. (2021b) |
| Spain | 2020 | Cherimoya | – | 44,081 | Anon. (2021a) |
| Taiwan | 2017 | Sugar apple | 5,146 | 3,446 | Tsai and Hsu (2017) |
| Thailand | 2013 | Sugar apple | 4,981 | 20,683 | Anon. (2014) |
| Vietnam | 2017 | Sugar apple | 6,500 | | Hoa *et al.* (2017) |

The three main states of soursop cultivation are the state of Nayarit with 2,456.94 ha with an average yield of 10.53 t $ha^{-1}$, the state of Michoacán with 423 ha and an average yield of 7.19 t $ha^{-1}$, and the state of Colima with 391.75 ha and an average yield of 10.06 t $ha^{-1}$. Sugar apple is mainly distributed in the provinces of Sierra Madre del Sur, Sierra Madre Occidental, Northern Gulf Coastal Plain and Yucatán peninsula. Mexico has a small area under sugar apple (91 ha) and cherimoya (60 ha) cultivation. Cherimoya is grown in Colima, State of Mexico, Guanajuato, Guerrero, Hidalgo, Jalisco, Michoacán, Morelos, Oaxaca, Puebla, Veracruz and Yucatán.

In Peru, approximately 20,000 t of cherimoya is produced annually (Anon., 2021c). The Lima region produces about 40% of the national production. Other production areas include Cajamarca, Piura, Junín and Apurimac. However, in many places in the highlands, cherimoya is commonly found in mixed orchards or as living fences. Peru exported 176 t of cherimoya between January and March 2023 to Chile (96.2%), the USA (2.2%) and other countries.

In Honduras, Costa Rica, Bolivia and Chile, there are very few organized orchards of cherimoya. In Costa Rica, cherimoya trees are planted on coffee farms for shading purposes. In Honduras, it is typically planted in backyards or small farms. Most of the harvested fruit are sold fresh in the open markets, or at fairs and roadside stands in the production area.

Although sugar apple production data are scare, India has a large area under cultivation. The species is grown commercially in India, Malaysia, Thailand, Vietnam, Taiwan, China, the West Indies, the Dominican Republic, the USA (Florida) and the Middle East.

Sugar apple is widely grown in tropical and subtropical situations in Vietnam. In southern Vietnam, it is grown mainly in the south-east, for example in Tây Ninh (3200 ha), Dong Nai (1228 ha), Bà Rịa–Vũng Tàu (1200 ha) and Ho Chi Minh City. Vietnam also grows about 1200 ha of soursop in the Mekong Delta (Hoa *et al.*, 2017).

In Thailand, sugar apple (Thai name: noi-na) is grown in tropical and dry regions, including the provinces of Nakhon Ratchasima, Lopburi and Saraburi. In 2013, the total sugar apple planting area in Thailand was 4,981 ha and production was approximately 20,683 t (Thanarut and Ko, 2017).

India has major share in sugar apple production in the world, producing 449,000 t from an area of 45,000 ha (NHB, 2021). The major growing states are Madhya Pradesh, Andhra Pradesh, Chhattisgarh, Rajasthan, Uttar Pradesh, Tamil Nadu, Bihar and West Bengal. India regularly exports sugar apple to Oman, United Arab Emirates, Kuwait, Qatar, the Maldives, Bahrain, Singapore, the UK, Bangladesh and Canada. The average international price per kilogram of cherimoya fruit is US$4. Peruvian cherimoya has a high demand in the international market., mainly from Chile, Spain, Canada, Italy, the USA and the Netherlands (Morales, 2019).

## References


Akah, P.A. (2023) *Annona senegalensis* Pers (Annonaceae): exploring the underutilized potential of an African medicinal plant. *Journal of Ethnopharmacology and Toxicology* 1, 16–26. DOI: 10.37446/jet/ra/1.1.2023.16-26

Alali, F.Q.., Liu, X.-X. and McLaughlin, J.L. (1999) Annonaceous acetogenins: recent progress. *Journal of Natural Products* 62, 504–540. DOI: 10.1021/np980406d

Anon. (2014) *Fruit Tree Database*. Department of Agricultural Extension, Bangkok, Thailand.

Anon. (2019) *Instituto Nacional de Innovacion Agropecuaria y Forestal*. Diagnostio de la Linea Base del Programa Nacional de Frulas 2019. Ministerio de Desarrollo Rural y Tierras, La Pa Z, Bolivia.

Anon. (2021a) Ministerio de Agriculture Pesca y Alimentación (MAPA), Spain. Available at: www.mapa.gob.es/es/ (accessed 21 August 2025).

Anon. (2021b) Direcao Regional de Estatistica da Madeira (DREM). Available at: https://estatistica.madeira.gov.pt/ (accessed 21 August 2025).

Anon. (2021c) Semana Nacional de la Chirimoya. Ministerio de Desarrollo Agrario y Riego, Peru. Available at: www.gob.pe/midagri (accessed 21 August 2025).

Anon. (2023) *Statistical Yearbook for Agricultural Production*. Government of Mexico, Mexico.
Arruda, H.S. and Pastore, G.M. (2019) Araticum (*Annona crassiflora* Mart.) as a source of nutrients and bioactive compounds for food and non-food purposes: a comprehensive review. *Food Research International* 123, 450–480. DOI: 10.1016/j.foodres.2019.05.011
Arruda, H.S., Borsoi, F.T., Andrade, A.C., Pastore, G.M. and Marostica Junior, M.R. (2023) Scientific advances in the last decade on the recovery, characterization and functionality of bioactive compounds from the araticum fruit (*Annona crassiflora* Mart.) *Plants* 12: 1536. DOI: 10.3390/plants12071536
Awan, J.A., Kar, A. and Udoudoh, P.J. (1980) Preliminary studies on the seeds of *Annona muricata* Linn. *Plant Food for Human Nutrition* 30, 163–168. DOI: 10.1007/BF01099054
Barbalho, S.M., Soares de Souza, M.D.S., Bueno, P.C.D.S., Guiguer, E.L., Farinazzi-Machado, F.M.V *et al.* (2012) *Annona montana* fruit and leaves improve the glycemic and lipid profiles of Wistar rats. *Journal of Medicinal Food* 15, 917–922. DOI: 10.1089/jmf.2012.0088
Bignell, G. (2022) *Final Report: New Custard Apple Varieties and Enhanced Industry Productivity (CU16002)*. Hort Innovation, Australia. Available at: www.horticulture.com.au/growers/help-your-business-grow/research-reports-publications-fact-sheets-and-more/cu16002/ (accessed 11 August 2025).
Bonavia, D., Ochoa, C.M., Tovar, Ó.S. and Palomino, R.C. (2004) Archaeological evidence of cherimoya (*Annona cherimolia* Mill.) and guanabana (*Annona muricata* L.) in Ancient Peru. *Economic Botany* 58, 509–522. DOI: 10.1663/0013-0001(2004)058[0509:AEOCAC]2.0.CO;2
Bourke, O.D. (1976) *Annona* spp. In: Garner, R.R. and Chaudhuri, S.S. (eds) *The Propagation of Tropical Fruit Trees*. Horticulture Commonwealth Bureau of Horticulture and Plantation Crops, East Malling, UK, pp. 223–248.
Bowden, W.M. (1945) A list of chromosome numbers in higher plants. I. Acanthaceae to Myrtaceae. *American Journal of Botany* 32, 81–92. DOI: 10.1002/j.1537-2197.1945.tb05090.x
Bowden, W.M. (1948) Chromosome numbers in the Annonaceae. *American Journal of Botany* 35, 377–381. DOI: 10.2307/2437937
Bridg, H. (2000) Micropropagation and determination of the *in vitro* stability of *Annona cherimola* Mill. and *Annona muricata* L. PhD thesis, Landwirtschaftlich Gärtnerische Fakultät der Humboldt-Universität Zu Berlin, Berlin.
Cavalcante, P.B. (1976) *Fruitas Comestiveis da Amazonia*. INPA, Belém do Pará, Brazil.
Chatrou, L.L. (1999) The Annonaceae and the Annonaceae project: a brief overview of the state of affairs. *Acta Horticulturae* 497, 43–58. DOI: 10.17660/ActaHortic.1999.497.2
Chatrou, L.W., Rainer, H. and Mass, P.J.M. (2004) *Annonaceae*. In: Smith, F.F., Mori, S.S., Stevenson, D.D. and Heald, S.S. (eds) *Flowering Plants of the Neotropics*. Princeton University Press, Princeton, New Jersey.
Chatrou, L.W., Pirie, M.D., Erkens, R.H.J., Couvreur, T.L.P., Neubig, K.M. *et al.* (2012) A new subfamilial and tribal classification of the pantropical flowering plant family Annonaceae informed by molecular phylogenetics. *Botanical Journal of the Linnean Society* 169, 5–40. DOI: 10.1111/j.1095-8339.2012.01235.x
Cota, L.G. Vieira, F.A., Melo Júnior, A.F., Brandao, M.M., Santana, K.N.O. *et al.* (2011) Genetic diversity of *Annona cassiflora* (Annonaceae) in northern Minas Gerais State. *Genetics and Molecular Research* 10, 2172–2180. DOI: 10.4238/vol10-3gmr1188
Couvreur, T.L., Pirie, M.D., Chatrou, L.W., Saunders, R.M., Su, Y.C. *et al.* (2011) Early evolutionary history of the flowering plant family *Annonaceae*: steady diversification and boreotropical geodispersal. *Journal of Biogeography* 38, 664–680. DOI: 10.1111/j.1365-2699.2010.02434.x
da Silva, E., Pinto, A.C.Q. and Azvedo, J.A. (1996) *Manejo da Irrigacao e Fertirrigacao na Cultura de Managueira.* Serie Documentos 61. Embrapa Cerrados, Brasília, Brazil.
Datta, P.C. and De, B. (1990) Karyology of some Indian Annonaceae. *Cytologia* 55, 187–196. DOI: 10.1508/cytologia.55.187
Duke, A.T. and du Cellier, J.L. (1993) Annonaceae. In: *CRC Handbook of Alternative Cash Crops*. CRC Press. Florida, pp. 33–46.
FAO (Food and Agriculture Organization of the United Nations) (1983) Food and fruit-bearing forest species. 1: examples from Eastern Africa. *FAO Forestry Paper* 4, 11–14.
FAO (Food and Agriculture Organization of the United Nations) (1994) Custard apples (*Annona* spp.). Bermejo, J.E.H. and Leon, J. (eds) In: *Neglected Crops: 1492 from a Different Perspective.* FAO Plant Production and Protection Series No. 26. pp. 85–92.
Fouqué, A. (1972) Especes fruitieres d'Amenque tropicale: famille des annonacées. *Fruits* 27, 62–72.
Fuentes, J.M.V., Tejacal, I.A., Agreda, F.J.M. and Fuentes, J.A. (2022) Soursop (*Annona muricata* L.). In: Cabrera, F.R. and Mota, D.D. (eds) *Postharvest of Underutilized Native Mexican Crops with High*

*Potential Marketing*. Metropolitan Autonomous University Iztapalapa Unit, Iztapalapa, Mexico. pp. 262–289.

Guo, X., Tang, C.C., Thomas, D.C., Couvreur, T.L.P. and Sanders, R.M.K. (2017) A mega-phylogeny of the *Annonaceae*: taxonomic placement of five enigmatic genera and support for a new tribe: Phoenicantheae. *Scientific Reports* 7: 7323. DOI: 10.1038/s41598-017-07252-2

Hoa, N.V., Dien, L.Q. Uyen, D.T.K. and Hang, N.T.N. (2017) Overview on guava, pineapple, wax apple, sugar apple research and production in Vietnam. *Acta Horticulturae* 1166, 15–23. DOI: 10.17660/ActaHortic.2017.1166.3

Hormaza, J.I., Carmona, E., Gonzalez-Padilla, I.M., Larranga, N., Lora, J. *et al.* (2020) Annona spp. atemoya, cherimoya, soursop and sugar apple. In: Litz, R.R., Pliego-Alfaro, F. and Hormaza, J.J. (eds) *Biotechnology of Fruit and Nut Crops*, 2nd edn. CAB International, Wallingford, UK, pp. 65–78.

Huber, H. (1985) Annonaceae. In: M.M. Dassanayake and F.F. Fosber (eds) *A Revised Handbook to the Flora of Ceylon, Vol.* 5. CRC Press, Boca Raton, Florida, pp. 1–75.

IBGE (2022) Instituto Brasileiro de Geografia e Estatistica. Available at: http://www.ibge.gov.br (accessed 11 August 2025).

Jamkhande, P.G. and Wattamwar, A.S. (2015) *Annona reticulata* Linn (Bullock's heart): plant profile, phytochemistry and pharmacological properties. *Journal of Traditional and Complementary Medicine* 5, 144–152.

Julián-Loaeza, A.P., Santos-Sánchez, N.F., Valdez-Blance, R., Sánchez-Guzmán, B.S. and Salas-Coronado, R. (2011) Chemical composition, color and antioxidant activity of three varieties of *Annona diversifolia* Safford fruits. *Industrial Crops and Products* 34, 1262–1268. DOI: 10.1016/j.indcrop.2010.06.012

Justino, A.B., Pereira, M.N., Peixoto, L.G., Vilela, D.D., Caixeta, D.C. *et al.* (2017) Hepatoprotective properties of polyphenol-enriched fraction from *Annona crassiflora* Mart. fruit peel against diabetes-induced oxidative and nitrosative stress. *Journal of Agricultural and Food Chemistry* 65, 4428–4438. DOI: 10.1021/acs.jafc.7b01355

Larranga, N., Albertazzi, F.J. and Hormaza, J.L. (2019) Phylogenetics of *Annona cherimola* (Annonaceae) and some of its close relatives. *Journal of Systemic Evolution* 57, 211–221. DOI: 10.1111/jse.12473

Leal, F. and Paull, R.R. (2023) The genus *Annona*: botanical characteristics, horticultural requirements and uses. *Crops Science* 63, 1030–1049. DOI: 10.1002/csc2.20833

Lefebvre, C., Segura, S. Carmona, A., Mathuriau, C., Barrios, S. *et al.* (2018) Linking local appreciation with conservation of an edible fruit species: the case study of llama (*Annona diversifolia* Saff.) in Tierra Caliente, Mexico. *Natural Resources* 9, 337–353. DOI: 10.4236/nr.2018.910021

Lemos, E.E.P. (2014) The production of annona fruits in Brazil. *Revista Brasileira de Fruticultura* 36, 77–85. DOI: 10.1590/S0100-29452014000500009

Leon, J. (1987) *Botanica de los Cultivos Tropicales*. IICA, San Jose, Costa Rica.

Leung, I., Veipaga, M.-L., Espinal, M., Zhang, W., Barnum, R. and Barbieri, M.A. (2022) Anti-lipid droplets accumulation effect of *Annona montana* (mountain soursop) leaves extract on differentiation of preadipocytes. *Biocell* 46, 567–578. DOI: 10.32604/biocell.2022.014009

Liaw, C.C., Wu, T.Y., Chang, F.R. and Wu, Y.C. (2010) Historic perspectives on Annonaceous acetogenins from the chemical bench to preclinical trials. *Plant Medica* 76, 1390–1404. DOI: 10.1055/s-0030-1250006

Lorenzi, H. (1998) *Arvores Brasileiras Manual de Identificacao e Cultivo de Palntas Arboreas Nativas do Brasil*, 2nd edn. Editora Plantarum, Nova Odessa, Brazil.

Love, K. and Paull, R.E. (2011) *Rollinia*. Fruit, Nut, and Beverage Crops Series; F_N-21. University of Hawaii at Manoa, Honolulu, Hawaii.

MAGAP (Ministry of Agriculture, Livestock, Aquaculture and Fisheries) (2019) *Agricultural Information System*. MAGAP, Ecuador.

Martin, C., Viruel, M.A., Lora, J. and Hormaza, J.I. (2019) Polyploidy in fruit tree crops of the genus *Annona* (Annonaceae). *Frontiers in Plant Science* 10: 99. DOI: 10.3389/fpls.2019.00099

Mass, P.J., Lubbert, Y.T., Westra, L.W.C., Brown, K.S. Jr, Mass, P.J.M. *et al.* (1992) *Rollinia*. Flora Neotropica Monograph 57. New York Botanical Garden Press Room Dept, New York.

Mass, P.J., Lubbert, Y.T., Westra, L.W.C., Verspagen, N. Rainer, H. *et al.* (2019) Twelve new and exciting Annonaceae from netropoics. *PhytoKeys* 126, 25–69. DOI: 10.3897/phytokeys.126.33913

Miege, J. (1960) Nombres chromosomiques de plants d'Afrique occidentale. *Revue de Cytologie et de Biologie Végétales* 21, 373–384.

Mootoo, B.S., Ali, A., Khan, A., Reynolds, W.F. and Mclean, S. (2000) Three novel monotetrahydrofuran annonaceous acetogenins from *Annona montana*. *Journal of Natural Products* 63, 807–811. DOI: 10.1021/np9903301

Morales, V. (2019) *Collection of Cumbe Cherimoya Grown in the Callejon de Conchucos (Ancash Region) for Fresh Export and Waste for Peruvian Agroindustry: Business Plan to Expand the Product Portfolio of the Company GEOAGRO SAC* (in Spanish). Available at: https://repositorio.esan.edu.pe/server/api/core/bitstreams/46acc5d9-866d-4175-bc54-3a71fdad16d6/content (accessed 11 August 2025).

Morawetz, W. (1986a) Remarks on karyological differentiation patterns in tropical woody plants. *Plant Systematic Evolution* 152, 49–100. DOI: 10.1007/BF00985351

Morawetz, W. (1986b) Systematics and karyoevolution in Magnoliidae: Tetrameranthus as compared with other Annonaceae genera of the same chromosome number. *Plant Systematic Evoloution* 154, 147–173. DOI: 10.1007/BF00984875

Morton, J.F. (1987) Annonaceae. In: *Fruits of Warm Climates*. Creative Resource Systems, Winterville, North Carolina, pp. 65–90.

Muriel, P. and Pitman, N. (2003) *Annona deceptrix*. In: *IUCN Red List of Threatened Species 2003*. Available at: www.iucnredlist.org/ (accessed 11 August 2025).

NHB (National Horticulture Board) (2021) *Horticulture Statistics at a Glance*. NHB, Ministry of Agriculture and Farmers Welfare, Gurugram, Haryana, India.

Ochse, J.J., Soule, M.J. Jr, Dijkman, M.J. and Wehlburg, C. (1974) Otros Cultivos Frutales. In: *Cultivo Y Mejoramiento de Plantas Tropicales Y Subtropicales*. Editorial Limusa, Mexico City, pp. 587–818.

Perrone, A., Youscfi, S., Salami, A., Papini, A. and Martinelli, F. (2022) Botanical, genetic, phytochemical and pharmaceutical aspects of *Annona cherimola* Mill. *Scientia Horticulturae* 296: 110896. DOI: 10.1016/j.scienta.2022.110896

Pico-Mendoza, J., Pinoragote, M., Madrid, L., Flor, J., Alva, J. *et al.* (2020) Germination capacity of *Annona deceptrix* (Westra) H. Rainer (Annonaceac) an endemic and endangered species in Manabi, Ecuador. *Indian Journal of Agricultural Research* 54, 329–335. DOI: 10.18805/IJARe.A-481

Pico-Mendoza, J., Madrid, L., Pinoargote, M., Limongi, R., Pena, G. *et al.* (2024) Simple sequence repeat markers (SSR) in *Annona deceptrix* Westra H. Rainer, an endangered species of the Ecuadorian coast. *Journal of Tropical Forest Science* 36, 132–138. DOI: 10.26525/jtfs2024.36.1.132

Pinto, A.C.de Q., Corderio, M.C.R., Andrade, S.R.M. de., Ferreira, F.R., de C. Filguieras, H.A. *et al.* (eds) (2005) *Annona Species*. International Centre for Underutilized Crops, University of Southampton, Southampton, UK.

Popenoe, W. (1921) The native home of cherimoya. *Journal of Heredity* 12, 331–336. DOI: 10.1093/oxfordjournals.jhered.a102116

Popenoe, W. (1939) *Importantes Frutas Tropicais*. Serie Agriculture 81–82. Uniao Panamericana, Depto de Cooperacao Agricola, Washington D.C. ().

Popenoe, W. (1974) *The Anonaceous Fruits*: *The Cherimoya*. In: *Manual of Tropical and Subtropical Fruits*, Facsimile of the 1920 edition. Hafner Press, London, pp. 161–189.

Purseglove, J.W. (1968) Other useful products: Annonaceae. In: *Tropical Crops: Dicotyledons*. Longman, London, UK. pp. 624–625.

Rainer, H. (2007) Monographic studies in the genus *Annona* L. (Annonaceae): inclusion of the genus *Rollinica* A. st. Hil. *Annalen des Naturhistorischen Museums in Wien: Serie B für Botanik und Zoologie* 108, 191–205.

Ramos, A.L.C., Silva, M.R., de Mendona, H. de O.P., Mazzinghy, A.C. do C., Silva, V.D.M. *et al.* (2022) Use of pulp, peel, and seed *of Annona crassiflora* Mart. in elaborating extracts for fingerprint analysis using paper spray mass spectrometry. *Food Research International* 160: 111687. DOI: 10.1016/j.foodres.2022.111687

Roesler, R. (2011) Effect of extracts from araticum (*Annona crassiflora*) of CC14 induced liver damage in rats. *Ciencia e Tecnologia de Alimentos* 31, 93–100. DOI: 10.1590/S0101-20612011000100012

Safford, W.E. (1913) Chelonocarpus, a new section of the genus *Annona*, with descriptions of *Annona scleroderma* and *Annona testudinea. Journal of the Washington Academy of Sciences* 3, 103–109.

Sanewski, G. (1988) *Growing Custard Apples*. Queensland Department of Primary Industries, Brisbane, Australia.

Silva Junior, M C. (2005) *100 Arvores do Cerrado, Guia de Campo.* Rede de Sementes do Cerrado, Brasilia, Brazil.

Sinchana, R., Mani, T.T., Pavithra, T. and Shiju, L. (2024) A review on a miracle plant *Annona glabra* Linn. *International Journal of Pharmacognosy* 11, 65–77. DOI: 10.13040/IJPSR.0975-8232.IJP.11(3).65-77

Sivrajan, V.C. and Sunil, C.N. (1993) *Annona glabra* L. (Annonaceae): a new record for India. *Rheedea* 3, 90–93. DOI: 10.22244/rheedea.1992.03.01.09

Sosa-Rueda, J., Peniche-Cardena, A., Mondragon-Vasquez, K., Dominguez-Chavez, J. and Cen-Pachero, F. (2023) Chemical study of *Annona scleroderma* and *Annona globiflora* plants as a source of cytotoxic compounds. *Revista Medica de la Universidad Veracruzana* 2022, 25–39. DOI: 10.25009/rmuv.2022.2.69

Thakur, D. and Singh, R. (1965) Studies on pollen morphology, pollination and fruit set in some annonas. *Indian Journal of Horticulture* 22, 10–18.

Thanarut, C. and Ko, J.-C. (2017) Guava, wax apple, sugar apple and pineapple production in Thailand. *Acta Horticulturae* 1166, 7–13. DOI: 10.17660/ActaHortic.2017.1166.2

Topete-Corona, C., Cuevas-Guzman, R., Sanchez-Rodriguez, E.V., Moreno-Hernandez, A., Morales-Arias, J.J. and Nunez-Lopez, N.N. (2020) Population structure and habitat of a tropical tree with, edible fruits, *Annona purpurea* (Annonaceae), in Western Mexico. *Revista de Biologia Tropical* 68, 1171–1184. DOI: 10.15517/rbt.v68i4.42195

Tsai, S.-J. and Hsu, Y.-T. (2017) The influence of population dynamics of spider mites on *Annona* spp. under pesticide stress. *Acta Horticulturae* 1166, 93–99. DOI: 10.17660/ActaHortic.2017.1166.13

Vidal-Lezama, E., Villegas-Monter, A., Vaquera-Huerta, H., Robledo-Paz, A. and Martinez-Palacios, A. (2019) *Annona purpurea* Moc. & Sesse ex Dunal native species of Mexico, underutilized. *Agro Productividad* 12, 9–15. DOI: 10.32854/agrop.v0i0.1326

Vogt, K. (1995) *A Field Worker's Guide to the Identification Propagation and Uses of Common Trees and Shrubs of Dryland Sudan.* SOS Sahel International, London.

Walker, J.J. (1972) Chromosome numbers, phylogeny, phytogeography of the Annonaceae and their bearing on the (original) basic chromosome number of angiosperms. *Taxon* 21, 57–65. DOI: 10.2307/1219222

Wu, Y., Chang, G., Ko, F. and Teng, C. (1995) Bioactive constituents from the stems of *Annona montana*. *Planta Medica* 61, 146–149. DOI: 10.1055/s-2006-958035

# 2 Ecology and Phenology

**Sisir Mitra***
*Former Professor, Faculty of Horticulture, Bidhan Chandra Krisi Viswavidyalaya (State Agricultural University), Mohanpur, West Bengal, India*

## 2.1 Introduction

Annonas are mostly subtropical fruits preferring a warm climate with a moderate winter and high humidity for a good cropping. The trees remain dormant during the cold season for a short period, but frost and prolonged cool weather can adversely affect their growth. Annonas can also tolerate adverse weather conditions, such as extreme temperature, high and low humidity, heavy and scanty rainfall, and desiccating winds, but at the cost of growth and yield. All *Annona* spp. are capable of growing in a wide range of soil types, from sandy soil to clay loams. although they differ in their soil preferences. Higher yields occur on more well-drained sandy to sandy-loam soils, except in the case of *Annona glabra* (pond apple), which can succeed in shallow ponds.

The plants show a rhythmic appearance of their vegetative and reproductive cycles over time. This phenomenon is termed phenology. Phenology is the study of periodic biological events, such as budbreak, flushing, flowering, fruit development, maturity and ripening, closely regulated by climate and seasonal changes. The accurate characterization of plant phenological stages has always been a challenge to the scientific community. Bleiholder *et al.* (1989) and Hess *et al.* (1997) developed the Biologische Bundesanstalt, Bundessortenamt und Chemische Industrie (BBHC) scale, a system for uniformly coding phenologically similar growth stages among all monocotyledonous and dicotyledonous plant species.

## 2.2 Climatic Factors

### 2.2.1 Altitude

Most *Annona* spp. are well adapted to altitudes below 1000 m and annual rainfall over 1000 mm with a dry period during flowering and an average annual temperature of 25–28°C (Leal and Paull, 2023).

Cherimoya (*Annona cherimola* Mill.) is reported to grow at altitudes between 900 and 2500 m above sea level (m.a.s.l.) (Popenoe, 1939; Belotto and Manica, 1994; Yaguana and Fernanda, 2018) on plateaus and in mountain valleys in subtropical areas in dry, cool climates. As well as high altitudes, it can grow also and yield well in the subtropics. It is cultivated in the Mediterranean region (Spain, Italy, Egypt, Israel) and southern coastal California and Portugal, as well as in South Africa, Argentina and Chile. In Spain, it is cultivated along the southern 'Sun Coast', especially in Malaga and Granada (Pinto *et al.*, 2005). In Ecuador, it grows between 700 and 2400 m.a.s.l. and better growth was observed

* Email: sisirm55@gmail.com

 *Annonas: Botany, Production and Uses* (ed. S. Mitra)
DOI: 10.1079/9781800627628.0002

between 1200 and 2000 m.a.s.l. (Yaguana and Fernanda, 2018).

Soursop (*Annona muricata* L.) is cultivated from sea level to an altitude of 1200 m (Pinto and Silva, 1994) and between latitudes of 27°N and 22.5°S. Its northern extremes include southern Florida (USA), Culiacán, Chiapas, Veracruz and Yucatan (Mexico), Cuba and south China, while its southern extreme is in central Brazil (Pinto *et al.*, 2005). The most suitable altitudinal range for optimum growth is 240–300 m.a.s.l. (Morton, 1987; Maheswari and Sinduja, 2020). However, a few high-yielding orchards were reported from Brazil at an altitude of about 1100 m.a.s.l. (Pinto and Silva, 1994).

Sugar apple (*Annona squamosa* L.) is usually cultivated in the lowlands, although in Cuba it was reported in cultivation up to 900 m.a.s.l. (Zayas, 1966). In Mexico, sugar apple is naturally distributed on the Pacific slope from Michoacán to Oaxaca, throughout the Yucatan Peninsula, between 0 and 1000 m.a.s.l. It is also frequently cultivated in semi-arid climates, such as north-eastern Brazil (Belotto and Manica, 1994). In India, its cultivation is spread over arid to semi-arid areas of Andhra Pradesh, Assam, Karnataka, Maharashtra, Madhya Pradesh, Odisha, Rajasthan and Tamil Nadu. It grows throughout the plains of India at an elevation not exceeding 1220 m.a.s.l. (Yadav *et al.*, 2023).

Custard apple (*Annona reticulata* L.) is a widely cultivated annona at low to medium elevations (0–1500 m) (Popenoe, 1974). It grows between latitudes of 25°N and 25°S and is reported to be found in almost all tropical areas of the world (Zayas, 1966; Nakasone and Paull, 1998).

Wild soursop (*Annona senegalensis* Pers.) is adapted to various altitudes, being cultivated from near sea level to 1800 m in Kenya and up to 2400 m in other parts of East Africa (FAO, 1983). Wild soursop is essentially grown in Africa, between the latitudes of 22.5°N and 22.5°S. Rollinia (*Annona delicosa* and *Annona mucosa*) grows from 100 to 1000 m.a.s.l. in Hawaii as long as it receives enough water (Love and Paull, 2011b).

### 2.2.2 Light

Solar radiation greatly affects the plant growth and productivity by influencing photosynthesis. Very little information is available on the growth response and physiological functions associated with different irradiance levels in Annonaceae fruit trees. Utsunomiya and Higuchi (1996) raised seedlings of cherimoya, sugar apple and soursop at 100%, 45%, 25% and 5% of full sunlight level to determine the influence of irradiance level on growth and photosynthesis. The shoot length and dry weight of whole plants in all three species decreased at 25% sunlight, but there was little difference between 45% and 100% sunlight. Total leaf area increased at 45% sunlight in sugar apple and soursop, while a decrease was noted in cherimoya and soursop at 25% sunlight. The leaf area ratio increased with decreasing light intensity in all three species. With decreasing light intensity, the leaves became thinner and the specific leaf weight decreased for all species studied. The chlorophyll content and photosynthetic rate markedly decreased at 25% sunlight. However, there were few differences in chlorophyll content and photosynthesis among these light intensities in cherimoya, indicating that cherimoya has a greater adaptability to low irradiance than soursop and sugar apple. Under low light conditions, stomatal conductance increased for all three species. At 5% sunlight, plant dry weight was less than about 90% compared with control (100%), and the photosynthetic rate decreased almost to the light compensation point.

### 2.2.3 Rainfall

Most *Annona* spp. grow well in areas with over 1000 mm annual rainfall with a dry period during flowering. Cherimoya requires an optimum level of rainfall that fluctuates between 675 and 1000 mm year$^{-1}$ (Xavier, 2018). In some countries, cherimoya is grown in subtropical areas with rainfall close to 1700 mm, which can affect fruit setting when the rainfall occurs during full bloom and increase disease incidence (Nakasone and Paull, 1998). Soursop can withstand high rainfall areas, and the mean annual rainfall suitable for its growth ranges from 1000 to 2500 mm, while a mean annual rainfall of under 1000 mm is unfavourable (Pinto *et al.*, 2005; Love and Paull, 2011a). In Sri Lanka, soursop is dominant in environments with an annual rainfall between 1900 and over 2500 mm (Ratnayake *et al.*, 2020).

Although soursop has been reported to be relatively tolerant to flooding/water-logged conditions compared with other *Annona* spp., heavy rainfall during peak flowering periods significantly reduces the efficiency of pollination (Nakasone and Paull, 1998). An annual rainfall of 600–800 mm is optimum for sugar apple (Satisha *et al*., 2016), although it can withstand higher rainfall (Yadav, 2018).

### 2.2.4 Flooding

Soil flooding is a common environmental stress in wetland areas and in areas with high rainfall, poor soil drainage and high water table fluctuations, including changes in soil physicochemical properties by alteration of soil structure, depletion of carbon dioxide ($CO_2$), anaerobic decomposition of organic matter, and reduction in the availability of iron and manganese (Kozlowski, 1997; Jackson and Colmer, 2005). Commercially grown annonas such as cherimoya, sugar apple and atemoya (*A. cherimola* × *A. squamosa*), when grown as seedling plants or on seedling rootstock, are susceptible to waterlogging (George *et al*., 1987; Schaffer *et al*., 2006). Soursop and pond apple have the potential to be used as flood-tolerant annona rootstocks (Núñez-Elisea *et al*., 1998). Pond apple is extremely flood tolerant and has been referred to as 'swamp-loving' (Popenoe, 1974; Zotz *et al*., 1997). Flooded pond apple seedlings develop hypertrophied stem lenticels and also exhibit thicker stems with reduced xylem density as a result of flooding (Núñez-Elisea *et al*., 1999). Núñez-Elisea *et al*. (1999) observed that seedlings of pond apple and soursop, and the scions of the atemoya cultivars '49-11' and 'Gefner' and custard apple grafted on to pond apple rootstock were flood tolerant based on their ability to survive flooded conditions. Scions of the normally flood-sensitive soursop and atemoya 'Gefner' and '49-11' tolerated root waterlogging when grafted on pond apple. In contrast, flooding of sugar apple seedlings and rootstocks and soursop rootstocks greatly reduced growth and net $CO_2$ assimilation rates, and resulted in 20–80% tree mortality. Seedlings of *Annona montana* Macfad. (mountain soursop) are also tolerant to continuous flooding and periodic waterlogging conditions (Núñez-Elisea *et al*., 1999).

Fu *et al*. (2012) studied the effects of continuous or periodical soil flooding on the tree performance of four different scion/rootstock combinations: (i) atemoya/hybrid atemoya × *A. reticulata*/pond apple (scion/interstock/ rootstock); (ii) hybrid atemoya × *A. reticulata*/pond apple (scion/rootstock); (iii) atemoya/hybrid atemoya × *A. reticulata*/*A. montana* (scion/interstock/ rootstock); and (iv) hybrid atemoya × *A. reticulata*/*A. montana* (scion/rootstock). They observed that flooding treatments reduced shoot extensions, leaf production, net photosynthetic rates and the maximum quantum efficiency of photosystem II ($F_v/F_m$) in plants of the atemoya cultivar 'African Pride' × *A. reticulata*/*A. montana*, which displayed wilting within 2 weeks of flooding. The wilted plants shed all leaves but remained alive and sprouted new but weak shoots after 16 weeks of flooding. Long-term flooding did not suppress but enhanced photosynthesis, as well as tree growth, in atemoya 'African Pride'/hybrid of atemoya 'African Pride' × *A. reticulata*/pond apple. They suggested using *A. glabra* (pond apple) instead of *A. montana* as the rootstock, and hybrid 'African Pride' × *A. reticulata* as an interstock to increase flood tolerance of commercial *Annona* spp.

### 2.2.5 Drought

Drought stress reduces plant growth by affecting photosynthesis, respiration and nutrient metabolism (Jaleel *et al*., 2008). The decrease in net photosynthesis results in inhibition of ribulose-1,5-bisphosphate carboxylase/oxygenase (Rubisco) enzyme activity and ATP synthesis (Dulai *et al*., 2006). Both atemoya and cherimoya are sensitive to water stress during fruit development. Soursop is more tolerant to water stress because it drastically reduces stomatal conductance and transpiration rate under water-stress situations (de Oliveira and Lameira, 2017). In a greenhouse study of sugar apple, Rodrigues *et al*. (2010) observed gas exchange sensitivity to high vapour pressure deficit (VPD). However, under water-deficit conditions, phytochemical efficiency dropped at midday and showed low tolerance to high VPD and light intensity under greenhouse conditions. In contrast, some chlorophyll fluorescence variables showed good photosystem II stability at the highest VPD hour (14:00 h). Carbohydrate metabolic changes showed an accumulation of soluble sugar and increases in the content of protein and free amino acids in water-deficit leaves.

Atemoya trees respond to adequate moisture during the growing season. In controlled-environment experiments, soil moisture stress (midday; 2.0 MPa) severely reduced flowering and fruit set (by >30%), particularly for trees growing under high temperature (28°C) (George and Nissen, 1998). Flowers and fruit were more sensitive to water stress than leaves. Under field conditions, trees may tolerate higher levels of stress than trees growing in containers due to diurnal and long-term osmotic adjustment, although the level of osmotic adjustment has been shown to be very small (George and Nissen, 1992). Kowitcharoen *et al.* (2017) noted lower photosynthesis rate, stomatal conductance and chlorophyll fluorescence ($F_v/F_m$) in sugar apple leaves in trees under drought conditions than the untreated control. Drought also decreased the relative water content in leaves, while proline concentrations in the leaves, fruit peel and pulp increased and proline accumulation was recorded as higher in the peel and pulp than in the leaves. The total sugar content in the peel and pulp under drought conditions was significantly higher than in the untreated control (watering of 30 l $day^{-1}$ per tree).

### 2.2.6 Wind

Atemoya trees are extremely susceptible to wind damage, especially when carrying a full crop. The fruit skin is easily damaged by rubbing and exposure to drying winds (Sanewski, 1991). In cherimoya, very strong winds can damage the shape of the tree crowns, break branches and even knock trees down (Morales Pérez, 2015). Strong winds during flowering and fruit setting periods can cause newly formed flowers and fruit to fall. It also affects the pollination and fertilization processes by drying out the stigmas due to the reduction in relative humidity. Strong wind during leaf fall, however, helps the mature leaves to fall, which allows new buds to develop and sprout (Viteri *et al.*, 2011). Soursop is susceptible to damage by strong winds as the tree has a soft wood and a relatively shallow root system. Strong winds also cause significant loss of flowers during the flowering period (Kome *et al.*, 2024).

### 2.2.7 Temperature

Cherimoya and atemoya thrive in the subtropics. They have a slightly greater cold tolerance and lower optimum temperature for fruit maturation than sugar apple (George and Nissen, 1986). Both atemoya and cherimoya are best grown in frost-free locations, as the young trees are killed at −1°C and mature trees at −3°C (Saavedra, 2022). Trees of atemoya and cherimoya are semi-deciduous and enter environmental-induced dormancy in the late winter and early spring. Cherimoya appears to initiate growth at a lower temperature (7°C) than atemoya (10°C) (George and Nissen, 1986, 1987). In cherimoya, when the day temperature reaches 33°C, the quality of the pollen and receptivity of the pistils decreased considerably, affecting the pollination and fertilization processes, and leaf burning and drop of young fruit can occur (Guirado *et al.*, 2003). The best temperature for cherimoya cultivation is 18–22°C in the summer and 5–18°C in the winter (Nakasone and Paull, 1998). Lora *et al.* (2012) studied the influence of temperature and moisture content on pollen performance of cherimoya in Malaga, Spain. The average daily temperature during the cherimoya flowering season in 2005 ranged from 20°C to 29°C and pollen germination ranged from 27% to 78%. Pollen germination showed a significant correlation (Pearson's correlation coefficient −0.372, $P = 0.21$), with the average temperature occurring 1 day prior to pollen shedding. Pollen germination was reduced at higher temperatures. In controlled-temperature chambers, the highest pollen germination was obtained in pollen from flowers maintained at 25°C and germinated at the same temperature. However, a significant decrease in pollen germination was observed in pollen from flowers kept at 30°C or 35°C and also in pollen germinated at 35°C or 15°C. The results indicated that temperatures higher than 30°C resulted in a decrease in pollen germination, whereas lower temperatures did not have a clear effect on pollen germination. Floral differentiation and development in cherimoya under warm (30/25°C) and cool (20/15°C) day/night temperatures were studied by Higuchi and Utsunomiya

(1999) in a temperature-controlled glass house. Floral differentiation was more favourable at 20/15°C than at 30/25°C. More flowers were produced and the flowering period was longer. The flowers were also larger at 20/15°C, but the growth rate of floral organs was faster at 30/25°C. The bud differentiation period was found to be the most temperature-sensitive stage to influence flower number and morphology.

Fruit-bearing atemoya trees can tolerate only light frosts; a temperature of −3°C causes tree damage, while young trees are even more sensitive, being damaged by −1°C. Temperatures of about 4°C for short periods can cause fruit splitting and skin discoloration. Frost-free sites are therefore essential. Temperatures of 25–28°C during the flowering period are favourable to good fruit set. At temperatures of about 28°C, atemoya trees produce fewer flowers but more growth, and drying of flower parts increases (Saneswki, 1988). High root temperatures increase shoot extension, leaf and flower production, and leaf and root dry weight in atemoya. George and Nissen (1987) observed a significant increase in root dry matter production at root temperatures between 17°C and 22°C in clonally propagated atemoya 'African Pride'.

Soursop is the most tropical annona (Popenoe, 1939; Nakasone and Paull, 1998), and is cultivated in damp tropical regions (Pinto *et al.*, 2005). Soursop is best cultivated between 21°C and 30°C as the fruit is susceptible to sudden changes in temperature, especially below 12°C (Pinto *et al.*, 2005). Successful pollination in soursop is achieved at temperatures around 25°C (Love and Paull, 2011a). In low mountainous ranges where the temperature varies between 15°C and 25°C, the tree produces very few fruit. A very low temperature and low humidity are detrimental for pollination and fruit set (Tripathi *et al.*, 2014). Jiménez-Zurita *et al.* (2023) observed the best growth (both vegetative and reproductive) of soursop in Compostela, Mexico, where the average temperature varied between 22°C and 28.2°C, with the average maximum (37°C) occurring in October and the average minimum (15°C) in December.

Sugar apple is more adaptable to low temperatures than soursop and is more tolerant of high temperatures than cherimoya (Belotto and Manica, 1994). It is more suited to warm, dry climates with seasonal rainfall typical of tropical regions. Optimum growth occurs in the temperature range of 15–25°C mean minimum and 25–35°C mean maximum. For optimum fruit maturation, sweetsop requires temperatures in the range of 17–21°C mean minimum and 25–30°C mean maximum (Sanewski, 1988). In Mexico, sugar apple is found in areas with annual temperatures in the range of 24–26°C in the Yucatan Peninsula, 22–24°C in the Gulf of Mexico and 26–28°C in the Pacific areas. Mature trees may tolerate short periods of −2°C without substantial injury but may be damaged or killed at temperatures of −2.2°C to −3.3°C (Crane *et al.*, 2019).

Custard apple cultivation is possible in both humid and semi-arid climates (Popenoe, 1974; Morton, 1987), although it is reported to prefer a humid climate (Fouque, 1972).

Pond apple is a tropical lowland species, requiring moist soil with regular inundations of fresh to brackish water, and can withstand periods of flooding with its roots submerged for several weeks continuously, but it does not appear to survive permanent inundation. It grows in tropical and subtropical climates and prefers full sun. It can grow in many natural environments including wetlands and sedgelands, rainforests (especially along the edges), along streams and riverbanks, alongside mangroves, in the high tide zones of beaches, and in agricultural fields, along fences and in farm drains (Morton, 1987).

Wild soursop can withstand a relative humidity as low as 44% at midday. The best temperature for wild soursop growth is between 16°C and 30°C (FAO, 1983). Rollinia (*A. deliciosa* and *A. mucosa*) is strictly a tropical tree, needing warm and humid climates. It does not tolerate cold (Love and Paull, 2011b).

## 2.3 Soil

In general, annonas are not too demanding about soil type (Nakasone and Paull, 1998), but grow well in fertile, well-aerated, well-drained, deep soil rich in organic matter (Zayas, 1966).

In Spain, cherimoya is cultivated in deep alluvial soil of metamorphic origin (Farré *et al.*, 1999). In Mexico, the best physio-chemical composition of soil for cherimoya cultivation is considered deep sandy clay soils with an organic matter content of 6.2%, 0.25% nitrogen,

2.8 p.p.m. phosphorus and potassium 0.79 mEq potassium 100 $g^{-1}$ of soil (Agustin and Angel, 1997). Atemoya prefers sandy and sandy-loam soils, but well-structured clay loams are also suitable. Although the main feeder roots of atemoya are relatively shallow, at least 1 m of well-drained soil is needed to avoid root rot and ensure good tree performance. If the topsoil is shallow (about 80 cm) or the internal drainage is marginal, atemoya can still be planted but the area will need to be mounted or drained (Sanewski, 1988).

Light or sandy textures such as loam, sandy loam, loamy sand and sandy clay loam constitute the optimal textural classes for soursop growth. Moderately suitable textural classes include clay loam, sandy clay, silty clay, silty clay loam and silt loam. Marginally suitable textural classes include clay and heavy clay soils (Kome *et al.*, 2024). Soils dominated by a heavy clay texture alongside high relative humidity are a major cause of root diseases of soursop (Pinto *et al.*, 2005). Sugar apple trees are well adapted to most well-drained soil types, including the sandy and limestone-based soils of south Florida. Trees in muck soils tend to grow more vigorously but produce fewer fruit due to high native nitrogen content. Sugar apple trees are intolerant of continuously wet or flooded soil (Crane *et al.*, 2019). In India, sugar apple is grown in different soil types, from sandy loam to saline, and on hill tracts or barren lands (Yadav *et al.*, 2023).

Rollinia grows best in deep, rich, organic soils and benefits from copious amounts of mulch as long as there is good drainage, but it tolerates poorer and highly acidic soils as long as there is sufficient water (Love and Paull, 2011b).

### 2.3.1 Soil pH

Cherimoya grows best in soil with a pH of 6.0–6.5 (Villachica, 1996). However, it tolerates alkalinity to a certain degree, growing well in soils with a pH of 7.5–8.5 and with carbonate content up to 29–30%, without showing major problems of iron chlorosis (Farré *et al.*, 1999). A soil pH of 6.0–6.5 appears to be most suitable for atemoya cultivation (Sanewski, 1988).

Soursop prefers a slightly acidic to neutral soil with a pH of 5.5–6.5 for optimal growth, while acidic soil conditions (pH $<5.0$) would be moderately or marginally suitable (Kome *et al.*, 2024). Custard apple can grow in soils with a pH of 5.0–8.0. Because of this high tolerance to variable soil types, custard apple is reported to be a good rootstock for cherimoya and soursop (Zayas, 1966; Popenoe, 1974).

### 2.3.2 Soil temperature

Soil temperature varies both seasonally and daily as a result of changes in radiant energy and energy changes taking place through the soil surface. It governs the soil physico-chemical and biological processes and also influences the interspheric processes of gas exchange between the atmosphere and the soil. Soil temperature affects soil water content, as well as its conductivity and availability to plants (Onwuka, 2016). There is an approximately 2% change in the various parameters present in the soil with every degree change in temperature (Frey *et al.*, 2013).

A high root temperature increases shoot extension, leaf and flower production, and leaf and root dry weight in *Annona* spp. With clonally propagated trees of the atemoya cultivar 'African Pride', significant increases in root dry matter production were recorded at root temperatures between 17°C and 22°C (George and Nissen, 1987). High root temperatures (27°C) were not detrimental to plant growth. Most cherimoya cultivars produced more dry matter at low root temperature (15°C) compared with the more tropical atemoya, sugar apple and pond apple (George and Nissen, 1987). Soursop is best adapted to a soil temperature of 25–35°C and pond apple to 20–25°C (Morton, 1987). Ojeda (2003) observed maximum growth of soursop at a soil temperature of 35°C in non-flooded conditions and at 25°C for flooded conditions. Ojeda (2003) stated that soursop may be a flood-tolerant rootstock only for areas subject to short-term flooding. The leaf chlorophyll index in pond apple, soursop and atemoya was found to be lowest at 5°C or 10°C and highest at 35°C soil temperature. The fresh and dry weight of roots, stems and leaves decreased almost linearly as the soil temperature decreased for the atemoya cultivar 'Gefner' on sugar apple and soursop rootstock (Ojeda, 2003). Soils with a hyperthermic and isohyperthermic temperature regime would be most suitable for the growth of soursop where the mean annual soil temperature is greater than 22°C. Soils with a thermic/isothermic temperature regime (15°C to <22°C) would be moderately suitable, while those with a mesic/isomesic regime (8°C to <15°C) would be marginally suitable. Soils with a frigid/isofrigid temperature regime (0°C to <8°C) would be severely limiting (Kome *et al.*, 2024).

### 2.3.3 Soil salinity

High salt concentrations in irrigation water and in soil cause serious stress in plants such as nutritional imbalance due to the accumulation of sodium, chlorine and sulfates inside the plant tissues, which reduces absorption of some essential nutrients such as nitrogen, phosphorus, calcium and potassium. In addition, the plants water absorption capacity is reduced due to decreased osmotic potential (Parihar *et al.*, 2015). Sugar apple plants under high salt concentrations have their gas exchange and chlorophyll reduced, showing their sensitivity to salinity stress (Sá *et al.*, 2015; Figueiredo *et al.*, 2019). Marler and Zozor (1996) studied the influence of salinity on photosynthetic characteristics, water relations and foliar mineral composition of sugar apple. They recorded inhibition of net $CO_2$ assimilation, stomatal conductance of $CO_2$ and transpiration within 2 weeks of initiating salinity treatments in pot-grown plants amended to 3 or 6 dS $m^{-1}$ with sea salt. The gas exchange continued until days 30–35. Salinity reduced $CO_2$, light energy and water-use efficiency. Root-zone salinity of 3 dS $m^{-1}$ administered for 52 days did not influence the foliar sodium concentration or the sodium:potassium ratio but increased chloride concentration and decreased nitrogen concentration. Salinity also reduced the osmotic potential of root tissue but did not influence foliar osmotic or pre-dawn xylem potential. These results indicated that sugar apple is sensitive to salinity stress. A high concentration of salt (3.0 dS $m^{-1}$) in the irrigation water reduced the number of fruit, fresh fruit mass, total number of seeds and mass of seeds in sugar apple (Silva *et al.*, 2022).

Soils with high base saturation (>60%) are highly suitable for soursop growth (Pinto and Silva, 1994), as they contain optimal quantities of plant nutrients such as calcium, magnesium and potassium, while soils with a low base saturation (<60%) are moderately suitable (Kome *et al.*, 2024).

Salinity is a limiting environmental factor for the establishment of wetland plant species because it causes physiological drought by producing a water deficit that reduces the osmotic potential of the soil (Salter *et al.*, 2007). Salinity negatively affected survival, biomass gain, leaf gain, and height and diameter growth of pond apple seedlings (Sánchez-García *et al.*, 2022).

## 2.4 Phenology

Plants show a rhythmic appearance of their vegetative and reproductive cycles over time. This phenomenon is termed phenology. The study of plant phenology provides knowledge about the patterns of plant growth and development, as well as the effects of environment and selective pressure on flowering and fruiting behaviour (Igboabuchi *et al.*, 2018). Phenological events are intimately linked to specific development phases of plant species such as reproductive bud development, flowering, fruit development and fruit maturation. These stages are described by the Biologische Bundesanstalt, Bundessortenamt und Chemische Industrie (BBCH) numerical scale, which is a system for uniformly coding phenologically similar growth stages among all monocotyledonous and dicotyledonous plant species (Bleiholder *et al.*, 1989; Hess *et al.*, 1997). The basic BBCH scale, represented by two digits, has primary and secondary scales, and each scale is subdivided into ten (0–9) clearly recognizable and distinguishable developmental phases. The primary scale describes the principal stages associated with the major developmental cycle of the plant.

### 2.4.1 Soursop

The phenological stages of soursop under various climatic conditions in Brazil (Nascimento *et al.*, 2002), Venezuela (Yamarte *et al.*, 2006) and Mexico (Jiménez-Zurita *et al.*, 2023) have been described. Jiménez-Zurita *et al.* (2023) studied these stages in two soursop-producing regions of Nayarit (Tepic and Compostela) in Mexico. Eight main stages of growth were typified using the extended BBCH scale: vegetative growth, which included the development of buds (stage 0), leaves (stage 1) and shoots (stage 3); flower growth, which included their emergence (stage 5) and opening stage (stage 6); fruit development, which included both fruit setting/growth (stage 7) and fruit maturity (stage 8); and lastly, the senescence of branches (implied by defoliation or the falling of leaves, stage 9) (Fig. 2.1).

In Compostela, the average temperature varied between 22°C and 28.2°C, while in Tepic it was between 19.8°C and 23.6°C. The differences in temperature influenced both vegetative and reproductive growth, especially in the leaves, shoots and flowers (phenological stages 110–119, 311–319 and 510–519, respectively). As a result, the duration of these stages was shorter in Compostela (25, 51 and 36 days) than in Tepic (47, 73 and 42 days). Fruit setting to physiological maturity took an average of 114 days in both situations. The researchers found a positive correlation between average temperature and the presence of floral structures from Compostela ($r = 0.56$, $P < 0.05$), suggesting that flowering might also be favoured by warm climates. Although defoliation occurs year-round, it was high between November and January, which was attributed to low temperatures and the dry season (lack of water). Senescence was negatively correlated with both lower precipitation ($r = 0.76$, $P < 0.01$) and relative humidity ($r = 0.74$, $P < 0.01$) (Jiménez-Zurita *et al.*, 2023).

### 2.4.2 Atemoya

The phenology and thermal requirements of atemoya 'Gefner' were studied by Mendes *et al.* (2019) in two agronomic seasons (August–April and May–January) in the Brazilian semi-arid region. The atemoya tree exhibited eight of the ten main stages according to the general BBCH scale (Fig. 2.2A–C). The onset was characterized by bud development (stage 0) and ended with beginning of the rest period (stage 9). The four vegetative stages included the development of buds (stage 0), leaf development (stage 1), shoots (buds) (stage 3) and the beginning of the rest period (stage 4). The reproductive stages corresponded to the emergence of the inflorescence (stage 5), flowering stage (stage 6), fruit growth (stage 7) and fruit maturation (stage 8).

In the first cycle, the duration and the thermal requirement for each subperiod were 24 days and 272 degree days (DD) from the closed leaf buds to the end of the leaf with 5 mm (CLB-EL5). It was 18 days and 207.2 DD from the first leaves separating until showing completely expanded leaves (FLS-FEL); around 221 days and 2545 DD from budding with 10% of its length to the maximum length (BIOL-BML); 29 days and 357 DD from the closed flowers with petals that form a long corolla (CFB-CFPL); 14 days and 163 DD from the beginning of the petal opening in the pre-female stage until the end of flowering (BPO-EF); 168 days and 1911 DD from fruiting at the beginning of the ovary growth to fruit at the ideal maturity size for harvest (FOG-IMH); and 65 days and 760 DD from the appearance of the ripening colour for harvest until fruit maturity (ARC-FMC). In the second cycle, with pruning performed on 16 May (autumn), the duration and the thermal requirements for each subperiod were 21 days and 248DD for CLB-EL5, 18 days and 197 DD for FLS-FEL, 218 days and 2474 DD for BIOL-BML, 31 and 326 DD for CFB-CFPL, 19 days and 207 DD for BPO-EP, 152 days and 1719 DD for FOG-IMH, and 46 days and 526 DD for ARC-FMC. The phenological cycle of the atemoya tree, from the stage of leaf buds closed and covered by brown scales until the complete physiological maturity of the fruit occurred in the periods between 217 and 206 days, with thermal requirements of 2469 and 2302 DD for the first and second agronomic seasons. These differences between the cycles were believed due to two climatic factors – rainfall and insolation (Mendes *et al.*, 2019).

### 2.4.3 Cherimoya

The phenological growth stages of the cherimoya tree were studied by Cautín and Agustí (2005) with two cultivars, 'Concha Lisa' and 'Bronceada' in Quillota, Chile, over two growing seasons (2000–2001 and 2001–2002), from December to October, using the extended BBCH scale. The researchers used seven of the ten principal stages, starting with shoot growth in cuttings or stumps (stage 0) and ending with initiation of a

**Fig. 2.1.** The eight main phenological stages of soursop (*Annona muricata*) as determined with the extended BBCH scale. The trees were cultivated in orchards located in Compostela and Tepic, in the state of Nayarit, Mexico. Photographs courtesy of Dr Iran Alia-Tejacal.

**Fig. 2.1.** Continued.

brief rest period (stage 9). Development of rosette leaves (stage 2) and vegetative harvestable parts (stage 4) were not considered because they do not apply in cherimoya. The fruit ripening stage (stage 8) was also not considered because the fruit does not change colour in cherimoya. The descriptions were considered valid for average climatic conditions for cherimoya growth (18–23° and 5–18°C average summer and winter temperatures, respectively), annual rainfall of 400–1500 mm and relative humidity higher than 80% during the fruit set stage.

The BBCH scale defined by Cautín and Agustí (2005) for cherimoya has several advantages. It defines accurately bud, leaf, shoot and fruit development and maturity. This is important for

**Fig. 2.2.** (A) Main phenological stages of the atemoya tree according to the general BBCH scale, for stages 00–35. 00: Dormancy: leaf buds closed and covered by brown scales. 01: Beginning of bud swelling, greenish-brown coloration. 03: End of the expansion of leaf buds; greenish-brown bud. 07: Beginning of budding; leaf ends visible with a greenish-brown colour. 09: End of greenish-brown leaf with 5 mm above the bud scale. 10: First leaves separating, with slightly brownish-green coloration. 11: First visible unfolded leaves of greenish-white colour. 13: Development of the third leaf of greenish-white colour, total unexpanded. 1.: Leaf development. 19: First fully expanded leaves, with leathery appearance. 32: Bud with 20% of final length. 35: Bud with 50% of final length. (B) Stages 37–65. 37: Bud with 70% of final length. 39: Bud reaching the maximum length of the species. 50: Floral buds enclosed by whitish scale. 51: Floral buds swelling. 54: Visible sepals and petals begin to lengthen. 55: Lengthening of petals. 57: Flowers still closed, petals lengthening. 59: Closed flowers, petals completely extended forming a long corolla. 60: Beginning of separation of petals: pre-female stage, pollen grain is not visible. 61: First flower at female stage: 10% of flowers open. 65: Full flowering, 50% of open flowers in completely male state, pollen grain released. (C) Stages 65–91. 65: Full flowering, at least 50% of flowers open and in completely male state, pollen grain released. 67: Fading of flower colour: majority of the petals fallen or dried. 69: End of flowering: visible fruiting, all petals fallen or dried. 71: onset of growth of green ovaries, sometimes with dry petals. 72: Fruit about 20% of final size. 74: Fruit about 40% of final size. 75: Fruit about half of final size. 77: Fruit about 70% of final size. 78: Fruit about 80% of final size. 79: Fruit have reached full size and are physiologically mature, with light green coloration and carpels distant from one another. 81: Appearance of fruit colour. 89: Fully mature fruit, detachable with ease. 91: End of growth of branches or buds, but the foliage remains green. Photographs taken in Janauba, Minas Gerais state, Brazil, courtesy of Dr Debora Souza Mendes.

**Fig. 2.2.** Continued.

the correct timing of orchard management, such as the effectiveness of fertilizer application, the specific time for application of plant growth regulators, and the application time for pest, disease and weed control.

### 2.4.4 Sugar apple

The phenological growth stages of sugar apple in dystrophic yellow latosol under the savannah conditions of Roraima, Brazil, were studied by de Moura *et al.* (2019). The study was carried out in four seasons of the year, the 2014 and 2015 rainy seasons and the 2014–2015 and 2015–2016 summer seasons. Production pruning was carried out in February and September in both 2014 and 2015. The period between production pruning and bud swelling was 6 days in 2014 and 9 days in 2015 when pruning was carried out in February, while it was 10 days and 7 days in 2014–2015 and 2015–2016, respectively. This was due to a reduction in rainfall and a decrease in relative humidity in the second cycle. The period between anthesis (pollination) and fruit harvest was 101 days for February pruning in both 2014 and 2015, while it was less for the September pruning (91 and 84 days in 2014–2015 and 2015–2016, respectively). The duration between production pruning and fruit harvest was 146 days for February pruning and 127 days for September pruning in 2014. After February pruning, the following 4-month period showed low rainfall when the plant was at full vegetative and reproductive development, causing the delay between the phenophases. The rainfall was regular for 3 months after the September pruning and the plants responded significantly better in less time. However, Liu *et al.* (2015) in Zhanijan, China, recorded a 215-day period between pruning and fruit harvest of sugar apple.

### 2.4.5 Other *Annona* spp.

The phenological growth stages of *Annona lutescens*, *A. reticulata* (custard apple), *A. macroprophyllata* and *A. purpurea* were studied in Chiapas, Mexico, in three areas of deciduous forests located at 16°31′3″N, 92°58′25″W; 16°28′82″N, 95°57′48″W; and

16°21′47″N, 92°58′47″N by González-Esquinca *et al.* (2016) using the extended BBCH scale. Observations over several years showed that *A. lutescens* and *A. reticulata* bear fruit at the end of dry season, whereas *A. macroprophyllata* and *A. purpurea* do so at the end of the rainy season and the beginning of the dry season. The dry season corresponded to the months of November–April when rainfall was below 46 mm and the maximum temperature reached 26°C. The rainy season occurred from June to September, with rainfall up to 265 mm and temperature under 25°C.

In all four species, the visibility of dormant buds was higher when the trees lost their leaves (stage 00), and leaf bud development in *A. lutcscens* and *A. reticulata* began during the transition from the dry season to the rainy season (stage 03). The branches produced mixed buds (vegetative and floral) in all species, and they were both terminal and axillary. Leaf flushing (branches with first leaves unfolded) in these trees began in May after the dry season and sometimes occurred before the rains, and the leaves emerged from the leaf buds after 3–4 days (stage 2). *A. lutescens* and *A. reticulata* flowered sporadically throughout the year, but most of the flowers were aborted. Full flowering occurred only during November and December. The flowers that emerged at the beginning of the dry season were the only ones that produced fruit. The flowering of *A. macroprophyllata* and *A. purpurea* only occurred in May and June. *A. lutescens* and *A. reticulata* fruited at the end of the dry season, but the other two species produced fruit during middle and at the end of the rainy season (González-Esquinca *et al.*, 2016). The flowers of all species exhibited dichogamy. With the fruit development, the petals fell in all species with the exception of *A. macroprophyllata*, whose petals tended to be more persistent (stage 70). The fruit reached half of its final size and full size at approximately 70 days (stage 75) and 120 days (stage 79) after pollination, respectively. The researchers identified the following stages of development: (i) fruit with incipient maturity that changed from green to yellow (*A. lutescens*) or from green-purple to red (*A. reticulata*); in *A. macroprophyllata* and *A. purpurea*, the fruit remained green or slightly brown (stage 81); (ii) fruit with firm flesh mature enough for harvest (stage 87); and (iii) fruit mature enough for consumption with an intense aroma and colour and a smooth texture (stage 89). The fruit of *A. macroprophyllata* split while still on the tree, showing a white- or pink-coloured pulp. Some development stages of the four *Annona* spp. are presented in Fig. 2.3. Dao *et al.* (2024) studied the phenology of *A. senegalensis* in western Burkina Faso at the sites of Dinderesso Classified Forest (11°12′38″N, 4°24′43″W) for the Sudanian climatic zone and at the sites of Toroba Classified Forest (12°29′14″N, 3°15′35″W) and Kousseri wasteland (12°33′09″N, 3°15′47″W) for the Sudano-Sahelian climatic zone. They observed leafing duration of 10 months in the Sudanian zone, and flowering and fruiting lasted for 3 months. In the Sudano-Sahelian zone, leafing lasted 8 months, while flowering and fruiting lasted 3 and 4 months, respectively. A total of 10,040 insects belonging to 48 species were collected in the two climatic zones. Forty-six species were found in Sudanian zone, while 25 species were recorded in the Sudano-Sahelian zone. The variations in plant phenology and the insect community were believed to be mainly due to the variations in rainfall across both climatic zones. The phenophases were correlated with climatic factors, in particular temperature and rainfall.

## 2.5 Conclusion

Annona fruit production is affected by many environmental factors such as air and soil temperatures, relative humidity, light, drought and flooding. Pond apples have potential for use as flood-tolerant rootstock for commercial *Annona* spp. because of their ability to survive and grow in flooded conditions. Both atemoya and cherimoya are sensitive to water stress during fruit development, while soursop is more tolerant of water stress. In general, annonas are not too demanding in terms of soil type but prefer fertile, well-aerated, well-drained, deep soil rich in organic matter.

Studying the developmental stages of different *Annona* spp. elucidates the relationship between its morphological and physiological characteristics and environmental factors, especially climatic factors. Understanding the phenology of a particular species helps to follow the correct management practices, and determine the optimum time for hand pollination, the prediction of likely pest outbreaks, the need for a specific fertilization, etc.

**Fig. 2.3.** Some development stages of *Annona* spp. (A–D) *Annona lutescens*. (A) Flowers in distinct development stages. (B) Male flower. (C) Fruit. (D) Flower bud. (E–H) *Annona macroprophyllata*. (E) Female and male flowers. (F) Male flower. (G) Developing fruit. (H) Leaf bud. (I–L) *Annona purpurea*. (I) Female flower. (J) The flower is used as chamber for pollination by Coleoptera spp. (K) Developing fruit. (L) Leaves buds. (M–P) *Annona reticulata.* (M) Male flower. (N) Female flower. (O) Fruit. (P) Branch with leaves and flowers fertilized. Photographs courtesy of Dr Alma Rosa González-Esquinca.

## References

Agustin, J.A. and Angel, D.N. (1997) In: Sao Jose, A.R., Vilas Bos, I., Morais, O.M. and Reboucas, T.N.H. (eds) *Anonaceas: Produces e Meracado*. Universidade Estadual do Sudoeste da Bahia, Bahia, Brazil, pp. 7–19.

Belotto, F.A. and Manica, I. (1994) Clima Sola. In: Manica, I. (ed.) *Fruticulturacultivo Das Anonaceas*. Parto Algre, pp. 13–17.

Bleiholder, H., Vanden Boom, T., Langeluddek, P. and Stauss, R. (1989) Einheitliche Codierung der phanologischen Stadien bei Kultur-und Schadptlazzen. *Gesunde Flanzen* 41, 381–384.

Cautín, R. and Agustí, M. (2005) Phenological growth stages of the cherimoya tree (*Annona cherimola* Mill.). *Scientia Horticulturae* 105, 491–497. DOI: 10.1016/j.scienta.2005.01.035

Crane, J.H., Balerdi, C.F. and Maguire, I. (2019) *Sugar Apple Growing in the Florida Home Landscape*, Vol. 38. University of Florida, Institute of Food and Agricultural Sciences Extension, Florida.

Dao, Z.A., Romba, R., Jaloux, B., Ouedraogo, A. and Gnankine, O. (2024) Effects of climate on the phenology of *Annona senegalensis* Pers. (Annonaceae) and the distribution of associated insects in Burkina Faso. *Ecology and Evolution* 14: e70154. DOI: 10.1002/ece3.70154

de Moura, E.A., Chagas, P.C., Chagas, E.A., de Oliveira, R.R., Araújo, W.F. *et al.* (2019) Phenological study of sugar apple (*Annona squamosa* L.) in dystrophic yellow latosol under the savanna conditions of Roraima. *Australian Journal of Crop Science* 13, 1467–1472. DOI: 10.21475/ajcs.19.13.09.p1557

de Oliveira, P.C. and Lameira, H.L.N. (2017) Physiological behaviour of *Annona muricata*, *Dipteryx odorata* and *Copaifera langsdorffii* in response to water and light stress. *Journal of Medicinal Plant Studies* 5, 193–199.

Dulai, S., Monar, I., Pronoy, J., Csernaka, A., Tarnai, R. and Molnár-Láng, M. (2006) Effect of drought on photosynthetic parameters and heat stability of PSH in wheat and in *Aegilops* species originating from dry habitats. *Acta Biologica Szegediensis* 50, 11–17.

FAO (Food and Agriculture Organization of the United Nations) (1983) Fruit and fruit breeding forest species 1: examples from Eastern Africa. *FAO Forestry Paper* 4, 11–14.

Farré, J.M., Hermoso, J.M., Guirado, E. and García-Tapia, J. (1999) Techniques of cherimoya cultivation in Spain. *Acta Horticulturae* 497, 91–118. DOI: 10.17660/ActaHortic.1999.497.5

Figueiredo, F.R.A., de Melo Gonçalves, A.C., Silva Ribeiro, J.E., da Silva, T.I., Nóbrega, J.S. *et al.* (2019) Gas exchanges in sugar apple (*Annona squamosa* l.) subjected to salinity stress and nitrogen fertilization. *Australian Journal of Crop Science* 139, 1959–1966. DOI: 10.21475/ajcs.19.13.12.p1754

Fouque, A. (1972) Especes fruitiers d Amerique tropicale. *Fruits* 27, 62–72.

Frey, S.D., Lee, J., Melillo, J.M. and Six, J. (2013) The temperature response of soil microbial efficiency and its feedback to climate. *Nature Climate Change* 3, 395–398. DOI: 10.1038/nclimate1796

Fu, X.Y., Peng, S.X., Yang, S., Chen, Y.H. and Zhang, J.Y. (2012) Effects of flooding on grafted annona plants of different scion, rootstock combinations. *Agricultural Science* 3, 249–256. DOI: 10.4236/as.2012.32029

George, A.P. and Nissen, R.J. (1986) Annonaceae. In: Page, P.E. (ed.) *Tropical Tree Fruit for Australia*. Queensland Department of Primary Industry, Brisbane, Australia, pp. 35–41.

George, A.P. and Nissen, R.J. (1987) The effects of root temperature on growth and dry matter production of *Annona* species. *Scientia Horticulturae* 31, 95–99. DOI: 10.1016/0304-4238(87)90111-7

George, A.P. and Nissen, R.J. (1992) Effect of environmental variables and cropping on plant water status of custard apple (*Annona cherimola* × *Annona squamosa*). *Journal of Horticultural Science* 67, 445–455. DOI: 10.1080/00221589.1992.11516270

George, A.P. and Nissen, R.J. (1998) The effect of temperature, vapour pressure deficit and soil moisture stress on growth, flowering and fruit set of custard apple (*Annona cherimola* × *Annona squamosa*) 'African pride'. *Scientia Horticulturae* 34, 183–191. DOI: 10.1016/0304-4238(88)90091-X

George, A.P., Nissen, R.J. and Brown, B.T. (1987) The custard apple. *Queensland Agricultural Journal* 113, 287–297.

González-Esquinca, A.R., De-la-Cruz-Chacón, I., Castro-Moreno, M. and Riley-Saldaña, C.A. (2016) Phenological strategies of *Annona* from deciduous forest of Chiapas, Mexico. *Botanical Sciences* 94, 531–541. DOI: 10.17129/botsci.645

Guirado, E., Hermoso, J., Perez, M. and Farré, M. (2003) *Experimental Farm the Nacla. Introduction to Cherimoya Cultivation*. Granada Rural Bank, Granada.

Hess, M., Barralis, G., Bleiholder, H., Buhr, L., Eggers, T.H. *et al.* (1997) Use of the extended BBCH scale – general for the descriptions of the growth stages of mono; and dicotyledonous weed species. *Weed Research* 37, 433–441. DOI: 10.1046/j.1365-3180.1997.d01-70.x

Higuchi, H. and Utsunomiya, N. (1999) Floral differentiation and development in cherimoya (*Annona cherimola* Mill.) under warm (30/25°C) and cool (20/15°C) day/night temperature. *Journal of the Japanese Society for Horticultural Science* 68, 707–716. DOI: 10.2503/jjshs.68.707

Igboabuchi, N.A., Echereme, C.B. and Ekwealor, K.U. (2018) Phenology in plant: concepts and uses. *International Journal of Science and Research Methodology* 11, 8–24.

Jackson, M.B. and Colmer, T.D. (2005) Response and adaptation by plants to flooding stress. *Annals of Botany* 96, 501–505. DOI: 10.1093/aob/mci205

Jaleel, C.A., Manivannan, P., Lakshmanan, G.M.A., Gomathinyagam, M. and Panneerselvam, R. (2008) Alterations in morphological parameters and photosynthetic pigments response of two varieties of *Catharanthus roseus* under soil water deficits. *Colloids and surfaces B: Bio-Interfaces* 61, 298–303. DOI: 10.1016/j.colsurfb.2007.09.008

Jiménez-Zurita, J.O., Alia-Tejacal, I., Balois-Morales, R., Villarrcal-Funtes, J.M., Nunez-Colin, C.A. *et al.* (2023) Phenological growth stages of soursop trees (*Annona muricata* L.) based on the extended BBCH-scale. *Revista Chapingo Serie Horticulturae* 29, 5–18. DOI: 10.5154/r.rchsh.2022.03.006

Kome, G.K., Enang, R.K., Silatsa, F.B.T., Yerima, B.P.K. and van Ranset, E. (2024) Baseline edaphic requirements of soursop (*Annona muricata* L.). *Tropical Plants* 3: e022. DOI: 10.48130/tp-0024-0023

Kowitcharoen, L., Wongs-Aree, C., Setha, S., Komkhuntod, R., Srilaong, V. *et al.* (2017) Physiological changes of fruit and C/N ratio in sugar apples (*Annona squamosa* L.) under drought conditions. *Acta Horticulturae* 1166, 195–202. DOI: 10.17660/ActaHortic.2017.1166.28

Kozlowski, T.T. (1997) Responses of woody plants to flooding and salinity. *Tree Physiology* 17, 490–490. DOI: 10.1093/treephys/17.7.490

Leal, F. and Paull, R.E. (2023) The genus *Annona*: botanical characteristics, horticultural requirements and uses. *Crop Science* 63, 1030–1049. DOI: 10.1002/csc2.20833

Liu, K., Li, H., Yuan, C., Huang, Y., Chen, Y. and Liu, J. (2015) Identification of phenological growth stages of sugar apple (*Annona squamosa* L.) using the extended BBCH-scale. *Scientia Horticulturae* 181, 76–80. DOI: 10.1016/j.scienta.2014.10.046

Lora, J., Hormaza, J.I. and Herrero, M. (2012) Pollen performance of *Annona cherimola* Mill. (Annonaceae) is affected by temperature and moisture content during the final stages of pollen development. *Acta Horticulturae* 932, 65–68. DOI: 10.17660/ActaHortic.2012.932.8

Love, K. and Paull, R.E. (2011a) *Soursop*. Fruit, Nut and Beverage Crops Series FM-22. University of Hawaii, Honolulu, Hawaii.

Love, K. and Paull, R.E. (2011b) *Rollinia*. Fruits and Nuts F_N-21. University of Hawaii at Manoa, Honolulu, Hawaii.

Maheswari, T.U. and Sinduja, S. (2020) Soursop: a promising fruit for cancer mitigation. *Plant Archives* 20, 1653–1656.

Marler, T.E. and Zozor, Y. (1996) Salinity influences photosynthetic characteristics, water relations, and foliar mineral composition of *Annona squamosa* L. *Journal of the American Society for Horticultural Science* 121, 243–248. DOI: 10.21273/JASHS.121.2.243

Mendes, D.S., Pereira, M.C.T., Nietsche, S., Xavier, H.R.A., Paixão, P.T.M. and Fonseca, M.P. (2019) Phenology and thermal requirements of the atemoya tree (*Annona cherimola* mill. × *Annona squamosa L.*). *Revista Ceres, Vicosa* 66, 200–209. DOI: 10.1590/0034-737x201966030006

Morales Pérez, A.A. (2015) [Application of resveratrol and 6-benzylaminopurine to increase postharvest life in custard apple]. Thesis, Universidad Autónoma del Estado de México, Toluca, Mexico. Available at: https://core.ac.uk/works/8621781/?t=49a336f3b47006e416d0de311a2c0a6b-8621781 (accessed 13 August 2025).

Morton, J.F. (1987) Soursop (*Annona muricata*). In: *Fruits of Warm Climates*. Purdue University, West Lafayette, Indiana, pp. 75–80.

Nakasone, H.Y. and Paull, R.E. (1998) Annonas. In: Nakasone, H.Y. and Paull, R.E. (eds) *Tropical Fruits*. Crop Production Science in Horticulture Series. CAB International, Wallingford, UK, pp. 45–75.

Nascimento, T.B., Gazel Filho, A.B. and Santos, J.A. (2002) Fenologia da gravioleira (*Annona muricata*) em área de Cerrado do Amapá, Brasil. *Acta Amazonica* 32, 367–367. DOI: 10.1590/1809-43922002323376

Núñez-Elisea, R., Schaffer, B., Fisher, J.B., Colls, A.M. and Crane, J.H. (1999) Influence of flooding on net $CO_2$ assimilation, growth and stem anatomy of *Annona* species. *Annals of Botany* 84, 771–780. DOI: 10.1006/anbo.1999.0977

Núñez-Elisea, R., Schaffer, B., Crane, J.H. and Colls, A.M. (1998) Impact of flooding on *Annona* species. *Proceedings of the Florida State Horticultural Society* 111, 317–319.

Ojeda, M. (2003) Flooding, soil temperature, iron nutrition, physiology and growth of *Annona* species. PhD thesis, University of Florida, Gainesville, Florida.

Onwuka, B.M. (2016) Effect of soil temperature on some soil properties and plant growth. *Scholarly Journal of Agricultural Science* 6, 89–93.

Parihar, P., Singh, S., Singh, R., Singh, V.P. and Prasad, S.M. (2015) Effect of salinity stress on plants and its tolerance strategies: a review. *Environmental Science and Pollution Research* 22, 4056–4075. DOI: 10.1007/s11356-014-3739-1

Pinto, A.C.Q. and Silva, E.M. (1994) *Soursop for Export: Technical Aspects of Production.* FRUPEX, Ministry of Agriculture, Supply and Agrarian Reform, Sec. of Rural Development Programme, Brazil.

Pinto, A.C.Q., Cordeiro, M.C.R., de Andrade, S.R.M., Ferreira, F.R., Filgueiras, H.A.C. *et al.* (2005) *Annona Species*. International Centre for Underutilized Crops, University of Southampton, Southampton, UK.

Popenoe, W. (1939) *Importance of Frutas Tropicais*. Uniao Panamericana, Depto de de Cooperacao, Agricola, Washington, DC.

Popenoe, W. (1974) *Manual of Tropical and Subtropical Fruits*. Hafner Press, New York.

Ratnayake, S.S., Kumar, L. and Kariyawasam, C.S. (2020) Neglected and underutized fruit species in Sri Lanka: prioritization and understanding the potential distribution under climate change. *Agronomy* 10: 34. DOI: 10.3390/agronomy10010034

Rodrigues, B.M., Souza, B.D., Nogueira, R.M. and Santos, M.G. (2010) Tolerance to water deficit in young trees of jackfruit and sugar apple. *Revista Ciência Agronômica* 41, 245–252. DOI: 10.1590/S1806-66902010000200011

Sá, F.V.S., Brito, M.E.B., Ferreira, I.B., Antônio Neto, P. and Silva, L.A. (2015) Balanco de sais e crescimento inicial de mudas de pinheira (*Annona squamosa* L.) sob substrates irrigados com agu salina. *Irriga* 20, 544–556. DOI: 10.15809/irriga.2015v20n3p544

Saavedra, E. (2022) Set and growth of *Annona cherimola* Mill. fruit obtained by hand-pollination and chemical treatments 1. *Journal of the American Society for Horticultural Science* 104, 668–673. DOI: 10.21273/JASHS.104.5.668

Salter, J., Morris, K., Bailey, P.C.E. and Boon, P.I. (2007) Interactive effects of salinity and water depth on the growth of *Melaleuca ericifolia* Sm. (swamp paperbark) seedlings. *Aquatic Botany* 86, 213–222. DOI: 10.1016/j.aquabot.2006.10.002

Sánchez-García, E.A., López-Rosas, H., Sosa, V.J., Linding-Cisneros, R. and Moreno-Casasola, P. (2022) Effect of water level and salinity on the growth of *Annona glabra* L. seedlings. *Wetlands Ecology and Management* 30, 579–593. DOI: 10.21.203/rs.3. rs-734846/V1

Sanewski, G.M. (1988) *Growing Custard Apples*. Department of Primary Industries,Queensland Government, Brisbane, Australia.

Sanewski, G.M. (1991) Custard apple – Cultivation and Crop Protection. Information Series Q 190031. Queensland Department of Primary Industry, Brisbane, Australia.

Satisha, J., Kurian, R.M. and Dinesh, M.R. (2016) *Production Technology of Tropical Fruits – A Handbook* (eds). IIHR, Bengaluru, India.

Schaffer, B., Davies, F.S. and Crane, J.H. (2006) Responses of subtropical and tropical fruit trees to flooding in calcareous soil. *HortScience* 41, 549–555. DOI: 10.21273/HORTSCI.41.3.549

Silva, A.A.R.D., Veloso, L.L.D.S.A., Lima, G.S.D., Gheyi, H.R., Sá, F.V.D.S. and Azevedo, C.A.V.D. (2022) Cultivation of custard-apple irrigated with saline water under combinations of nitrogen, phosphorus and potassium. *Revista Caatinga* 35, 181–190. DOI: 10.1590/1983-21252022v35n118rc

Tripathi, P.C., Sankar, V. and Senthil Kumar, R. (2014) *Soursop: An Emerging Fruit of the Future*. Technical Bulletin, Vol. 5. ICAR-IIHR, Chettalli, India.

Utsunomiya, N. and Higuchi, H. (1996) Effects of irradiance level on the growth and photosynthesis of cherimoya, sugar apple and soursop seedlings. *Environment Control in Biology* 34, 201–207. DOI: 10.2525/ecb1963.34.201

Villachica, V. (1996) *Frutales y Hortalizas Promisorios de La Amazonia*. Secretaria Pro-Tempore, Lima, Peru.

Viteri, P., Leon, J., Soria, N. and Vasquez, W. (2011) Advances in management of forced production in cherimoya to improve and/or alter the physiological processes of plants and plan fruit harvests. In: *III International Seminar on Cherimoya and II on Guanaba*. Faculty of Agronomy, University of Cuenca, Cuenca, Ecuador, pp. 20–28.

Xavier, B.V.S. (2018) Effect of 1-methylcyclopropene on the inhibition of ethylene in the ripening of cherimoya (*Annona cherimola* Mill). Thesis, Universidad Central del Ecuador, Ecuador.

Yadav, V. (2018) Cultivation of custard apple fetches more. *Indian Horticulture* 63, 42–46.

Yadav, Y., Singh, A.K., Hiwale, S.S., Singh, S. and Sharma, B.D. (2023) Thar Amrit: a new high-yielding custard apple variety for dry land. *Indian Horticulture* 68, 11–13.

Yaguana, F. and Fernanda, P. (2018) [Morphological characterization of native populations of cherimoya (*Annona cherimola* Mill.) for use in the province of Loja]. Available at: https://dspace.unl.edu.ec/items/9ef64062-52f0-4293-a104-c98a89eb684c (accessed 13 August 2025).

Yamarte, M., Marin, M., Bautista, D. and Avilan, L. (2006) Caracteristicas del crecimiento de las ramas del guanabano (*Annona muricata* L.) bajo las condiciones de un bosque muy seco tropical. *Revista de La Facultad de Agronomica* 23, 1–16.

Zayas, J.C. (1966) *Las Frutas Anonaceas*. Ediciones Fruticuba, Havana, Cuba.

Zotz, G., Tyree, M.T. and Patiño, S. (1997) Hydraulic architecture and water relations of a flood-tolerant tropical tree, *Annona glabra*. *Tree Physiology* 17, 359–365. DOI: 10.1093/treephys/17.6.359

# 3 Composition and Uses

**Gabriela Aguilar-Hernández[1], Libier Meza-Espinoza[2] and Efigenia Montalvo-González[3*]**

[1]*Division of Agricultural Sciences and Engineering, Centro Universitario de los Altos (CUALTOS)-University of Guadalajara, Guadalajara, Jalisco, Mexico;* [2]*Department of Agricultural Sciences, Technological University of the Coast, Santiago, Ixcuintla, Nayarit, Mexico;* [3]*Integral Laboratory of Food Research, National Technologic of Mexico/Technological Institute of Tepic, Tepic, Nayarit, Mexico*

## 3.1 Introduction

The term 'annona' is etymologically derived from the Latin *annona* meaning 'annual harvest'. *Annona* spp. belong to the family Annonaceae, which is native to Central America and the West Indies and is one of the most prominent tropical and subtropical families of trees, shrubs and lianas (Escribano *et al.*, 2007). The important cultivated *Annona* spp. are *A. muricata* L., *A. cherimola* Mill., *A. squamosa* L. and *A. reticulata* L. (Pinto *et al.*, 2005; Pareek *et al.*, 2011; Anuragi *et al.*, 2016). The edible part of these fruits has a sweet taste, a creamy consistency and an exquisite smell, so it is mainly used for fresh consumption or processed into juices, nectars, yogurts, jams, ice cream and confectionery, among other products. Annona pulp is a source of vitamins (A, B, C and D), minerals (iron, calcium, phosphorus and potassium), fibre, carbohydrates and fatty acids (Pareek *et al.*, 2011; Anuragi *et al.*, 2016; Egydio and Santos, 2016).

The remaining parts of the fruit, such as the peel, columella and seeds, are not used for daily purposes and are mainly discarded, so they can be used to obtain secondary metabolites. *Annona* spp. seeds constitute about 23–40% of the total weight of the fruit. They are an important source of oil rich in fatty acids, amino acids and bioactive compounds such as acetogenins (ACGs) (Pareek *et al.*, 2011; Anuragi *et al.*, 2016; Egydio and Santos, 2016). In addition, the different parts of the *Annona* spp. tree are commonly prepared by concentration, maceration or infusion and used for medicinal purposes to treat health problems such as fever, diarrhoea, dysentery, haematuria, urethritis, asthma, rheumatism, and parasitic and hepatic diseases. The extracts also exert antispasmodic, laxative, anti-emetic, antisudorific, antitussive, anticold and antidepressant effects (Egydio and Santos, 2016; Anaya-Esparza *et al.*, 2020). These same extracts have a high potential for pharmaceutical applications because they contain various bioactive compounds, such as alkaloids, essential oils, phenolic compounds and ACGs, which have been attributed to a range of pharmacological activities, including antibacterial, antifungal, anticancer, antidepressant and antioxidant effects (Kazman *et al.*, 2022; López-Romero *et al.*, 2022; Aguilar-Hernández *et al.*, 2023). The aim of this chapter is to investigate the nutritional composition of the four *Annona* spp. mentioned above and their main traditional and pharmacological uses, as well as their composition, analysis of the different ACGs present in their seeds, and their main biological activities and mechanisms of action.

* E-mail: emontalvo@ittepic.edu.mx


DOI: 10.1079/9781800627628.0003

## 3.2 General Description of the Fruits

The *A. muricata* (soursop) tree is 5–10 m tall with a 15–83 cm diameter trunk, low branches and edible fruit. The fruit are irregularly shaped (typically heart-shaped), dark green when unripe and pale green when ripe. The fruit are 20–25 cm long and 10–12 cm in diameter, weigh 0.25–10 kg, and have 50–200 seeds or more. In addition, the fruit consists of a 67% edible fraction. The remaining 33% is made up of the peel (20%), seeds (9–10%) and columella (4%) (Pinto *et al.*, 2005; Badrie and Schauss, 2010; Aguilar-Hernández *et al.*, 2020, 2023; López-Romero *et al.*, 2022).

The fruit of *A. cherimola* Mill. (cherimoya) is usually heart-shaped, oval and conical with dimensions of approximately 7.5–12.5 cm and weights of 90–700 g. The surface of the fruit is covered with small conical protuberances on the carpels, which are greenish-yellow when ripe. The flesh is edible, white and has numerous seeds (21–41 seeds per fruit), approximately 1.5–2 cm long and 1 cm wide (Pinto *et al.*, 2005; Kazman *et al.*, 2022).

The fruit of *A. squamosa* L. (sugar apple) is rounded, heart-shaped, ovate or conical, approximately 6–10 cm long, weighing 120–330 g, and containing 35–45 black seeds that are 1.5–2 cm long and 0.6–0.8 cm wide. These fruit grow on 3–8 m high trees, with large branches with brownish or light brown bark and thin leaves (Pinto *et al.*, 2005; Padmanabhan and Paliyath, 2016; Kazman *et al.*, 2022).

The tree of *A. reticulata* L. (custard apple) reaches heights of approximately 6–7.5 m. Its fruit are edible, heart-shaped, rough and yellow when unripe and yellowish-red when ripe. They have abundant, smooth, blackish seeds, and are of 10–12 cm in size and weigh 0.1–1 kg (Pinto *et al.*, 2005; Jamkhande and Wattamwar, 2015). The edible fraction constitutes 45% of the total weight of the fruit, which, when ripe, is creamy-yellow, sweet and low in acidity (Pareek *et al.*, 2011).

## 3.3 Nutritional Composition

The nutritional composition of *A. muricata*, *A. cherimola*, *A. squamosa* and *A. reticulata* vary considerably depending on their stage of development, edaphic and climatic conditions, time of year, soil type, environmental practices and postharvest conditions (Albuquerque *et al.*, 2016; Anuragi *et al.*, 2016). The nutritional composition of the edible fraction of these four *Annona* spp. is presented in Table 3.1. The pulp has a low caloric content (56–114 Kcal 100 $g^{-1}$ of pulp) and mainly comprises water (65–85.1%), carbohydrates (11.7–26.2%), dietary fibre (0.42–5.32%), proteins (0.43–3.84%) and lipids (0.1–1.25%). In addition, the pulp of these fruits is a rich source of vitamin C and minerals such as potassium, phosphorus and calcium.

### 3.3.1 Minerals

Macrominerals, also known as macroelements, are minerals that the human body needs and are part of the biological structure of humans, playing an essential role in almost all metabolic processes (Ali, 2023). Microelements, which occur at trace levels, are as important as macroelements, which are usually present in higher concentrations. Biochemical processes can only occur with these essential elements and at appropriate proportions. Although minerals comprise only 4–6% of our body weight, they are crucial to our daily diet. Some minerals are required in quantities greater than 100 mg $day^{-1}$, while microelements are needed in quantities less than 100 mg $day^{-1}$ (Bhattacherjee and Tandon, 2021; Ali, 2023; Bhattacherjee, 2023). Macro- and microelements are classified as follows:

- Macroelements: calcium ($Ca^{2+}$), phosphorus ($PO_4^{3-}$), magnesium ($Mg^{2+}$), sodium ($Na^+$), potassium ($K^+$) and chloride ($Cl^-$).
- Microelements (trace elements): zinc ($Zn^{2+}$), manganese ($Mn^{2+}$), iron ($Fe^{2+/3+}$), iodine ($I^-$), selenium ($Se^{2-}$), cobalt ($Co^{2+/3+}$), chromium ($Cr^{3+}$), copper ($Cu^{1+/2+}$), fluorine ($F^-$) and molybdenum (Mo), among others.

*Annona* spp. fruits are a rich source of minerals such as potassium, phosphorus and calcium, although magnesium and iron are also found in lower concentrations (Table 3.1). Potassium content varies depending on the species, with *A. reticulata* having the highest content (382 mg 100 $g^{-1}$ of edible fraction), followed by *A. cherimola* (287 mg 100 $g^{-1}$), *A. muricata* (278 mg 100 $g^{-1}$) and *A. squamosa* (247 mg 100 $g^{-1}$).

**Table 3.1.** Nutritional composition of *Annona muricata* L., *A. squamosa* L., *A. cherimola* Mill. and *A. reticulata* L. fruits (100 $g^{-1}$ of edible pulp). Data from Badrie and Schauss (2010); Pareek *et al.* (2011); Albuquerque *et al.* (2016); Anuragi *et al.* (2016); Egydio and Santos (2016); Padmanabhan and Paliyath (2016); Chang *et al.* (2018) and Neta *et al.* (2019).

| Nutritional composition | *A. muricata* L. | *A. squamosa* L. | *A. cherimola* Mill. | *A. reticulata* L. |
|---|---|---|---|---|
| Energy (kcal) | 64–71 | 80–114 | 56–102 | 75 |
| Water (g) | 81–85.1 | 65–81 | 71.5–79.26 | 73–75.8 |
| Total sugars (g) | 7.59–8.92 | 16.62 | 14.41 | 14.89 |
| Carbohydrates (g) | 13.41–18.2 | 14–26.2 | 11.7–22 | 18.7–22.71 |
| Proteins (g) | 0.43–1.15 | 0.7–3.84 | 1.36–1.96 | 1.85 |
| Lipids (g) | 0.3–0.97 | 0.1–1.25 | 0.103–0.5 | 0.35 |
| Dietary fibre (g) | 0.42–2.34 | 0.9–3.28 | 1.0–5.32 | 2.18–2.55 |
| Ash (g) | 0.4–1.69 | 0.6–1.36 | 0.37–1.56 | 0.95–1.48 |
| Iron (mg) | 0.6–0.82 | 0.28–1.8 | 0.4–0.8 | 1.0 |
| Magnesium (mg) | 21 | 21 | 17 | 18 |
| Calcium (mg) | 8.8–15 | 17–53.7 | 21.7–34 | 24 |
| Phosphorus (mg) | 27.1–29 | 23.6–55.3 | 30.2–47 | 26 |
| Potassium (mg) | 278 | 247 | 287 | 382 |
| Vitamin A (mg) | 8.9–20 | 0.004–0.007 | – | Trace |
| Vitamin $B_1$ (mg) | 0.06–0.11 | 0.10–0.11 | 0.06–0.12 | 0.07 |
| Vitamin $B_5$ (mg) | 0.89–1.52 | 0.65–1.28 | 0.6–1.02 | 0.7 |
| Vitamin $B_{12}$ (mg) | 0.05–0.12 | 0.057–0.167 | 0.11–0.14 | 0.12 |
| Vitamin C (mg) | 16.4–39.24 | 17.65–42.2 | 0.115–19.60 | 23.36–30 |
| Vitamin E (mg) | – | – | 0.071–0.23 | – |
| Total phenolic content (mg) | 0.16–160.28 | 0.3 | 0.35–12 | 0.41 |

Potassium is the major cation inside cells and is vital in intra- and extracellular water distribution. It is involved in maintaining membrane potential, has a role in cellular osmolarity and acid–base balance, and favours the transmission of neuronal stimuli and the regulation of cardiac and muscular function (Olechno *et al.*, 2021; Yamada and Inaba, 2021; Ali, 2023). Phosphorus is another macronutrient present in *A. squamosa* (23.6–55.3 mg 100 $g^{-1}$ of edible fraction), *A. cherimola* (30.2–47 mg 100 $g^{-1}$ of edible fraction), *A. muricata* (27.1–29 mg 100 $g^{-1}$) and *A. reticulata* (26 mg 100 $g^{-1}$) (Table 3.1). This component is involved in various physiological processes, and approximately 85% of this element is found in bones and teeth, 14% in soft tissues and only 1% in extracellular fluid. Other functions include regulating acid–base balance, cell signalling and the cell energy cycle (EFSA, 2019; Olechno *et al.*, 2021). The daily requirement for phosphorus is 550 mg, and *Annona* spp. fruits provide between 4.29% and 10.05% 100 $g^{-1}$ of fresh pulp (EFSA, 2019).

*Annona* spp. fruits are also a source of calcium, which plays a crucial role in vital human functions, as it is essential for the formation, growth and preservation of bones and teeth and the stability of the cellular cytoskeleton. Calcium is the most abundant cation in the body, present at approximately 1200 g, representing about 2% of body weight. This component is also responsible for transmitting nerve signals, muscle contraction, blood coagulation, intracellular signals (found inside the cell) and stabilization of cell membranes. Bone tissue is responsible for most of the calcium transport processes in the body (Beto, 2015; Ciosek *et al.*, 2021). Calcium has been found in concentrations ranging from 8.8 to 53.7 mg 100 $g^{-1}$ of edible fraction of the fruits (Table 3.1), with lower concentrations in *A. muricata* (8.8–15 mg) and higher concentrations in *A. squamosa* (17–53.7 mg). However, the mineral content of the fruits will always vary depending on soil nutrients, type of fruit, cultivation method, environmental pollution, production processes, altitude above sea level and climatic conditions, among other factors (Olechno *et al.*, 2021).

### 3.3.2 Sugars

The content, types and proportion of sugars are vital in determining fruit flavour. Sugars are the product of starch hydrolysis; among the main sugars in fruits are sucrose, fructose, glucose and sorbitol, and their composition and concentration are essential factors in determining fruit quality and flavour. In plants, sugars serve crucial functions as structural and storage substances, respiratory substrates and intermediate metabolites involved in various biochemical processes that maintain normal metabolic functions. In addition, sugars act as signalling molecules that govern flowering, fruit ripening and the response to abiotic stress (Du *et al.*, 2024). *Annona* spp. fruits have a high content (58% of dry mass) of sugars including fructose (3.5%), glucose (2.97–5.1%), sucrose (1.02–3.4%) and oligosaccharides (1.2–2.5%), which are part of the main carbohydrates (Badrie and Schauss, 2010; Pareek *et al.*, 2011; Padmanabhan and Paliyath, 2016). The sugar content varies depending on the species (7.59–16.62 g 100 $g^{-1}$ of edible fraction). The lowest total sugar content is found in *A. muricata* (7.59–8.92 g 100 $g^{-1}$), with higher content in *A. cherimola* (14.41 g 100 $g^{-1}$), *A. reticulata* (14.89 g 100 $g^{-1}$) and *A. squamosa* (16.62 g 100 $g^{-1}$) (Table 3.1). In most fruits, including *Annona* spp. fruits, the sugar content varies depending on the stage of maturity, and this content becomes higher as the fruit advanced towards maturity. Fruit maturity can be observed with the decomposition of starch, which produces sugar and organic acid and initiates fruit ripening (Pareek *et al.*, 2011).

### 3.3.3 Dietary fibre

Dietary fibre is the edible portion of plants or their carbohydrate analogues that resists absorption and digestion in the smaller intestine and complete or partial fermentation in the large intestine (AACC, 2001). This definition includes polysaccharides, oligosaccharides, lignin and other plant substances. Dietary fibre promotes beneficial physiological effects, including laxation, improved intestinal transit time, glycaemic control, cholesterol reduction, weight maintenance, reduced caloric intake, increased satiety and increased mineral absorption, and some compounds serve as a prebiotic for the intestinal microbiota (Miller, 2014; Cruz-Requena *et al.*, 2019).

*Annona* spp. fruits have a significant content of dietary fibre per 100 g of edible fraction, which varies in each species; *A. muricata* (0.42–2.34 g) and *A. reticulata* (2.18–2.55 g) have the lowest content, while *A. squamosa* (0.9–3.28 g) and *A. cherimola* (1.0–5.32 g) have the highest concentrations. According to the US Food and Drug Administration, the required daily intake of dietary fibre is 25 g for a 2000 kcal $day^{-1}$ diet or 30 g for a 2500 kcal $day^{-1}$ diet. In the USA, information on the amount of dietary fibre per serving must appear on the nutrition facts panel of food products unless the product contains less than 1 g of fibre (FDA, 2018; Cruz-Requena *et al.*, 2019). Permitted claims in the USA are as follows:

- Source of fibre: products containing at least 10% of the recommended daily value or 2.5 g of fibre.
- High fibre: products containing at least 20% or 5 g of fibre.

Thus, according to these statements, the fruits of the *Annona* spp. discussed in this chapter can be considered an essential source with high dietary fibre content, contributing about 1.68–21.28 g 100 $g^{-1}$ of fruit for a 2000 kcal $day^{-1}$ diet or 1.4–17.73 g 100 $g^{-1}$ for a 2500 kcal $day^{-1}$ diet.

### 3.3.4 Proteins and fats

Although the edible fraction of *Annona* spp. fruits has a low protein and fat content, the seeds have a high content of these components. The fruits have a fat content of around 0.3–1.25 g 100 $g^{-1}$ of edible fraction (see Table 3.1), with *A. muricata* (0.3–0.97 g 100 $g^{-1}$) and *A. squamosa* (0.1–1.25 g 100 $g^{-1}$) having the highest concentration, while *A. cherimola* (0.1–0.5 g 100 $g^{-1}$) and *A. reticulata* (0.35 g 100 $g^{-1}$) have a lower content.

The protein content in these fruits ranged from 0.43 to 3.84 g 100 $g^{-1}$ of the edible fraction (Table 3.1), with the highest content found in *A. squamosa* (0.7–3.84 g 100 $g^{-1}$) and a lower concentration in the other species (0.43–1.96 g 100 $g^{-1}$). Proteins provide amino acids, precursors of volatile compounds formed during the ripening process, giving the typical aroma to the fruit.

In addition, they can also provide nitrogen that could be reassimilated to form nitrogenous compounds such as polyamines, which increase during ripening (Cordeiro *et al.*, 2013). Nineteen amino acids (Table 3.2) were found in *A. muricata* seeds, with a total of nine essential amino acids in concentrations of 1.9–51.5 mg kg$^{-1}$, with higher concentrations of leucine (29.3–51.5 mg kg$^{-1}$), valine (15.3–28.0 mg kg$^{-1}$), lysine (11.4–23.6 mg kg$^{-1}$), isoleucine (13.1–23.4 mg kg$^{-1}$) and phenylalanine (16.8–23.2 mg kg$^{-1}$), and lower concentrations of tryptophan (1.9–3.4 mg kg$^{-1}$), methionine (7.4–12.6 mg kg$^{-1}$), histidine (12.2–15.0 mg kg$^{-1}$) and threonine (7.5–19.6 mg kg$^{-1}$). In addition, ten non-essential amino acids were also found at concentrations of 7.7–134.5 mg kg$^{-1}$, such as glutamic acid (60.4–134.5 mg kg$^{-1}$), arginine (31.4–81.6 mg kg$^{-1}$) and aspartic acid (31.2–68.1 mg/kg) with the highest concentrations. Cysteine (7.7–16.9 mg kg$^{-1}$), tyrosine (16.6–24.5 mg kg$^{-1}$), serine (13.4–30.4 mg kg$^{-1}$), proline (15.5–36.9 mg kg$^{-1}$), alanine (18.1–38.4 mg kg$^{-1}$) and glycine (20.8–39.7 mg kg$^{-1}$) were identified at lower concentrations. Eight essential amino acids were identified in *A. squamosa* at concentrations of 2.52–3.90 mg kg$^{-1}$, with phenylalanine (3.72 mg kg$^{-1}$) and lysine (3.90 mg kg$^{-1}$) at higher concentrations, and nine non-essential amino acids at concentrations of 1.60–13.50 mg kg$^{-1}$. In *A. cherimola*, seven essential amino acids were identified (0.03–1.55 mg kg$^{-1}$), with phenylalanine having the highest concentration (1.55 mg kg$^{-1}$), while of the eight non-essential amino acids found at concentrations of 0.1–18.47 mg kg$^{-1}$, the most abundant being arginine (18.47 mg kg$^{-1}$).

Sixteen fatty acids were identified in the seed oil of *Annona* spp. fruits (Table 3.3). Nine fatty acids were classified as saturated (palmitic, lauric, behenic, lignoceric, capric, myristic, stearic, pentadecyl and margaric) and seven as unsaturated (palmitoleic, oleic, linoleic, linolenic, arachidonic, gadoleic and sapienic), of which oleic acid (39.18–56.50%) and linoleic acid (20.49–35.90%) were reported at higher concentrations in *A. muricata, A. squamosa* and *A. reticulata*, while in *A. cherimola* the most predominant fatty acid was palmitic acid (9.71–31.26%) and linoleic acid (6.60–24.66%).

**Table 3.2.** Content of amino acids found in *Annona muricata* L., *A. squamosa* L. and *A. cherimola* Mill. seeds (mg kg$^{-1}$). Data from Egydio *et al.* (2013); Abdualrahman *et al.* (2019); Villacís-Chiriboga *et al.* (2023).

| Amino acids | *A. muricata* L. | *A. squamosa* L. | *A. cherimola* Mill. |
|---|---|---|---|
| Essential amino acids | | | |
| Methionine | 7.4–12.6 | 2.52 | 0.03 |
| Threonine | 7.5–19.6 | 2.83 | 0.38 |
| Valine | 15.3–28.0 | 3.65 | – |
| Isoleucine | 13.1–23.4 | 3.50 | 0.03 |
| Leucine | 29.3–51.5 | 3.50 | 0.12 |
| Phenylalanine | 16.8–23.2 | 3.72 | 1.55 |
| Lysine | 11.4–23.6 | 3.90 | 0.22 |
| Histidine | 12.2–15.0 | 2.90 | 0.29 |
| Tryptophan | 1.9–3.4 | – | – |
| Non-essential amino acids | | | |
| Glutamine | – | – | 2.43 |
| Glutamic acid | 60.4–134.5 | 13.50 | 0.40 |
| Aspartic acid | 31.2–68.1 | 7.90 | 0.21 |
| Glycine | 20.8–39.7 | 4.26 | 0.10 |
| Serine | 13.4–30.4 | 2.70 | 0.13 |
| Alanine | 18.1–38.4 | 4.82 | 0.42 |
| Proline | 15.5–36.9 | 3.27 | – |
| Cysteine | 7.7–16.9 | 1.60 | – |
| Arginine | 31.4–81.6 | 6.08 | 18.47 |
| Tyrosine | 16.6–24.5 | 3.11 | 0.17 |

**Table 3.3.** Fatty acids found in *Annona muricata* L., *A. squamosa* L., *A. cherimola* Mill. and *A. reticulata* L. seed oils (% of total fatty acids). Data from Azam *et al.* (2005); Cordeiro *et al.* (2013); Chen *et al.* (2016); Mariod *et al.* (2017); Pinto *et al.* (2018); Abdualrahman *et al.* (2019); Panadare *et al.* (2020) and Solís-Fuentes *et al.* (2020).

| Seed oil composition | *A. muricata* L. | *A. squamosa* L. | *A. cherimola* Mill. | *A. reticulata* L. |
|---|---|---|---|---|
| Saturated fatty acids | | | | |
| Palmitic acid | 19–25.50 | 9.92–17.79 | 9.71–31.26 | 17.20 |
| Lauric acid | – | 0.08 | – | – |
| Behenic acid | 0.08–0.10 | 2.01 | 0.11–1.23 | – |
| Lignoceric acid | 0.10–0.15 | – | 0.76–9.17 | – |
| Capric acid | – | 0.17–0.30 | – | – |
| Myristic acid | 0.06 | 0.60–0.70 | – | – |
| Stearic acid | 3.3–7.90 | 4.29–10.60 | 1.73–5.13 | 7.50 |
| Pentadecylic acid | 0.40 | – | 5.49–8.84 | – |
| Margaric acid | 0.07–0.11 | | 0.40–1.05 | – |
| Unsaturated fatty acids | | | | |
| Palmitoleic acid | 1.34–17.29 | – | 0.07–0.29 | 4.20 |
| Oleic acid | 39.18–44.00 | 39.72–56.50 | 4.11–8.59 | 48.40 |
| Linoleic acid | 21.74–35.90 | 20.49–29.13 | 6.60–24.66 | 21.70 |
| Linolenic acid | 1.16 | 1.37–2.18 | 5.34–18.12 | – |
| Arachidonic acid | 0.42–1.19 | 1.06 | – | – |
| Gadoleic acid | 0.07–0.16 | – | – | – |
| Sapienic acid | – | 1.43 | – | – |

### 3.3.5 Vitamins

Vitamins are a group of organic compounds that improve the nutritional status of humans and play an essential role in many physiological functions. However, they cannot be synthesized in humans due to mutations in the L-gluconolactone oxidase gene, which catalyses the synthesis of L-ascorbic acid; thus, these compounds are essential in the human diet. However, these compounds are susceptible to degradation and their stability can be affected by many factors, including storage time, storage conditions (temperature, time, oxygen, light and soil moisture), and the nature of the food matrix, which can lead to the degradation of vitamins by oxidation and isomerization (Granger and Eck, 2018; Yang *et al.*, 2024). Generally, there are two forms of vitamins: fat soluble (A, D, E and K) and water soluble (B complex vitamins and vitamin C), with fruits being a rich source of vitamins (Bhattacherjee, 2023). *Annona* spp. fruits have a significant vitamin C content, contributing 24.5–52.75 mg 100 $g^{-1}$ of fruit of the recommended daily intake (80 mg $day^{-1}$). Vitamin C is an essential nutrient known as an antioxidant that helps prevent scurvy in humans and is involved in the primary prevention of common and complex diseases such as anaemia, the common cold, bleeding disorders, coronary heart disease, stroke and cancer. It also contributes to forming connective tissues such as teeth, blood vessels and bones, and aids in wound healing (Granger and Eck, 2018; Bhattacherjee and Tandon, 2021). This vitamin exists in two primary forms: ascorbic acid and dehydroascorbic acid. Ascorbic acid helps to reduce extracellular oxidants, increase endothelium-dependent vasodilation and reduce oxidation of low-density lipoproteins (Granger and Eck, 2018). Under undesirable conditions, vitamin C is prone to decomposition by aerobic (during processing) and anaerobic (during storage) pathways, which is dependent on oxygen, light, heat, storage temperature and time (Yang *et al.*, 2024).

The vitamin C content of fruits varies among the *Annona* spp. (Table 3.1), with the lowest content found in *A. cherimola* and *A. reticulata* (0.115–19.60 and 23.36–30 mg 100 $g^{-1}$ edible fraction, respectively), while the highest concentration was reported in *A. muricata* (16.4–39.24 mg 100 $g^{-1}$) and *A. squamosa* (17.65–42.2 mg 100 $g^{-1}$). Badrie and Schauss (2010) reported an ascorbic acid content of 29.6 mg 100 $g^{-1}$ of edible fraction in the fruit of *A. squamosa* from

Havana, Cuba, and Anuragi *et al.* (2016) found an ascorbic acid content of 39.24 mg $100^{-1}$ g of edible fraction in fruit grown in Gir Forest, Gujarat, India. Abdualrahman *et al.* (2019) showed that the pulp of *A. squamosa* harvested in the region of Darfur, Sudan, had a significant amount of ascorbic acid content (31.22 mg 100 $ml^{-1}$). The variations in ascorbic acid content in the different *Annona* spp. fruits depends on their stage of development, climatic conditions, soil type, place of cultivation and postharvest conditions (Albuquerque *et al.*, 2016; Anuragi *et al.*, 2016).

Among the four *Annona* spp. reviewed in this chapter, only *A. muricata* (8.9–20 mg 100 $g^{-1}$ of edible fraction) showed significant amounts of vitamin A, while *A. squamosa* (0.004–0.007 mg 100 $g^{-1}$) and *A. reticulata* (trace amounts) reported very low quantities (Table 3.1). Vitamin A deficiency is a public health problem in developing countries as it is an essential compound in a wide range of physiological processes, including immune function, vision, cell proliferation, intercellular communication and reproduction, and its deficiency can lead to increased morbidity and mortality from infectious diseases, xerophthalmia and night blindness (Campbell *et al.*, 2009; Debelo *et al.*, 2017). Therefore, according to the data provided in Table 3.1, *A. muricata* fruit could contribute to the recommended daily intake of vitamin A (300 μg $day^{-1}$ in children, 750 μg $day^{-1}$ in women and 900 μg $day^{-1}$ for men), providing 890–2000 μg $day^{-1}$ from 100 g of fruit (Debelo *et al.*, 2017).

Vitamin E is found in cell membranes, both intracellularly and extracellularly. Its principal function is to prevent membrane damage and regulate the redox balance in the body. In addition, it has essential health benefits, including anti-allergic, anticancer, anticardiovascular, antidiabetic, antilipidemic, antihypertensive, anti-inflammatory and neuroprotective effects (Al-Sowayan and Almarzougi, 2024). However, the presence of vitamin E in *Annona* spp. fruits has yet to be reported, except in *A. cherimola* pulp, where concentrations of 0.071–0.23 mg 100 $g^{-1}$ of edible fraction were found (Table 3.1). Likewise, vitamin $B_1$ and $B_{12}$ from *Annona* spp. fruits were reported in relatively low concentrations, ranging from 0.05 to 0.167 mg 100 $g^{-1}$ of edible fraction for both vitamins. Vitamin $B_5$ concentrations were found in the range of 0.6–1.28 mg 100 $g^{-1}$ of the edible fraction for *A. cherimola*, *A. reticulata* and *A. squamosa*; however, in *A. muricata*, concentrations of 0.89–1.52 mg 100 $g^{-1}$ of edible fraction were reported (Pareek *et al.*, 2011).

## 3.4 Acetogenins

ACGs are the main bioactive compounds isolated from different parts of the plant or fruit of members of the family Annonaceae. The seeds have the highest content, and more ACGs have been identified in seeds (Durán-Ruiz *et al.*, 2019; Aguilar-Hernández *et al.*, 2020; Neske *et al.*, 2020; Aguilar-Hernández *et al.*, 2022; López-Romero *et al.*, 2022). ACGs are lipophilic compounds consisting of a long, linear aliphatic chain with 35–37 carbon atoms, one, two or three tetrahydrofuran (THF) or tetrahydropyran rings in its central region, a terminal α-β-unsaturated γ-lactone, and various functional groups such as hydroxyl groups, ketones or epoxides (Bermejo *et al.*, 2005; Bonneau *et al.*, 2016; Aguilar-Hernández *et al.*, 2022).

### 3.4.1 Acetogenins found in the seeds of *A. muricata*, *A. squamosa*, *A. cherimola* and *A. reticulata*

A broad spectrum of ACGs have been identified from methanol, ethanol, aqueous and other organic solvent extracts from the seeds of *A. muricata*, *A. squamosa*, *A. cherimola* and *A. reticulata* and a number of biological activities have been attributed to them. In total, 127 ACGs have been identified in the seeds of the fruits of these species (Table 3.4). Thirty-five ACGs have been reported in *A. muricata*, 68 in *A. squamosa*, 15 in *A. cherimola* and nine in *A. reticulata*. The first ACGs isolated from these seeds were identified in the 1990s from *A. muricata*, where the presence of murisolin, 15-oleylsolamin, corossolone (Myint *et al.*, 1991), muricin G, muricatin C (Cortes *et al.*, 1991a,b,c), solamin (Myint *et al.*, 1991), diepomuricanin A (Laprévote *et al.*, 1992), corepoxylone (Gromek *et al.*, 1993), gigantetrocin B, muricatetrocins A + B and muricatin A (Rieser *et al.*, 1993). In *A. squamosa*, the first detection of ACGs was reported in the late 1980s and early 1990s with the pioneer Fujimoto *et al.* (1988)

**Table 3.4.** Acetogenins isolated from the seeds of *Annona muricata* L., *A. squamosa* L., *A. cherimola* Mill. and *Annona reticulata* L.

| Species | Acetogenins | Molecular formula | Reference(s) |
|---|---|---|---|
| *Annona muricata* L. | Murihexol | $C_{35}H_{66}O_8$ | Yu *et al.* (1998) |
| | Muricatenol | $C_{37}H_{68}O_6$ | Li *et al.* (2000) |
| | Diepomuricanin A | $C_{35}H_{62}O_4$ | Laprévote *et al.* (1992) |
| | Corepoxylone | $C_{35}H_{60}O_5$ | Gromek *et al.* (1993) |
| | Solamin | $C_{35}H_{64}O_5$ | Myint *et al.* (1991) |
| | 15-Palmitoylsolamin | $C_{51}H_{94}O_6$ | Gleye *et al.* (1998) |
| | 15-Oleylsolamin | $C_{53}H_{96}O_6$ | Gleye *et al.* (1998) |
| | Murisolin | $C_{35}H_{64}O_6$ | Cortes *et al.* (1991a,b,c); Gleye *et al.* (1998) |
| | Corossolin | $C_{35}H_{64}O_6$ | Cortes *et al.* (1991a,b,c); Gleye *et al.* (1998); Kim *et al.* (2001a,b) |
| | Corossolone | $C_{35}H_{62}O_6$ | |
| | Muricin G or annomocherin | $C_{35}H_{62}O_7$ | Cortes *et al.* (1991a,b,c); Li *et al.* (1994); Kim *et al.* (2001a,b); |
| | Muricatin C | $C_{35}H_{62}O_8$ | Cortes *et al.* (1991a,b,c); Li *et al.* (1994); Gui and Yu (1996); Kim *et al.* (2001a,b) |
| | Muricatatin C | $C_{35}H_{66}O_8$ | |
| | Muricatin A (muricatatin A) | $C_{35}H_{64}O_8$ | Rieser *et al.* (1993); Li *et al.* (1994); Gui and Yu (1996) |
| | Muricatin B (muricatatin B) | $C_{35}H_{64}O_8$ | Li *et al.* (1994) |
| | Muricatalicin | $C_{35}H_{64}O_8$ | Gui and Yu (1996) |
| | Gigantetrocin B | $C_{35}H_{64}O_7$ | Rieser *et al.* (1993) |
| | Muricatetrocins A + B | $C_{35}H_{64}O_7$ | |
| | Muricatalin | $C_{35}H_{64}O_8$ | Ghi *et al.* (1995) |
| | Annopentocin A + B | $C_{35}H_{64}O_8$ | Zeng *et al.* (1996a,b,c) |
| | Annomuricinone D | $C_{35}H_{64}O_8$ | |
| | Annopentocin C | $C_{35}H_{64}O_8$ | |
| | Muricin A | $C_{35}H_{64}O_7$ | Chang and Wu (2001) |
| | Longifolicin | $C_{37}H_{68}O_6$ | |
| | Muricin B | $C_{35}H_{64}O_7$ | |
| | Muricin C | $C_{35}H_{64}O_7$ | |
| | Muricin D | $C_{33}H_{60}O_7$ | |
| | Muricin E | $C_{33}H_{60}O_7$ | |
| | Muricin F | $C_{35}H_{62}O_7$ | |
| | Annocatalin | $C_{35}H_{64}O_7$ | Liaw *et al.* (2002) |
| | Muricin H | $C_{35}H_{64}O_6$ | |
| | Muricin I | $C_{37}H_{66}O_6$ | |
| | *cis*-Annomontacin | $C_{37}H_{66}O_7$ | Chang *et al.* (2003) |
| | Annocatacin A | $C_{35}H_{62}O_6$ | |
| | Annocatacin B | $C_{35}H_{62}O_6$ | |
| | Robustocin | $C_{35}H_{62}O_5$ | Gleye *et al.* (2000a,b) |
| | Annonamuricin A, B, C, D | $C_{35}H_{64}O_8$ | Sun *et al.* (2017) |
| | Reticulatamol | $C_{35}H_{66}O_3$ | Tam *et al.* (1994b) |
| *Annona squamosa* L. | Neo-epoxyrolin | $C_{36}H_{64}O_3$ | Li *et al.* (2010) |
| | Epoxyrolin B | $C_{36}H_{66}O_3$ | |
| | Dieposabadelin | $C_{35}H_{62}O_4$ | Bajin ba Ndob *et al.* (2009) |
| | Squamocenin A | $C_{35}H_{60}O_4$ | |
| | Dotistenin | $C_{35}H_{62}O_5$ | |
| | Lepirenin | $C_{35}H_{62}O_5$ | |

*Continued*

**Table 3.4.** Continued.

| Species | Acetogenins | Molecular formula | Reference(s) |
|---|---|---|---|
| | *cis*-Annotemoyin 1 | $C_{35}H_{64}O_5$ | Liaw *et al.* (2008) |
| | Squadiolin A | $C_{37}H_{68}O_8$ | |
| | Squadiolin B | $C_{37}H_{68}O_8$ | |
| | Squadiolin C | $C_{35}H_{64}O_7$ | |
| | Squafosacin B | $C_{37}H_{68}O_7$ | |
| | Squafosacin C | $C_{35}H_{64}O_6$ | |
| | Squafosacin F | $C_{35}H_{62}O_5$ | |
| | Squafosacin G | $C_{37}H_{68}O_5$ | |
| | Squamocenin B | $C_{35}H_{62}O_5$ | Yu *et al.* (2005) |
| | Squamoxinone D | $C_{37}H_{66}O_6$ | Miao *et al.* (2016b) |
| | Squamocin I | $C_{37}H_{66}O_7$ | |
| | Squamocin II | $C_{37}H_{66}O_7$ | |
| | Squamocin III | $C_{37}H_{66}O_7$ | |
| | Squamoxinone E | $C_{37}H_{66}O_7$ | Miao *et al.* (2016a) |
| | Squamocin IV | $C_{37}H_{66}O_7$ | |
| | Squamocin V | $C_{35}H_{64}O_6$ | |
| | Annonacin A | $C_{35}H_{64}O_7$ | Lieb *et al.* (1990) |
| | Squamosten B | $C_{35}H_{64}O_7$ | Ma *et al.* (2019) |
| | Muricin O | $C_{35}H_{64}O_7$ | |
| | Squamotin A | $C_{37}H_{66}O_7$ | |
| | Squamotin B | $C_{37}H_{66}O_6$ | |
| | Squamotin C | $C_{37}H_{66}O_6$ | |
| | Squamotin D | $C_{37}H_{66}O_8$ | |
| | Annosquatin IV | $C_{37}H_{66}O_8$ | |
| | Annosquatin V | $C_{37}H_{66}O_7$ | |
| | Neoannonin | $C_{35}H_{62}O_6$ | Kawazu *et al.* (1989) |
| | Squamocin I | $C_{35}H_{62}O_6$ | Fujimoto *et al.* (1994) |
| | Squamocin K | $C_{35}H_{62}O_6$ | |
| | Squamocin N | $C_{37}H_{66}O_6$ | |
| | Squamostatin E | $C_{37}H_{66}O_7$ | |
| | Annosquacin A | $C_{35}H_{62}O_6$ | Chen *et al.* (2012) |
| | Annosquacin B | $C_{37}H_{66}O_6$ | |
| | Annosquacin C | $C_{37}H_{66}O_7$ | |
| | Annosquacin D | $C_{37}H_{66}O_6$ | |
| | Annosquatin A | $C_{37}H_{66}O_7$ | |
| | Annosquacin B | $C_{37}H_{66}O_8$ | |
| | Annosquacin I | $C_{37}H_{66}O_7$ | Chen *et al.* (2011) |
| | Annosquacin II | $C_{37}H_{66}O_8$ | |
| | Annosquatin I | $C_{37}H_{66}O_8$ | |
| | Annonareticin | $C_{37}H_{66}O_7$ | Li *et al.* (2010) |
| | Squamostatin C | $C_{37}H_{66}O_8$ | |
| | 4-Hydroxy-squamocin Y | $C_{37}H_{66}O_8$ | Li *et al.* (2017) |
| | 28-Hydroxy-squamocin L | $C_{37}H_{66}O_7$ | |
| | Squamostatin A | $C_{37}H_{66}O_8$ | Fujimoto *et al.* (1990) |
| | Squamocin | $C_{37}H_{66}O_7$ | Fujimoto *et al.* (1988) |
| | Annonin III | $C_{35}H_{66}O_7$ | Londershausen *et al.* (1991) |
| | Squamocin B | $C_{35}H_{62}O_7$ | Sahai *et al.* (1994) |
| | Squamocin D | $C_{37}H_{66}O_7$ | |
| | Squamocin F | $C_{37}H_{66}O_7$ | |
| | Squamocin O1 | $C_{37}H_{66}O_8$ | Araya *et al.* (2002) |
| | Squamocin O2 | $C_{37}H_{66}O_8$ | |
| | Squamocin P | $C_{37}H_{66}O_8$ | Ma *et al.* (2017) |
| | Annosquatin III | $C_{37}H_{66}O_7$ | |

*Continued*

**Table 3.4.** Continued.

| Species | Acetogenins | Molecular formula | Reference(s) |
|---|---|---|---|
| | Annonsilin A | $C_{37}H_{66}O_7$ | Yang *et al.* (1995) |
| | Squamostatin D | $C_{37}H_{66}O_7$ | Hirayama *et al.* (1993) |
| | Squamostanin A | $C_{37}H_{66}O_8$ | Yang *et al.* (2009) |
| | Squamostanin B | $C_{37}H_{66}O_8$ | |
| | Squamostanin C | $C_{37}H_{66}O_8$ | |
| | Squamostanin D | $C_{37}H_{66}O_8$ | |
| | Annotemoyin L | $C_{35}H_{64}O_5$ | Ma *et al.* (2024) |
| | Annotemoyin Y | $C_{35}H_{64}O_5$ | |
| | Annotemoyin X | $C_{35}H_{64}O_5$ | |
| *Annona cherimola* Mill. | Cherimolin I | $C_{37}H_{66}O_8$ | Cortes *et al.* (1993) |
| | Cherimolin II | $C_{37}H_{66}O_8$ | |
| | Squamostatin A | $C_{37}H_{66}O_8$ | |
| | Otivarin | $C_{37}H_{68}O_8$ | |
| | Jetein | $C_{35}H_{66}O_7$ | |
| | Laherradurin | $C_{37}H_{68}O_7$ | Ríos *et al.* (1989) |
| | Itrabin | $C_{35}H_{64}O_7$ | Cortes *et al.* (1991a,b) |
| | Molvizarin | $C_{35}H_{62}O_7$ | |
| | Motrilin | $C_{37}H_{66}O_7$ | |
| | Tucumanin | $C_{37}H_{68}O_7$ | Barrachina *et al.* (2004) |
| | Annocherimolin | $C_{37}H_{66}O_7$ | Kim *et al.* (2001a) |
| | Annomolin | $C_{35}H_{64}O_7$ | Kim *et al.* (2001a) |
| | Annocherinone | $C_{35}H_{62}O_7$ | Woo *et al.* (1999) |
| | Annocherin | $C_{35}H_{62}O_7$ | |
| | Annomolon A | $C_{35}H_{62}O_7$ | Son *et al.* (2003) |
| | Annomolon B | $C_{35}H_{62}O_8$ | |
| *Annona reticulata* L. | Reticulatamone | $C_{35}H_{64}O_3$ | Tam *et al.* (1995) |
| | Reticulatain 1 | $C_{37}H_{68}O_5$ | |
| | Reticulatain 2 | $C_{37}H_{68}O_5$ | |
| | Dieporeticanin 1 | $C_{37}H_{66}O_4$ | Tam *et al.* (1994a) |
| | Dieporeticanin 2 | $C_{37}H_{66}O_4$ | |
| | Dieporeticenin | $C_{37}H_{64}O_4$ | |
| | Isomurisolenin | $C_{35}H_{62}O_6$ | Chang *et al.* (1998) |
| | 10-OH-asimicin | $C_{37}H_{66}O_8$ | Araya *et al.* (1994a,b) |
| | Squamostolide | $C_{22}H_{36}O_5$ | Xie *et al.* (2003) |

reporting the presence of squamocin in the seeds of this fruit; the presence of neoannonin (Kawazu *et al.*, 1989), annonacin A (Lieb *et al.*, 1990), squamostatin A (Fujimoto *et al.*, 1988) and annonin III (Londershausen *et al.*, 1991) were also reported. The first ACG identification in *A. cherimola* was made by Ríos *et al.* (1989), who reported the presence of laherradurin in the fruit seeds. A year later, Cortes *et al.* (1991a,b) also identified the presence of itrabin, motrilin and molvizarin. These same authors subsequently reported the presence of cherimolin I and II, squamostatin A, otivarin and jetein (Cortes *et al.*, 1993). The first ACGs reported in *A. reticulata* seeds were dieporeticanin 1 and 2, dieporeticenin, reticulatain 1 and 2, reticulatamone (Tam *et al.*, 1994a, 1995), 10-OH-asimicin (Araya *et al.*, 1994a,b), isomurisolenin (Chang *et al.*, 1998) and squamostolide (Xie *et al.*, 2003).

Due to the broad spectrum of ACGs found in all existing Annonaceae species and the different structures that have been identified, these compounds have awakened a critical interest at the pharmaceutical level as several biological activities have been attributed to them, with the antimicrobial, antiproliferative and cytotoxic properties being the most studied (Zafra-Polo *et al.*, 1998; Alali *et al.*, 1999; Liaw *et al.*, 2016; Miao *et al.*, 2016a; Ma *et al.*, 2019; López-Romero *et al.*, 2022; Aguilar-Hernández *et al.*, 2023). The main mechanisms of action of ACGs are discussed below.

### 3.4.2 General mechanism of action of acetogenins

Different mechanisms have been proposed for ACGs, including their effect on various biological activities. ACGs have shown potency in inhibiting NADH ubiquinone oxide reductase of mitochondrial complex I in the electron transporter chain and NADH oxidase of the plasma membrane, causing ATP reduction in cells and leading to apoptosis (Bermejo *et al.*, 2005). Inhibition of these enzymes alters the flow of negatively charged ions through the respiratory chain, inhibiting the proton cascades that affect cellular respiration and directly causing cell death through autophagy, apoptosis or cell-cycle disruption (Abe *et al.*, 2008; Pinto *et al.*, 2017).

However, it has been reported that the biological effects of ACGs depend mainly on their chemical structure, which possesses the following characteristics: three OH groups flanking the THF system and a THF ring linked to the lactonic ring by an alkyl spacer (whose optimal length is 13 carbon atoms), which provides greater inhibitory potency (Abe *et al.*, 2005, 2008). In this regard, Shimada *et al.* (1998a,b) proposed that the THF rings serve as a hydrophilic anchor to the membrane, enhancing the alignment and configuration of the functional groups of the ACGs, and the terminal lactone is considered a component of the spacer domain, which is the segment of the molecule that is expected to directly couple with the enzyme active site in mitochondrial complex I. This interaction facilitates lateral diffusion within the inner mitochondrial membrane.

Another mechanism involving the THF segments of ACGs and their adjacent OH groups is independent interactions with bivalent cations such as $Ca^{2+}$ and $Mg^{2+}$. The bivalent cations can form complexes with the ACG molecules, allowing the THF ring cores and neighbouring OH groups to bind to the ions and form chelating complexes. The creation of chelating complexes involving ACGs and $Ca^{2+}$ can alter the intracellular calcium balance by sequestering these ions within ACG–$Ca^{2+}$ complexes. This process facilitates the movement of ions across the mitochondrial membrane, leading to elevated $Ca^{2+}$ levels within the cell, specifically within the mitochondria, resulting in a decrease in mitochondrial membrane potential and the initiation of pro-apoptotic signalling (Liaw *et al.*, 2011; Juang *et al.*, 2016).

Other studies have suggested that the bioactivity of ACGs also depends on their purity, the concentration evaluated and their structure; for example, ACGs with two THF rings are more active than those with a single THF ring as inhibitors of the mitochondrial respiratory chain (Neske *et al.*, 2020; Aguilar-Hernández *et al.*, 2023). Therefore, in the following section, we will analyse the main biological activities of ACGs from the seeds of the fruits studied in this chapter.

### 3.4.3 Main biological activities of acetogenins from the seeds of *A. muricata*, *A. squamosa*, *A. cherimola* and *A. reticulata*

Various investigations have reported that the ACGs of different *Annona* spp. possess numerous biological properties, including anti-inflammatory, antioxidant, antiprotozoal, antipyretic, anti-ulcer, antihyperglycemic, antimalarial, antidiarrhoeal, antihelminthic and antileishmanial. This section will address the antimicrobial and antiproliferative/cytotoxic activities of crude extracts and fractions of ACGs isolated from the *Annona* spp. fruits discussed in this chapter.

#### *3.4.3.1 Antimicrobial activity of ACGs*

The antimicrobial effect of ACGs has been little studied, but here we present the most relevant studies on the effect of these compounds as antibacterials (against Gram-positive and Gram-negative bacteria) and antifungals.

A study by Rahman *et al.* (2005) showed that the ACGs annotemoyin 1 and 2 from *A. squamosa* seeds showed a more significant effect (13 mm inhibition) against *Aspergillus flavus* compared with nystatin (8 mm inhibition) using concentrations of 200 μg per disc. Dang *et al.* (2011) conducted a study using a methanolic extract of *A. squamosa* seeds consisting of squamocin, squamocin G and squamostatin A. These authors demonstrated that the ACGs present in this extract (0.006–0.048 μg $ml^{-1}$) showed potent antifungal activity *in vitro* and *in vivo* against *Phytophthora infestans*, which causes tomato late blight. In addition, squamostatin A effectively controlled the development of wheat leaf rust caused by *Puccinia recondita*. Taechowisan *et al.* (2016) reported that squamocin showed a minimum inhibitory concentration (MIC) of 32 μg $ml^{-1}$ and a minimum bactericidal concentration (MBC) of 128 μg $ml^{-1}$ against *Salmonella typhi* and *Bacillus cereus* with inhibition zone diameters of 27.5 and 28.2 mm, respectively. Aqueous and methanolic seed extracts of *A. squamosa* have been reported to show activity against *Staphylococcus aureus* with an MIC of 50 mg $ml^{-1}$ and an MBC of 100 mg $ml^{-1}$ (Nasser *et al.*, 2017). Likewise, a study by López-Romero *et al.* (2022) demonstrated that isolated ACGs from *A. muricata* seeds had a significant antifungal effect against *Candida albicans*, *C. glabrata*, *C. tropicalis* and *C. krusei* at concentrations of 0.0125–4 mg $ml^{-1}$. However, the most significant effect was observed at 0.8 mg $ml^{-1}$, and the order of inhibition was as follows: *C. albicans* (15.50 mm) > *C. tropicalis* (14 mm) > *C. krusei* (13.50 mm) > *C. glabrata* (8.50 mm). These authors demonstrated that ACGs could be a potent antifungal inhibitor compared with ketoconazole (500 mg $ml^{-1}$), which only inhibited *C. albicans* (14.75 mm). The effect of ACGs against this type of yeast is attributed to the inhibition of NADH dehydrogenase or NADH:ubiquinone-6-oxidoreductase type II enzymes present in yeast and fungi. Thus, cell death is induced by blocking these enzymes, although this depends on the type of microorganism. In addition, microbial inhibition can be caused by adhesion of the ACGs to the cell surface or by their diffusion inside the cell, preventing the formation of nucleic acids (DNA and RNA) and inactivating the ability of the genetic material to function (Rizqilah *et al.*, 2019).

Another study by Aguilar-Hernández *et al.* (2023) demonstrated that ACGs isolated from *A. muricata* seeds (used at 0.0125–4 mg $ml^{-1}$) showed significant inhibitory effects (7–18 mm) when tested against Gram-negative bacteria (*Streptococcus mitis*, *Listeria monocytogenes*, *Staphylococcus aureus*, *Enterococcus faecalis*, *Streptococcus mutans* and *Streptococcus salivaris*) and Gram-positive (*Aeromonas hydrophila*, *Escherichia coli*, *Salmonella choleraasius*, *Burkholderia cenocepacia*, *Salmonella* Paratyphi and *Klebsiella pneumoniae*) bacteria. The results were similar to those obtained with the application of ampicillin (500 mg $ml^{-1}$) against *L. monocytogenes* and *Staphylococcus aureus*; however, for Gram-negative bacteria (except for *E. coli* and *Salmonella* Paratyphi), this inhibition was more significant (8–15 mm) compared with ampicillin, which showed no inhibition against the rest of the bacteria. The more substantial effect on Gram-negative bacteria is attributed to their composition, as these bacteria have an outer membrane, lipoproteins, lipopolysaccharides and a single peptidoglycan layer, which makes them more susceptible to damage by ACGs compared with Gram-positive bacteria, which have a thick cell wall surrounded by multiple layers of peptidoglycan and surface glycopolymers such as teichoic acids, making them less vulnerable to damage. Furthermore, bacterial membranes function as barriers preventing drugs and bioactive substances (such as ACGs) from entering the cell and keeping intracellular proteins intact (Pinto *et al.*, 2017). ACGs damage the bacterial membrane by increasing permeability and leaking nucleotides, inhibiting bacterial growth and leading to cell death. A process of insertion into the cell membrane similar to that of fatty acids affects the viability and functionality of bacteria (Salinas-Salazar *et al.*, 2016).

The antimicrobial action of the extracts varies depending on the type of microorganism, the concentration, their purity and the amount of unsaturation in the aliphatic chain of the ACGs. Thus, ACGs from the seeds of *Annona* spp. should be tested against various microorganisms to evaluate their antimicrobial potential.

#### *3.4.3.2 Cytotoxic/antiproliferative activity of ACGs from the seeds of* Annona *spp.*

Cell toxicity or cytotoxicity is defined as adverse effects that result from interference with the cellular structure, cellular processes or both that occur in all cells and are essential for cell function, survival and proliferation. Therefore, a cytotoxic or antiproliferative agent is a drug or bioactive substance, such as ACGs, that inhibits cell proliferation, preventing cell growth and the ability of the cell to divide and multiply, which in turn inactivates signal transduction pathways and causes cell-cycle arrest with destructive actions on cells, causing irreparable damage to DNA, inhibiting its synthesis or interfering with cell division mechanisms, for example blocking topoisomerases or microtubule binding (Casado-Hernández *et al.*, 2016; Muñoz and Cuca, 2016; Rady *et al.*, 2018). Several studies have shown that ACGs can significantly inhibit cell proliferation by inhibiting critical enzymes involved in DNA replication and interfering with metabolic processes necessary for cell growth, directly inducing programmed cell death (Alali *et al.*, 1999; Lima *et al.*, 2022; Onnis, 2023).

A study carried out by Chang and Wu (2001) revealed that muricins C and D, muricatetrocins A and B, longifolicin, corossolin and corossolone (ACGs isolated from the seeds of *A. muricata*) showed higher cytotoxicity against Hep-G2 (half-maximal inhibitory concentration ($IC_{50}$) of $6.6 \times 10^{-4}$–$4.8 \times 10^{-1}$ μg $ml^{-1}$) and Hep 2,2,15 ($IC_{50}$ of $4.83 \times 10^{-3}$–$2.84 \times 10^{-1}$ μg $ml^{-1}$) cells compared with adriamycin, which showed an $IC_{50}$ of 0.24 and 0.45 μg $ml^{-1}$, respectively, for these cell lines. These authors showed that the cytotoxic effect of ACGs depends on their relative structure and the spacer fraction, which provides maximum potency. In the same way, Liaw *et al.* (2002) used ACGs isolated from *A. muricata* seeds (muricin H, muricin I, *cis*-annomontacin, *cis*-corossolone and annocatalin) to evaluate their effect against Hep-G2 ($IC_{50}$ of 0.095–5.7 μg $ml^{-1}$) and Hep 2,2,15 ($IC_{50}$ of 0.0034–0.22 μg $ml^{-1}$) cells. The results indicated that all the ACGs evaluated showed a cytotoxic effect against both cell lines; however, annocatalin showed a more significant effect and high selectivity against Hep 2,2,15 cells with an $IC_{50}$ of 0.0034 μg $ml^{-1}$ compared with adriamycin, which had an $IC_{50}$ of 0.45 μg $ml^{-1}$. Yuan *et al.* (2003) demonstrated the apoptotic activity of annonacin (with an $IC_{50}$ <0.5 μg $ml^{-1}$) isolated from *A. reticulata* seeds in ovarian cancer (PA-1 and SKOV3), cervical cancer (HeLa and HeLa S3), breast cancer (MCF-7), bladder cancer (T24) and skin cancer (BCC) cells. These authors observed that annonacin induced p21 expression more rapidly in T24 cells lacking p53, suggesting that these ACGs potentiated p21 activity and $G_1$ phase arrest in a p53-independent pathway. Furthermore, a decrease in procaspase-3 expression levels was observed at 8 h and at later time points. A study by Pardhasaradhi *et al.* (2005) evaluated the effect of organic (2–5%) and aqueous (5–25%) extracts of *A. squamosa* seeds against breast cancer (MCF-7), erythroleukemia (K-562) and colon cancer (COLO-205) cell lines. The authors demonstrated that organic and aqueous extracts induced apoptosis in MCF-7 and K-562 cells, as assessed by annexin-V–FITC staining, while no cell death was observed in COLO-205 cells. In addition, DNA fragmentation was observed in MCF-7 and K-562 cells but not in COLO-205 cells, which showed no nuclear damage. Miao *et al.* (2016a) demonstrated that ACGs from *A. squamosa* seeds (squamocin IV, squamocin V, and squamoxinone E) showed cytotoxicity against liver cancer (Hep-G2, SMMC-7721, BEL-7402), stomach cancer (BGC-830) and lung cancer (H460) cell lines. The results showed that squamoxinone E had the highest cytotoxic effect, with $IC_{50}$ values of 0.103, 0.687, 4.19, 0.43 and 6.56 μg $ml^{-1}$, respectively, for the cell lines, respectively, compared with fluorouracil, which showed a lower effect with $IC_{50}$ values of 0.742, 12.5, 4.85, 30.7 and 18.8 μg $ml^{-1}$, respectively. Likewise, squamocin IV and V showed selective cytotoxic activity against H460 cells with $IC_{50}$ values of 0.049 and 0.288 μg $ml^{-1}$, respectively. The number of THF rings and OH groups present in the structures of ACGs directly affects the functionality of these compounds; furthermore, the antiproliferative activity of ACGs is considered selective, as some *in vitro* studies have shown that acetogenic extracts are more toxic to cancer cells than to normal cells (Daddiouaissa *et al.*, 2019). The antiproliferative effect of annonamuricins A, B, C and D and annonacin at concentrations of 20 μg $ml^{-1}$ against prostate cancer cells (PC-3) was demonstrated by Sun *et al.* (2017). From the results obtained, it was observed that all ACGs showed inhibition; however, this effect was more significant in annonamuricin C (97.7%) > annonacin (96.9%) > annonamuricin B (96%) > annonamuricin D (96%) > annonamuricin A (86%). These authors concluded that although all ACGs inhibited cell growth, this inhibition depended on the number of hydroxyl groups and the positions of these groups. Meanwhile, Yap *et al.* (2017) reported that annonacin had an antiproliferative effect on endometrial cancer cell lines (ECC-1 and HEC-1A) and primary cells (EC6-ept and EC14-ept) with $IC_{50}$ values ranging from 4.62 to 4.92 μg $ml^{-1}$.

Another study conducted by Ma *et al.* (2019) evaluated the effect of nine ACGs from *A. squamosa* seeds (squamotin A–D, annosquatin IV and V, muricin O, squamosten B and gigantecin against liver cancer SMMC 7721/Adr, breast cancer MCF-7/Adr and lung cancer A549/T cell) lines. These authors demonstrated that ACGs were much more cytotoxic ($IC_{50}$ values of 2.56, 8.54, 14.10, 1.43, 5.68, 7.88, 5.43, 5.40 and 15.45 μmol $l^{-1}$, respectively, for each ACG tested) than adriamycin ($IC_{50}$ 1815 μmol $l^{-1}$) against SMMC 7721/Adr cells. Likewise, it was observed that squamotin D, annosquatin IV and V, muricin O, squamosten B and gigantecin showed a selective effect against MCF-7/Adr cells with $IC_{50}$ values of 2.55–5.78 μmol $l^{-1}$, whereas A549/T cells were inhibited mainly by annosquatin V, squamosten B and annosquatin IV with $IC_{50}$ values of 3.27, 5.23 and 6.13 μmol $l^{-1}$, respectively, compared with adriamycin (7.96 μmol $l^{-1}$). Likewise, Fuel *et al.* (2021) evaluated the antiproliferative effect of three different extracts of *Moringa oleifera*, *Tropaeolum tuberosum* and *A. cherimola* seeds against colon cancer cell lines: T-84 (4 $\times 10^3$ cells per well), HCT-15 (5 $\times 10^3$ cells per well), SW480 (5 $\times$ $10^3$ cells per well) and CCD18 (4 $\times 10^3$ cells per well). These authors showed that the ethanolic extract of *A. cherimola* showed higher proliferative inhibition (66–90%) in HCT15 cells ($IC_{50}$ of 46.35 and 123.6 μg $ml^{-1}$, respectively) compared with *M. oleifera* (57–76%) with $IC_{50}$ values of 36.9 and 98.4 μg $ml^{-1}$and *T. tuberosum* (54–76%) with $IC_{50}$ values of 62.1 and 165.6 μg $ml^{-1}$, respectively.

ACGs thus represent a promising class of compounds with cytotoxic potential, and could be an effective, attractive and low-dose alternative to destroy cancer cells selectively compared with conventional drugs, which have been shown to affect both cancer cells and normal cells with a high proliferative index generating collateral damage in patients (Jacobo-Herrera *et al.*, 2019); however, to replace the use of drugs by bioactive compounds such as ACGs, many studies are required to evaluate their toxicity, efficacy and efficiency, and to further understand their mechanism of action in cancer treatments.

## 3.5 Main Uses of *Annona* Spp. Fruits

### 3.5.1 Food uses

*Annona* spp. fruits are widely cultivated and distributed in tropical and subtropical regions worldwide. They are commonly known in Ecuador, Peru, Asia, Brazil, South and Central America, Africa, India, Egypt and Mexico (Jamkhande and Wattamwar, 2015; Kalidindi *et al.*, 2015; Vasarri *et al.*, 2020). The edible part of the fruits is mainly consumed fresh or used semi-processed or processed for the production of soft drinks, mixed with wine and milk to make milkshakes, and processed to obtain juices, nectars, syrups, purees, ice creams, sorbets, yogurts, jellies, and jams for baking in cookies, cakes and similar foods. They are also used as raw materials for producing powders, sticks and flakes (Pinto *et al.*, 2005; Pareek *et al.*, 2011; Coria-Téllez *et al.*, 2018a,b; Anaya-Esparza *et al.*, 2020).

### 3.5.2 Pharmacological uses

Various parts of the plant or fruits, including the pulp, leaves, root, bark, peel and seeds of *Annona* spp., have been prepared by concentration, maceration, infusion or pastes and have been the subject of countless medicinal uses. These uses are attributed to the presence of various secondary metabolites and bioactive compounds, including alkaloids, terpenoids, saponins, flavonoids, coumarins, phenols, tannins, anthraquinones and, mainly, ACGs (Badrie and Schauss, 2010; Coria-Téllez *et al.*, 2018a).

Extracts obtained from the various components of the plant or fruits have been used to treat problems associated with colds, flu, coughs, pain, fever, diarrhoea, dysentery, asthma, psoriasis, parasitic diseases, heart, liver, arthritis, rheumatism and neuralgia. They are also used as anti-inflammatories, antispasmodics, laxatives, anti-emetics and antidepressants (Amoo *et al.*, 2008; Jamkhande and Wattamwar, 2015; Padmanabhan and Paliyath, 2016; Calzada *et al.*, 2017a,b; Coria-Téllez *et al.*, 2018a; Rady *et al.*, 2018; Vasarri *et al.*, 2020). In addition, these species have vasodilator, cardiodepressive, antispasmodic, antimutagenic, anticonvulsant, insecticidal, antiviral, antidiabetic, antihypertensive, antimicrobial, antioxidant, cytotoxic and antitumor effects (Badrie and Schauss, 2010; Nirmal *et al.*, 2010; Zaman, 2013; Nasser *et al.*, 2017; Coria-Téllez *et al.*, 2018a; Ahmed and Mariod, 2019; Fuel *et al.*, 2021;

Kazman *et al.*, 2022; López-Romero *et al.*, 2022; Aguilar-Hernández *et al.*, 2023).

Various studies have been carried out on the pharmacological activities of extracts or bioactive compounds obtained from the different components of *Annona* spp. trees or fruits. Chavan *et al.* (2012) evaluated the analgesic and anti-inflammatory effect of a fraction of sesquiterpene from *A. reticulata* stem used at 12.5 and 25 mg $kg^{-1}$. In the acetic acid-induced writhing response in mice, the sesquiterpene fraction and standard aspirin (100 mg $kg^{-1}$) similarly reduced the amount of writhing, with 56–66% and 82% inhibition, respectively; likewise, in carrageenan-induced paw oedema, the sesquiterpene fraction and aspirin showed significant inhibition of inflammatory oedema in the first and second hours after carrageenan treatment (65–74%). Jamkhande *et al.* (2014) evaluated the antimicrobial effect of methanol extracts of *A. reticulata* root, finding a more significant effect against *Bacillus subtilis* (23.6 mm), *E. coli* (25.6 mm) and *Salmonella* Typhi (20.3 mm) compared with penicillin (19.66, 12 and 19.6 mm, respectively). The cytotoxicity of ACGs present in the fruit of *A. muricata* was tested against prostate cancer cell lines (PC-3) at concentrations of 20 μg $ml^{-1}$, with inhibition ranging from 70% to approximately 95% (Sun *et al.*, 2014). In the same vein, Chen *et al.* (2016) showed that *A. squamosa* seed oil inhibited the growth of H22 tumour cells in mice by up to 53.54% when administered orally at a dose of 1 ml $kg^{-1}$ $day^{-1}$. The authors attributed its effect to the presence of unsaturated fatty acids and ACGs through the IL-6/JAK/STAT3 pathway.

Bhoir *et al.* (2018) evaluated the antipsoriatic effect of *A. squamosa* seed oil (20 μl) applied to the ears of BALB/c mice. The authors demonstrated that the seed oil reduced (82%) the thickness of the ears to a similar degree to that of the positive control (98%; tenovate 0.05%). Rizwana *et al.* (2021) investigated the antifungal effect of aqueous and methanol extracts of *A. muricata* pulp and seeds against *Alternaria alternata*, a pathogenic fungus of tomato. The results showed that both methanol and aqueous extracts showed significant inhibition of radial mycelial growth as a function of concentration (0.5–6%), with the maximum inhibition of mycelial growth found with the methanol extracts of the seeds (90.93%) at a concentration of 6% and with greater effectiveness compared with the pulp (79.53%), which was attributed to the presence of various bioactive compounds such as polyphenols, flavonoids, terpenoids, alkaloids and saponins.

## 3.6 Conclusion

The fruits of the *Annona* spp. discussed in this chapter are widely consumed naturally or processed for their content of bioactive compounds, to which various pharmacological properties have been attributed. These fruits and their constituents have an important vitamin content, mainly vitamin C, as well as minerals such as iron, magnesium, calcium, phosphorus and potassium, and dietary fibre that contribute to the recommended daily intake. In addition, the seeds are rich in fatty acids, amino acids and various bioactive compounds such as ACGs; however, this content varies depending on factors such as species, the part of the fruit or plant, the place of cultivation, the type of soil, climatic conditions and postharvest processing. Both the edible fraction and the various parts of the tree or fruits of *Annona* spp. have traditionally been used to obtain infusions, extracts or pastes used for medicinal purposes to treat different health conditions such as cough, asthma, nasal congestion, fever, diarrhoea, pain and problems associated with psoriasis, constipation, inflammation, diabetes and cancer, among others. The seeds are a promising source of fatty acids and ACGs with potential effects on various biological activities, such as antimicrobial, antiproliferative, antihypertensive, anticancer and anti-inflammatory, among others, so they could be a viable alternative to replace the use of drugs that may cause adverse health effects. However, it is essential to mention that a broad spectrum of research is required to demonstrate that the components present in the extracts and oils from the seeds do not have adverse health effects and to understand the mechanisms of action of the ACGs present in this matrix.

## References

AACC (American Association of Cereal Chemists) (2001) The definition of dietary fiber. Report of the dietary fiber definition committee to the board of directors of the American Association of Cereal Chemists. *Cereal Foods World* 46, 112–126.

Abdualrahman, M.A.Y., Ma, H., Zhou, C., Yagoub, A.E.A., Ali, A.O. *et al.* (2019) Postharvest physicochemical properties of the pulp and seed oil from *Annona squamosa* L. (Gishta) fruit grown in Darfur region, Sudan. *Arabian Journal of Chemistry* 12, 4514–4521. DOI: 10.1016/j.arabjc.2016.07.008

Abe, M., Murai, M., Ichimaru, N., Kenmochi, A., Yoshida, T. *et al.* (2005) Dynamic function of the alkyl spacer of acetogenins in their inhibitory action with mitochondrial complex I (NADH-ubiquinone oxidoreductase). *Biochemistry* 44, 14898–14906. DOI: 10.1021/bi051568t

Abe, M., Kubo, A., Yamamoto, S., Hatoh, Y., Murai, M. *et al.* (2008) Dynamic function of the spacer region of acetogenins in the inhibition of bovine mitochondrial NADH-ubiquinone oxidoreductase (complex I). *Biochemistry* 47, 6260–6266. DOI: 10.1021/bi800506s

Aguilar-Hernández, G., Vivar-Vera, M. de los Á., García-Magaña, M. de L., González-Silva, N., Pérez-Larios, A. and Montalvo-González (2020) Ultrasound-assisted extraction of total acetogenins from the soursop fruit by response surface methodology. *Molecules* 25: 1139. DOI: 10.3390/molecules25051139

Aguilar-Hernández, G., Zepeda-Vallejo, L.G., García-Magaña, M.L., López-García, U.M., Aguilera-Aguirre, S. *et al.* (2022) Acetogenin content from soursop fruit (*Annona muricata* l.) seeds: evaluation of different extraction conditions. *Biotecnia* 24, 12–19.

Aguilar-Hernández, G., López-Romero, B.A., Pérez-Larios, A., Ruvalcaba-Gómez, J.M., Castellanos-Huerta, I. *et al.* (2023) Antibacterial activity of crude extract and purified acetogenins from *Annona muricata* seeds. *Applied Sciences* 13: 558. DOI: 10.3390/app13010558

Ahmed, R.H.A. and Mariod, A.A. (2019) *Annona squamosa*: phytochemical constituents, bioactive compounds, traditional and medicinal uses. In: Mariod, A.A. (ed.) *Wild Fruits: Composition, Nutritional Value and Products*. Springer, Berlin/Heidelberg, pp. 143–155.

Alali, F.Q., Liu, X.X. and McLaughlin, J.L. (1999) Annonaceous acetogenins: recent progress. *Journal Natural Product* 62, 504–540. DOI: 10.1021/np980406d

Albuquerque, T.G., Santos, F., Sanches-Silva, A., Beatriz Oliveira, M., Bento, A.C. *et al.* (2016) Nutritional and phytochemical composition of *Annona cherimola* Mill. fruits and by-products: POTENTIAL health benefits. *Food Chemistry* 193, 187–195. DOI: 10.1016/j.foodchem.2014.06.044

Ali, A.A.H. (2023) Overview of the vital roles of macro minerals in the human body. *Journal of Trace Elements and Minerals* 4: 100076. DOI: 10.1016/j.jtemin.2023.100076

Al-Sowayan, N.S. and Almarzougi, R.A. (2024) Vitamin E reduces oxidative stress in brains of male albino male rats undergoing immobilization. *Saudi Journal of Biological Sciences* 31: 103900. DOI: 10.1016/j.sjbs.2023.103900

Amoo, I.A., Emenike, A.E. and Akpambang, V.O.E. (2008) Compositional evaluation of *Annona cherimoya* (custard apple) fruit. *Trends in Applied Sciences Research* 3, 216–220. DOI: 10.3923/tasr.2008.216.220

Anaya-Esparza, L.M., García-Magaña, M. de L., Abraham Domínguez-Ávila, J., Yahia, E.M., Salazar-López, N.J. *et al.* (2020) Annonas: underutilized species as a potential source of bioactive compounds. *Food Research International* 138: 109775. DOI: 10.1016/j.foodres.2020.109775

Anuragi, H., Dhaduk, H.L., Kumar, S., Dhruve, J.J., Parekh, M.J. *et al.* (2016) Molecular diversity of *Annona* species and proximate fruit composition of selected genotypes. *3 Biotech* 6: 204. DOI: 10.1007/s13205-016-0520-9

Araya, H., Fujimoto, Y. and Hirayama, K. (1994a) Structural elucidation of tetrahydrofuranic acetogenins by means of precursor-ion scanning method. *Journal of Synthetic Organic Chemistry* 52, 765–777. DOI: 10.5059/yukigoseikyokaishi.52.765

Araya, H., Hara, N., Fujimoto, Y., Srivastava, A. and Sahai, M. (1994b) Squamosten-A, a novel mono-tetrahydrofuranic acetogenin with a double bond in the hydrocarbon chain, from *Annona aquamosa* L. *Chemical and Pharmaceutical Bulletin* 42, 388–391. DOI: 10.1248/cpb.42.388

Araya, H., Sahai, M., Singh, S., Singh, A.K., Yoshida, M. *et al.* (2002) Squamocin-$O_1$ and squamocin-$O_2$, new adjacent bis-tetrahydrofuran acetogenins from the seeds of *Annona squamosa*. *Phytochemistry* 61, 999–1004. DOI: 10.1016/s0031-9422(02)00351-5

Azam, M.M., Waris, A. and Nahar, N. (2005) Prospects and potential of fatty acid methyl esters of some non-traditional seed oils for use as biodiesel in India. *Biomass and Bioenergy* 29, 293–302. DOI: 10.1016/j.biombioe.2005.05.001

Badrie, N. and Schauss, A.G. (2010) Soursop (*Annona muricata* l.): composition, nutritional value, medicinal uses, and toxicology. In: Watson, R.R. and Preedy, V.R. (eds) *Bioactive Foods in Promoting Health: Fruits and Vegetables*, Academic Press, London, pp. 621–643. DOI: 10.1016/B978-0-12-374628-3.00039-6

Bajin ba Ndob, I., Champy, P., Gleye, C., Lewin, G. and Akendengué, B. (2009) Annonaceous acetogenins: precursors from the seeds of *Annona squamosa*. *Phytochemistry Letters* 2, 72–76. DOI: 10.1016/j.phytol.2008.11.006

Barrachina, I., Neske, A., Granell, S., Bermejo, A., Chahboune, N. *et al.* (2004) Tucumanin, a β-hydroxy-γ-lactone bistetrahydrofuranic acetogenin from *Annona cherimolia*, is a potent inhibitor of mitochondrial complex I. *Planta Medica* 70, 866–868. DOI: 10.1055/s-2004-827237

Bermejo, A., Figadere, B., Zafra-Polo, M.-C., Barrachina, I., Estornell, E. and Cortes, D. (2005) Acetogenins from Annonaceae: recent progress in isolation, synthesis and mechanisms of action. *Natural Product Reports* 22, 269–303. DOI: 10.1039/b500186m

Beto, J.A. (2015) The role of calcium in human aging. *Clinical Nutrition Research* 4, 1–8. DOI: 10.7762/cnr.2015.4.1.1

Bhattacherjee, A.K. (2023) Composition and uses. In: Mitra, S. (ed.) *Jackfruit: Botany. Production and Uses*. CAB International, Wallingford, UK, pp. 20–54.

Bhattacherjee, A.K. and Tandon, D.K. (2021) Composition and processing. In: Mitra, S. (ed.) *Guava: Botany, Production and Uses*. CAB International, Wallingford, UK, pp. 33–63.

Bhoir, S.S., Vishwapathi, V. and Singh, K.K. (2018) Antipsoriatic potential of *Annona squamosa* seed oil: an *in vitro* and *in vivo* evaluation. *Phytomedicine* 54, 265–277. DOI: 10.1016/j.phymed.2018.07.003

Bonneau, N., Cynober, T., Jullian, J.C. and Champy, P. (2016) $^{1}$H qNMR quantification of anonaceous acetogenins in crude extracts of *Annona muricata* L. fruit pulp. *Phytochemical Analysis* 28, 251–256. DOI: 10.1002/pca.2668

Calzada, F., Correa-Basurto, J., Barbosa, E., Mendez-Luna, D. and Yepez-Mulia, L. (2017a) Antiprotozoal constituents from *Annona cherimola* Miller, a plant used in Mexican traditional medicine for the treatment of diarrhea and dysentery. *Pharmacognosy Magazine* 13: 148. DOI: 10.4103/0973-1296.197636

Calzada, F., Solares-Pascasio, J., Ordoñez-Razo, R., Velazquez, C., Barbosa, E. *et al.* (2017b) Antihyperglycemic activity of the leaves from *Annona cherimola* Miller and rutin on alloxan-induced diabetic rats. *Pharmacognosy Research* 9, 1–6. DOI: 10.4103/0974-8490.199781

Campbell, A.A., Thorne-Lyman, A., Sun, K., de Pee, S., Kraemer, K. *et al.* (2009) Indonesian women of childbearing age are at greater risk of clinical vitamin A deficiency in families that spend more on rice and less on fruits/vegetables and animal-based foods. *Nutrition Research* 29, 75–81. DOI: 10.1016/j.nutres.2008.12.004

Casado-Hernández, I., Mora-González, N., Ferrer-Carmenates, G., Fernández-Torres, S. and Pino Blanco, D. (2016) Citotoxicidad *in vitro* y potencialidades de los compuestos quinoides como agentes antitumorales. *Revista Cubana de Hematología, Inmunología y Hemoterapia* 32, 30–42.

Chang, F.R. and Wu, Y.C. (2001) Novel cytotoxic annonaceous acetogenins from *Annona muricata*. *Journal of Natural Products* 64, 925–931. DOI: 10.1021/np010035s

Chang, F.R., Chen, J.L., Chiu, H.F., Wu, M.J. and Wu, Y.C. (1998) Acetogenins from seeds of *Annona reticulata*. *Phytochemistry* 47, 1057–1061. DOI: 10.1016/s0031-9422(97)00675-4

Chang, F.R., Liaw, C.C., Lin, C.Y., Chou, C.J., Chiu, H.F. and Wu, Y.-C. (2003) New adjacent bis-tetrahydrofuran annonaceous acetogenins from *Annona muricata*. *Planta Medica* 69, 241–246. DOI: 10.1055/s-2003-38485

Chang, L.S., Karim, R., Mohammed, A.S and Mohd Ghazali, H. (2018) Characterization of enzyme-liquefied soursop (*Annona muricata* L.) puree. *LWT – Food Science and Technology* 94, 40–49. DOI: 10.1016/j.lwt.2018.04.027

Chavan, M.J., Wakte, P.S. and Shinde, D.B. (2012) Analgesic and anti-inflammatory activities of the sesquiterpene fraction from *Annona reticulata* L. bark. *Natural Product Research* 26, 1515–1518. DOI: 10.1080/14786419.2011.564583

Chen, Y., Chen, J. and Li, X. (2011) Cytotoxic bistetrahydrofuran annonaceous acetogenins from the seeds of *Annona squamosa*. *Journal of Natural Products* 74, 2477–2481. DOI: 10.1021/np200708q

Chen, Y., Chen, J., Wang, Y., Xu, S. and Li, X. (2012) Six cytotoxic annonaceous acetogenins from *Annona squamosa* seeds. *Food Chemistry* 135, 960–966. DOI: 10.1016/j.foodchem.2012.05.041

Chen, Y., Chen, Y., Shi, Y., Ma, C., Wang, X. *et al.* (2016) Antitumor activity of *Annona squamosa* seed oil. *Journal of Ethnopharmacology* 193, 362–367. DOI: 10.1016/j.jep.2016.08.036

Ma, C.-Y., M., Lu, L., Li, L., Liu, L. and Chen, J.-W. (2019) Eight new cytotoxic annonaceous acetogenins from the seeds of *Annona squamosa*. *Chinese Journal of Natural Medicines* 17, 291–297. DOI: 10.1016/S1875-5364(19)30032-9

Ciosek, Ż., Kot, K., Kosik-Bogacka, D., Łanocha-Arendarczyk, N. and Rotter, I. (2021) The effects of calcium, magnesium, phosphorus, fluoride, and lead on bone tissue. *Biomolecules* 11: 506. DOI: 10.3390/biom11040506

Cordeiro, N., Sousa, L., Freitas, N. and Gouveia, M. (2013) Changes in the mesocarp of *Annona cherimola* Mill. 'Madeira' during postharvest ripening. *Postharvest Biology and Technology* 85, 179–184. DOI: 10.1016/j.postharvbio.2013.05.014

Coria-Téllez, A.V., Montalvo-Gónzalez, E., Yahia, E.M. and Obledo-Vázquez, E.N. (2018a) *Annona muricata*: a comprehensive review on its traditional medicinal uses, phytochemicals, pharmacological activities, mechanisms of action and toxicity. *Arabian Journal of Chemistry* 11, 662–691. DOI: 10.1016/j.arabjc.2016.01.004

Coria-Téllez, A.V., Montalvo-González, E. and Obledo-Vázquez, E.N. (2018b) Soursop (*Annona muricata*). *Fruit and Vegetable Phytochemicals* 2, 1243–1252. DOI: 10.1002/9781119158042

Cortes, D., Myint, S.H. and Hocquemiller, R. (1991a) Molvizarin and motrilin: two novel cytotoxic bis-tetrahydro-furanic γ-lactone acetogenins from *Annona cherimolia*. *Tetrahedron* 47, 8195–8202. DOI: 10.1016/S0040-4020(01)91014-2

Cortes, D., Myint, S.H., Laurens, A., Hocquemiller, R., Leboeuf, M. and Cavé, A. (1991b) Corossolone et corossoline, deux nouvelles γ-lactones mono-tétrahydrofuraniques cytotoxiques. *Canadian Journal of Chemistry* 69, 8–11. DOI: 10.1139/v91-002

Cortes, D., Myint, S.H., Leboeuf, M. and Cavé, A. (1991c) A new type of cytotoxic acetogenins: the tetrahydrofuranic β-hydroxy methyl γ-lactones. *Tetrahedron Letters* 32, 6133–6134. DOI: 10.1016/0040-4039(91)80771-W

Cortes, D., Myint, S.H., Dupont, B. and Davoust, D. (1993) Bioactive acetogenins from seeds of *Annona cherimolia*. *Phytochemistry* 32, 1475–1482. DOI: 10.1016/0031-9422(93)85162-K

Cruz-Requena, M., Escobedo-García, S., Salas-Tovar, J.A., Mora-Cura, Y. and Chávez-González, M.L. (2019) Definitions and regulatory perspectives of dietary fibers. In: Galanakis, C.M. (ed.) *Dietary Fiber: Properties, Recovery, and Applications*. Academic Press, pp. 1–25. DOI: 10.1016/B978-0-12-816495-2.00001-0

Daddiouaissa, D., Amid, A., Kabbashi, N.A., Fuad, F.A.A., Elnour, A.M. and Epandy, M.A.K.M.S. (2019) Antiproliferative activity of ionic liquid-graviola fruit extract against human breast cancer (MCF-7) cell lines using flow cytometry techniques. *Journal of Ethnopharmacology* 236, 466–473. DOI: 10.1016/j.jep.2019.03.003

Dang, Q.L., Kim, W.K., Nguyen, C.M, Choi, Y.H., Choi, G-J. *et al.* (2011) Nematicidal and antifungal activities of annonaceous acetogenins from *Annona squamosa* against various plant pathogens. *Journal of Agricultural and Food Chemistry* 59, 11160–11167. DOI: 10.1021/jf203017f

Debelo, H., Novotny, J.A. and Ferruzzi, M.G. (2017) Vitamin A. *Advances in Nutrition Journal* 8, 992–994. DOI: 10.3945/an.116.014720

Du, M., Zhu, Y., Nan, H., Zhou, Y. and Pan, X. (2024) Regulation of sugar metabolism in fruits. *Scientia Horticulturae* 326: 112712. DOI: 10.1016/j.scienta.2023.112712

Durán-Ruiz, C.A., Cruz-Ortega, R., Zaldívar-Riverón, A., Zavaleta-Mancera, H.A., De-la-Cruz-Chacón, I. and González-Esquinca, A.R. (2019) Ontogenic synchronization of *Bephratelloides cubensis*, *Annona macroprophyllata* seeds and acetogenins from Annonaceae. *Journal of Plant Research* 132, 81–91. DOI: 10.1007/s10265-018-01078-3

Egydio, B.A.P.M. and Santos, D.Y.A.C. (2016) Nutritional value of the pulp of different sugar apple cultivars (*Annona squamosa* L.). *Nutritional Composition of Fruit Cultivars* 195–214. DOI: 10.1016/B978-0-12-408117-8.00009-X

Egydio, A.P.M., Catarina, C.S., Floh, E.I.S. and Santos, D.Y.A.C. dos (2013) Free amino acid composition of *Annona* (Annonaceae) fruit species of economic interest. *Industrial Crops and Products* 45, 373–376. DOI: 10.1016/j.indcrop.2012.12.033

Escribano, P., Viruel, M.A. and Hormaza, J.I. (2007) Molecular analysis of genetic diversity and geographic origin within an *ex situ* germplasm collection of cherimoya by using SSRs. *Journal of the American Society for Horticultural Science* 132, 357–367. DOI: 10.21273/JASHS.132.3.357

EFSA (European Food Safety Authority) (2019) Dietary reference values for nutrients summary report. Technical Report. *EFSA Supporting Publications* 14: e15121E. DOI: 10.2903/sp.efsa.2017.e15121

FDA (US Food and Drug Administration) (2018) *Guidance for Industry: The Declaration of Certain Isolated or Synthetic Non-digestible Carbohydrates as Dietary Fiber on Nutrition and Supplement Facts Labels*. FDA-2018-D-1323. FDA, Silver Spring, Maryland. Available at: www.fda.gov/regulatory-information/search-fda-guidance-documents/guidance-industry-declaration-certain-isolated-or-synthetic-non-digestible-carbohydrates-dietary (accessed 15 August 2025).

Fuel, M., Mesas, C., Martínez, R., Ortiz, R., Quiñonero, F. *et al.* (2021) Antioxidant and antiproliferative potential of ethanolic extracts from *Moringa oleifera*, *Tropaeolum tuberosum* and *Annona cherimola* in colorectal cancer cells. *Biomedicine & Pharmacotherapy* 143: 112248. DOI: 10.1016/j.biopha.2021.112248

Fujimoto, Y., Eguchi, T., Kakinuma, K., Ikekawa, N., Sahai, M. and Gupta, Y.K. (1988) Squamocin, a new cytotoxic bis-tetrahydrofuran containing acetogenin from *Annona squamosa*. *Chemical and Pharmaceutical Bulletin* 36, 4802–4806. DOI: 10.1248/cpb.36.4802

Fujimoto, Y., Murasaki, C., Kakinuma, K., Eguchi, T., Ikekawa, N. *et al.* (1990) Squamostatin-A: unprecedented bis-tetrahydrofuran acetogenin from *Annona squamosa*. *Tetrahedron Letters* 31, 535–538. DOI: 10.1016/0040-4039(90)87027-W

Fujimoto, Y., Murasaki, C., Shimada, H., Nishioka, S., Kakinuma, K. *et al.* (1994) Annonaceous acetogenins from the seeds of *Annona squamosa*. Non-adjacent bis-tetrahydrofuranic acetogenins. *Chemical and Pharmaceutical Bulletin* 42, 1175–1184. DOI: 10.1248/cpb.42.1175

Ghi, H.Q., Yu, J.G. and Yu, Z.L. (1995) Muricatalin, a new polyketide from *Annona muricata* (Annonaceae). *Chinese Chemistry Letters* 6, 45–48.

Gleye, C., Laurens, A., Hocquemiller, R., Faucheur, N., Serani, L. and Laprévote, O. (1998) Mass spectrometry and structure elucidation of 15-palmitoyl- and 15-oleylsolamin, the first known fatty acid esters of acetogenins from *Annona muricata* L. *Rapid Communications in Mass Spectrometry* 12, 1051–1056. DOI: 10.1002/(SICI)1097-0231(19980831)12:16<1051::AID-RCM283>3.0.CO;2-2

Gleye, C., Rafidiarison, N., Duret, P., Laurens, A. and Hocquemiller, R. (2000a) Robustocin, a new acetogenin from the seeds of *Annona muricata*. *Natural Product Letters* 14, 239–245. DOI: 10.1080/10575630008041238

Gleye, C., Raynaud, S., Fourneau, C., Laurens, A., Laprévote, O. *et al.* (2000b) Cohibins C and D, two important metabolites in the biogenesis of acetogenins from *Annona muricata* and *Annona nutans*. *Journal of Natural Products* 63, 1192–1196. DOI: 10.1021/np000061a

Granger, M. and Eck, P. (2018) Dietary vitamin C in human health. *Advances in Food and Nutrition Research* 83, 281–310. DOI: 10.1016/bs.afnr.2017.11.006

Gromek, D., Figadère, B., Hocquemiller, R., Cavé, A. and Cortes, D. (1993) Corepoxylone, a possible precursor of mono-tetrahydrofuran γ-lactone acetogenins: biomimetic synthesis of corossolone. *Tetrahedron* 49, 5247–5252. DOI: 10.1016/S0040-4020(01)82374-7

Gui, H.Q. and Yu, J.G. (1996) Muricatalicin, a new penta-hydroxy derivative of acetogenin from *Annona muricata* (Annonaceae). *Chinese Chemistry Letters* 7, 561–564.

Hirayama, K., Akashi, S., Yuji, R., Niitsu, U. and Fujimoto, Y. (1993) Structural studies of polyhydroxy-bis(tetrahydrofuran)acetogenins from *Annona squamosa* using the combination of chemical derivatization and precursor-ion scanning mass spectrometry. *Organic Mass Spectrometry* 28, 1516–1524. DOI: 10.1002/oms.1210281225

Jacobo-Herrera, N., Pérez-Plasencia, C., Castro-Torres, V.A., Martínez-Vázquez, M., González-Esquinca, A.R. and Zentella-Dehesa, A. (2019) Selective acetogenins and their potential as anticancer agents. *Frontiers in Pharmacology* 10: 783. DOI: 10.3389/fphar.2019.00783

Jamkhande, P.G. and Wattamwar, A.S. (2015) *Annona reticulata* Linn. (bullock's heart): plant profile, phytochemistry and pharmacological properties. *Journal of Traditional and Complementary Medicine* 5, 144–152. DOI: 10.1016/j.jtcme.2015.04.001

Jamkhande, P.G., Wattamwar, A.S., Pekamwar, S.S. and Chandak, P.G. (2014) Antioxidant, antimicrobial activity and *in silico* PASS prediction of *Annona reticulata* Linn. root extract. *Beni-Suef University Journal of Basic and Applied Sciences* 3, 140–148. DOI: 10.1016/j.bjbas.2014.05.008

Juang, S.-H., Chiang, C.-Y., Liang, F.-P., Chan, H.-H., Yang, J.-S. *et al.* (2016) Mechanistic study of tetrahydrofuran- acetogenins in triggering endoplasmic reticulum stress response-apotoposis in human nasopharyngeal carcinoma. *Scientific Reports* 6: 39251. DOI: 10.1038/srep39251

Kalidindi, N., Thimmaiah, N.V., Jagadeesh, N.V., Nandeep, R., Swetha, S. and Kalidindi, B. (2015) Antifungal and antioxidant activities of organic and aqueous extracts of *Annona squamosa* Linn. leaves. *Journal of Food and Drug Analysis* 23, 795–802. DOI: 10.1016/j.jfda.2015.04.012

Kawazu, K., Alcantara, J.P. and Kobayashi, A. (1989) Isolation and structure of neoannonin, a novel insecticidal compound from the seeds of *Annona squamosa*. *Agricultural and Biological Chemistry* 53, 2719–2722. DOI: 10.1271/bbb1961.53.2719

Kazman, B.S.M., Harnett, J.E. and Hanrahan, J.R. (2022) Usos tradicionales, fitoquímica y actividades farmacológicas de Annonacae. *Moléculas* 27: 3462. DOI: 10.3390/molecules27113462

Kim, D.H., Ma, E.S., Suk, K.D., Son, J.K., Lee, J.S. and Woo, M.H. (2001a) Annomolin and annocherimolin, new cytotoxic annonaceous acetogenins from *Annona cherimolia* seeds. *Journal of Natural Products* 64, 502–506. DOI: 10.1021/np000335u

Kim, D.H., Son, J.K. and Woo, M.H. (2001b) Annomocherin, annonacin and annomontacin: a novel and two known bioactive mono-tetrahydrofuran annonaceous acetogenins from *Annona cherimolia* seeds. *Archives of Pharmacal Research* 24, 300–306. DOI: 10.1007/BF02975096

Laprévote, O., Girard, C., Das, B.C., Laugel, T., Roblot, F. *et al.* (1992) Location of epoxy rings in a long chain acetogenin by fast-atom bombardment and linked scan (*B/E*) mass spectrometry of lithium-cationized molecules. *Rapid Communications in Mass Spectrometry* 6, 352–355. DOI: 10.1002/rcm.1290060509

Li, C.M., Mu, Q.X., Hao, J., Sun, H.D., Zheng, H.L. and Wu, Y.C. (1994) Three new bioactive annonaceous acetogenins from *Annona muricata*. *Chinese Chemistry Letters* 5, 747–750.

Li, D.-Y., Yu, J.G., Luo, X.Z., Sun, L. and Yang, S.L. (2000) Muricatenol, a linear acetogenin from *Annona muricata* (Annonaceae). *Chinese Chemical Letters* 11, 239–242.

Li, X., Chen, X.L., Chen, J.W. and Sun, D.D. (2010) Annonaceous acetogenins from the seeds of *Annona squamosa*. *Chemistry of Natural Compounds* 46, 101–105. DOI: 10.1007/s10600-010-9538-0

Li, Y., Miao, Y.J., Ma, C.Y. and Chen, J.W. (2017) Two new annonaceous acetogenins from seeds of *Annona squamosa*. *Chinese Traditional Herbal Drugs* 48, 2375–2381.

Liaw, C.-C., Chang, F.-R., Lin, C.-Y., Chou, C.-J., Chiu, H.-F. *et al.* (2002) New cytotoxic monotetrahydrofuran annonaceous acetogenins from *Annona muricata*. *Journal of Natural Products* 65, 470–475. DOI: 10.1021/np0105578

Liaw, C.-C., Yang, Y.-L., Chen, M., Chang, F.-R., Chen, S.-L. *et al.* (2008) mono-tetrahydrofuran annonaceous acetogenins from *Annona squamosa* as cytotoxic agents and calcium ion Chelators. *Journal of Natural Products* 71, 764–771. DOI: 10.1021/np0704957

Liaw, C.-C., Liao, W.-Y., Chen, C.-S., Jao, S.-C., Wu, Y.-C. *et al.* (2011) The calcium-chelating capability of tetrahydrofuranic moieties modulates the cytotoxicity of annonaceous acetogenins. *Angewandte Chemie* 123, 8031–8037. DOI: 10.1002/anie.201100717

Liaw, C.-C., Liou, J.-R., Wu, T.-Y., Chang, F.-R. and Wu, Y.-C. (2016) Acetogenins from Annonaceae. *Progress in the Chemistry of Organic Natural Products* 101, 113–230. DOI: 10.1007/978-3-319-22692-7_2

Lieb, F., Nonfon, M., Wachendorff-Neumann, U. and Wendisch, D. (1990) Annonacine und Annonastatin aus *Annona squamosa*. *Medical Plant* 56, 317–319. DOI: 10.1055/s-2006-960968

Lima, N.N. das C., Faustino, D.C., Allahdadi, K.J., França, L.S. de A. and Pinto, L.C. (2022) Acetogenins from annonaceae plants: potent antitumor and neurotoxic compounds. *PharmaNutrition* 20: 100295. DOI: 10.1016/j.phanu.2022.100295

Londershausen, M., Leicht, W., Lieb, F., Moeschler, H. and Weiss, H. (1991) Molecular mode of action of annonins. *Pesticide Science* 33, 427–438. DOI: 10.1002/ps.2780330405

López-Romero, B.A., Luna-Bárcenas, G., García-Magaña, M. de L., Anaya-Esparza, L.M., Zepeda-Vallejo, L.G. *et al.* (2022) Extraction of acetogenins using thermosonication-assisted extraction from *Annona muricata* seeds and their antifungal activity. *Molecules* 27: 6045. DOI: 10.3390/molecules27186045

Ma, C., Wang, Q., Shi, Y., Li, Y., Wang, X. *et al.* (2017) Three new antitumor annonaceous acetogenins from the seedsof*Annonasquamosa*.*NaturalProductResearch*31,2085–2090.DOI:10.1080/14786419.2016.1274897

Ma, C.-Y., M., Lu, L., Li, L., Liu, L. and Chen, J.-W. (2019) Eight new cytotoxic annonaceous acetogenins from the seeds of *Annona squamosa*. *Chinese Journal of Natural Medicines* 17, 291–297. DOI: 10.1016/S1875-5364(19)30032-9

Ma, C., Li, Y., Lu, J., Wang, M., Li, X. *et al.* (2024) Three new cytotoxic annonaceous acetogenins from the seeds of *Annona squamosa Natural Product Research* 38, 1135–1139. DOI: 10.1080/14786419.2022.2134362

Mariod, A.A., Mirghani, M.E.S. and Hussein, I. (2017) *Annona squamosa* L. sugar apple seed oil. *Unconventional Oil Seeds and Oil Sources* 25, 145–150. DOI: 10.1016/B978-0-12-809435-8.00025-1

Miao, Y.-J., Shi, Y.-Y., Xu, X.-F., Chen, Y., Chen, J.-W. and Li, X. (2016a) Three cytotoxic Annonaceous acetogenins from the seeds of *Annona squamosa*. *Phytochemistry Letters* 16, 92–96. DOI: 10.1016/j.phytol.2016.03.006

Miao, Y., Xu, X., Yuan, F., Shi, Y., Chen, Y. *et al.* (2016b) Four cytotoxic annonaceous acetogenins from the seeds of *Annona squamosa*. *Natural Product Research* 30, 1273–1279. DOI: 10.1080/14786419.2015.1055490

Muñoz, C.D.R. and Cuca, S.L.E. (2016) Plant-derived cytotoxic compounds and their relationship with inhibitor of apoptosis (IAP) proteins. *Revista Colombiana de Cancerología* 20, 124–134. DOI: 10.1016/j.rccan.2015.10.002

Myint, S.H., Cortes, D., Laurens, A., Hocquemiller, R., Lebęuf, M. *et al.* (1991) Solamin, a cytotoxic mono-tetrahydrofuranic γ-lactone acetogenin from *Annona muricata* seeds. *Phytochemistry* 30, 3335–3338. DOI: 10.1016/0031-9422(91)83204-X

Nasser, M., Majid, E.-M., Falah, A.-S., Layla, C.., Akram, H. *et al.* (2017) Antibacterial, antioxidant and antiproliferative activities of the hydroalcoholic extract of the Lebanese *Annona squamosa* L. seeds. *International Research Journal of Pharmacy* 8, 1–7. DOI: 10.7897/2230-8407.0712141

Neske, A., Ruiz Hidalgo, J., Cabedo, N. and Cortes, D. (2020) Acetogenins from annonaceae family. Their potential biological applications. *Phytochemistry* 174: 112332. DOI: 10.1016/j.phytochem.2020.112332

Neta, M.T.S.L., de Jesus, M.S., da Silva, J.L.A., Araujo, H.C.S., Sandes, R.D.D. *et al.* (2019) Effect of spray drying on bioactive and volatile compounds in soursop (*Annona muricata*) fruit pulp. *Food Research International* 124, 70–77. DOI: 10.1016/j.foodres.2018.09.039

Nirmal, S.A., Gaikwad, S.B., Dhasade, V.V., Dhikale, R.S., Kotkar, P.V. and Dighe, S. (2010) Anthelmintic activity of *Annona reticulata* leaves. *Pharmaceutical Biological Chemical Sciences* 1, 115–118.

Olechno, E., Puścion-Jakubik, A., Socha, K. and Zujko, M.E. (2021) Coffee brews: are they a source of macroelements in human nutrition *Foods* 10: 1328. DOI: 10.3390/foods10061328

Onnis, V. (2023) Nuevos agentes antiproliferativos. *Pharmaceuticals* 16: 1437. DOI: 10.3390/ph16101437

Padmanabhan, P. and Paliyath, G. (2016) Annonaceous fruits. In: Caballero, B., Finglas, P. and Toldrá, F. (eds) *Encyclopedia of Food and Health*. Academic Press, Kidlington, UK, pp. 169–173. DOI: 10.1016/B978-0-12-384947-2.00031-3

Panadare, D.C., Gondaliya, A. and Rathod, V.K. (2020) Comparative study of ultrasonic pretreatment and ultrasound assisted three phase partitioning for extraction of custard apple seed oil. *Ultrasonics Sonochemistry* 61: 104821. DOI: 10.1016/j.ultsonch.2019.104821

Pardhasaradhi, B.V., Reddy, M., Ali, A.M., Kumari, A.L. and Khar, A. (2005) Differential cytotoxic effects of *Annona squamosa* seed extracts on human tumour cell lines: role of reactive oxygen species and glutathione. *Journal Bioscience* 30, 237–244. DOI: 10.1007/BF02703704

Pareek, S., Yahia, E.M., Pareek, O.P. and Kaushik, R.A. (2011) Postharvest physiology and technology of *Annona* fruits. *Food Research International* 44, 1741–1751. DOI: 10.1016/j.foodres.2011.02.016

Pinto, A.C.Q., Cordeiro, M.C.R., Andrade, S.R.M., Ferreira, F.R. and Filgueiras, H.A.C. *et al.* (2005) Taxonomy and Botany. In: *Annona Species*. International Centre for Underutilised Crops, University of Southampton, Southampton, UK, pp. 3–18.

Pinto, L.C., de Souza, C.O., de Souza, S.A., da Silva, H.B., da Silva, R.R. *et al.* (2018) Potential of *Annona muricata* L. seed oil: phytochemical and nutritional characterization associated with non-toxicity. *Grasas y Aceites* 69: e234. DOI: 10.3989/gya.0777171

Pinto, N. de C.C., Campos, L.M., Evangelista, A.C.S., Lemos, A.S.O., Silva, T.P. *et al.* (2017) Antimicrobial *Annona muricata* L. (soursop) extract targets the cell membranes of Gram-positive and Gram-negative bacteria. *Industrial Crops and Products* 107, 332–340. DOI: 10.1016/j.indcrop.2017.05.054

Rady, I., Bloch, M.B., Chamcheu, R.C.N., Banang, M.S. and Anwar, M.R. *et al.* (2018) Anticancer properties of graviola (*Annona muricata*): a comprehensive mechanistic review. *Oxidative Medicine and Cellular Longevity* 2018: 1826170. DOI: 10.1155/2018/1826170

Rahman, M., Parvin, S., Ekramul Haque, M., Ekramul Islam, M. and Mosaddik, M.A. (2005) Antimicrobial and cytotoxic constituents from the seeds of *Annona squamosa*. *Fitoterapia* 76, 484–489. DOI: 10.1016/j.fitote.2005.04.002

Rieser, M.J., Fang, X.P., Anderson, J.E., Miesbauer, L.R., Smith, D.L. and McLaughlin, J.L. (1993) Muricatetrocins A and B and gigantetrocin B: three new cytotoxic monotetrahydrofuran-ring acetogenins from *Annona muricata*. *Helvetica Chimica Acta* 76, 2433–2444. DOI: 10.1002/hlca.19930760703

Rizqilah, R., Apriyanto, D.R. and Mulyaningsih, R.E. (2019) Inhibitory of soursop leaves (*Annona muricata* L.) extract against *Malassezia furfur* growth. *Proceedings of International Conference on Applied Science and Health* 4, 229–233.

Rizwana, H., Bokahri, N.A., Alsahli, S.A., Al Showiman, A.S., Alzahrani, R.M. and Aldehaish, H.A. (2021) Postharvest disease management of *Alternaria* spots on tomato fruit by *Annona muricata* fruit extracts. *Saudi Journal of Biological Sciences* 28, 2236–2244. DOI: 10.1016/j.sjbs.2021.01.014

Ríos, J., Cortes, D. and Valverde, S. (1989) Acetogenins, aporphinoids, and azaanthraquinone from *Annona cherimolia* seeds. *Planta Medica* 55, 321–323. DOI: 10.1055/s-2006-962020

Sahai, M., Singh, S., Singh, M., Gupta, Y.K., Akashi, S. *et al.* (1994) Annonaceous acetogenins from the seeds of *Annona squamosa*. Adjacent bis-tetrahydrofuranic acetogenins. *Chemical and Pharmaceutical Bulletin* 42, 1163–1174. DOI: 10.1248/cpb.42.1163

Salinas-Salazar, C., Hernández-Brenes, C., Rodríguez-Sánchez, D.G., Castillo, E.C., Navarro-Silva, J.M. and Pacheco, A. (2017) Inhibitory activity of avocado seed fatty acid derivatives (acetogenins) against *Listeria monocytogenes*. *Journal of Food Science* 82, 134–144. DOI: 10.1111/1750-3841.13553

Shimada, H., Grutzner, J.B., Kozlowski, J.F. and McLaughlin, J.L. (1998a) Membrane conformations and their relation to cytotoxicity of asimicin and its analogues. *Biochemistry* 37, 854–866. DOI: 10.1021/bi972348l

Shimada, H., Kozlowski, J.F. and McLaughlin, J.L. (1998b) The localisations in liposomal membranes of the tetrahydrofuran ring moieties of the annonaceous acetogenins, annonacin and sylvaticin, as determined by $^1$H NMR spectroscopy. *Pharmacological Research* 37, 357–364. DOI: 10.1006/phrs.1998.0314

Solís-Fuentes, J.A., Hernández-Medel, M.R. and Durán-de-Bazúa, M.C. (2020) Soursop seed: soursop (*Annona muricata* L.) seed, therapeutic, and possible food potential. In: Preedy, V.R. and Watson, R.R. (eds) *Nuts and Seeds in Health and Disease Prevention*, 2nd edn, pp. 15–25. DOI: 10.1016/B978-0-12-818553-7.00002-4

Son, J.K., Kim, D.H. and Woo, M.H. (2003) Two new epimeric pairs of acetogenins bearing a carbonyl group from *Annona cherimolia* seeds. *Journal of Natural Products* 66, 1369–1372. DOI: 10.1021/np0301487

Sun, S., Liu, J., Kadouh, H., Sun, X. and Zhou, K. (2014) Three new anti-proliferative annonaceous acetogenins with mono-tetrahydrofuran ring from graviola fruit (*Annona muricata*). *Bioorganic & Medicinal Chemistry Letters* 24, 2773–2776. DOI: 10.1016/j.bmcl.2014.03.099

Sun, S., Liu, J., Sun, X., Zhu, W., Yang, F. *et al.* (2017) Novel annonaceous acetogenins from Graviola (*Annona muricata*) fruits with strong anti-proliferative activity. *Tetrahedron Letters* 58, 1895–1899. DOI: 10.1016/j.tetlet.2017.04.016

Taechowisan, T., Singtotong, C. and Phutdhawong, W.S. (2016) Antibacterial and antioxidant activities of acetogenins from *Streptomyces* sp. VE2; an endophyte in *Vernonia cinerea* (L.) less. *Journal of Applied Pharmaceutical Science* 6, 67–72. DOI: 10.7324/JAPS.2016.60810

Tam, V.T., Chi Hieu, P.Q., Chappe, B., Roblot, F. and Laprevote, O. *et al.* (1994a) Four new acetogenins from the seeds of *Annona reticulata* 1. *Natural Products Letters* 4, 255–262. DOI: 10.1080/10575639408043915

Tam, V.T., Chaboche, C., Figadère, B., Chappe, B., Hieu, B.C. and Cavé, A. (1994b) First synthesis of a new acetogenin of Annonaceae, reticulatamol: activated tin hydride with enhanced reducing ability. *Tetrahedron Letters* 35, 883–886. DOI: 10.1016/S0040-4039(00)75989-2

Tam, V.T., Hieu, P.Q.C., Chappe, B., Roblot, F., Figadere, B. and Cavé, A. (1995) Reticulatain-1 and -2 with reticulatamone: three new polyketides from the seeds of *Annona reticulata*. *Bulletin Societe Chimique France* 132, 324–329.

Vasarri, M., Barletta, E., Vinci, S., Ramazzotti, M., Francesconi, A. *et al.* (2020) *Annona cherimola* Miller fruit as a promising candidate against diabetic complications: An *in vitro* study and preliminary clinical results. *Foods* 9: 1350. DOI: 10.3390/foods9101350

Villacís-Chiriboga, J., Prandi, B., Ruales, J., van Camp, J., Sforza, S. and Elst, K. (2023) Valorization of soursop (*Annona muricata*) seeds as alternative oil and protein source using novel de-oiling and protein extraction techniques. *LWT – Food Science and Technology* 182: 114777. DOI: 10.1016/j.lwt.2023.114777

Woo, M.H., Kim, D.H., Fotopoulos, S.S. and McLaughlin, J.L. (1999) Annocherin and (2,4)-*cis* - and *trans*-annocherinones, monotetrahydrofuran annonaceous acetogenins with a C-7 carbonyl group from *Annona cherimolia* seeds. *Journal Natural Products* 62, 1250–1255. DOI: 10.1021/np990135m

Xie, H.H., Wei, X.Y., Wang, J.D., Liu, M.F. and Yang, R.Z. (2003) A new cytotoxic acetogenin from the seeds of *Annona squamosa*. *Chinese Chemical Letters* 14, 588–590.

Yamada, S. and Inaba, M. (2021) Potassium metabolism and management in patients with CKD. *Nutrients* 13: 1751. DOI: 10.3390/nu13061751

Yang, H., Hou, L., Sun, H.M. and Ye, S.H. (2024) Degradation kinetics of vitamins in different enteral feeding formulas during storage at different temperatures. *Heliyon* 10: e29653. DOI: 10.1016/j.heliyon.2024.e29653

Yang, H.J., Li, X., Zhang, N., Chen, J.W. and Wang, M.Y. (2009) Two new cytotoxic acetogenins from *Annona squamosa*. *Journal of Asian Natural Products Research* 11, 250–256. DOI: 10.1080/10286020802682916

Yang, R., Zheng, X., Wu, S. and Qin, G. (1995) Annonsilin A – a novel seco-tris-tetrahydrofuranyl annonaceous acetogenin. *Journal Integrative Plant Biology* 37, 492–495.

Yap, C.V., Subramaniam, K.S., Khor, S.W. and Chung, I. (2017) Annonacin exerts antitumor activity through induction of apoptosis and extracellular signal-regulated kinase inhibition. *Pharmacognosy Research* 9: 378. DOI: 10.4103/pr.pr_19_17

Yu, J.-G., Luo, X.-Z., Sun, L., Li, D.-Y., Huang, W.-H. and Liu, C.Y. (2005) [Chemical constituents from the seeds of *Annona squamosa*]. *Yao Xue Xue Bao* 40, 153–158 (in Chinese).

Yu, J.-G., Gui, H.-Q., Luo, X.-Z. and Sun, L. (1998) Murihexol, a linear acetogenin from *Annona muricata*. *Phytochemistry* 49, 1689–1692. DOI: 10.1016/s0031-9422(98)00224-6

Yuan, S.-S.F., Chang, H.-L., Chen, H.-W., Yeh, Y.-T., Kao, Y.-H. *et al.* (2003) Annonacin, a mono-tetrahydrofuran acetogenin, arrests cancer cells at the $G_1$ phase and causes cytotoxicity in a Bax- and caspase-3-related pathway. *Life Sciences* 72, 2853–2861. DOI: 10.1016/s0024-3205(03)00190-5

Zafra-Polo, M.C., Figadère, B., Gallardo, T., Tormo, J.R. and Cortes, D. (1998) Natural acetogenins from Annonaceae, synthesis and mechanisms of action. *Phytochemistry* 48, 1087–1117. DOI: 10.1016/S0031-9422(97)00917-5

Zaman, K. (2013) Pharmacognostical and phytochemical studies on the leaf and stem bark of *Annona reticulata* Linn. *Pharmacognosy Phytochemical* 1, 1–7.

Zeng, L., Ye, Q., Oberlies, N.H., Shi, G., Gu, Z.M. *et al.* (1996a) Recent advances in annonaceous acetogenins. *Natural Product Reports* 13: 275. DOI: 10.1039/np9961300275

Zeng, L., Zhang, Y., Ye, Q., Shi, G., He, K. *et al.* (1996b) *Cis*-gigantrionenin and 4-acetyl gigantetrocin A, two new bioactive annonaceous acetogenins from *Goniothalamus giganteus*, and the stereochemistries of acetogenin 1,2,5-triols. *Bioorganic & Medicinal Chemistry* 4, 1271–1279. DOI: 10.1016/0968-0896(96)00111-3

Zeng, L., Wu, F.E., Oberlies, N.H., McLaughlin, J.L. and Sastrodihadjo, S. (1996c) Five new monotetrahydrofuran ring acetogenins from the leaves of *Annona muricata*. *Journal of Natural Products* 59, 1035–1042. DOI: 10.1021/np960447e

# 4 Processing and Products

**Alka Sharma[1] and V. Eyarkai Nambi[2*]**

*[1]Institute Technology Management Unit, ICAR-Central Institute of Post Harvest Engineering and Technology, Ludhiana, Punjab, India; [2]Department of Food Plant Operations Incubation and Entrepreneurship, National Institute of Food Technology Entrepreneurship and Management-Thanjavur, Tamil Nadu, India*

## 4.1 Introduction

Annona fruits are mainly consumed fresh. However, some of them, such as soursop (*Annona muricata*), sugar apple (*Annona squamosa*), cherimoya (*Annona cherimola*) and atemoya (*A. cherimola* × *A. squamosa*), can be processed and used in the preparation of juice, nectars, sherbets, ice cream, syrup and cakes (Pinto *et al.*, 2005). Soursop is the most suitable for processing, not only because of its high sugar content and delicate aromatic flavour but also because its pulp does not oxidize like that of sugar apple and cherimoya (Villachica *et al.*, 1996; Mororo *et al.*, 1997).

Value addition enables the creation of a wide range of sugar apple-based products. These include pulp, nectar, juice, ice cream, yogurt, desserts, sweets and even alcoholic beverages such as wine and liqueur. Diversifying product offerings caters to different consumer preferences and opens up new market segments, thereby increasing overall demand.

Proper processing techniques ensure the retention of essential nutrients present in annona fruits. Technologies such as minimal processing, freeze-drying or vacuum packaging help preserve vitamins, minerals and antioxidants, maintaining the nutritional integrity in the developed products. Processed products offer convenience and accessibility to consumers, as they are available in convenient forms that require minimal preparation. Ready-to-eat snacks, beverages or pre-packaged desserts are more accessible to busy consumers, thereby expanding the consumer base and market reach. Value-added products typically command higher prices compared with fresh fruit, translating into increased profitability for growers and processors. By adding value through processing, businesses can capture a larger share of the market and enhance their competitiveness in the industry. Processed products also have better export potential compared with fresh fruit due to their longer shelf life and reduced susceptibility to damage during transportation. Exporting value-added products opens up international markets, increasing revenue opportunities for growers and processors. In addition, processing not only utilizes the fruit but also allows the use of by-products such as peel, seeds and pulp. These by-products can be repurposed for animal feed, composting or extraction of bioactive compounds, further enhancing the sustainability and economic viability of processing operations.

* Email: eyarkainambi@iifpt.edu.in


DOI: 10.1079/9781800627628.0004

## 4.2 Processing and Product Development

### 4.2.1 Soursop

#### *4.2.1.1 Extraction of pulp*

Soursop pulp has a distinctive and highly pleasing flavour and aroma and is widely used in the preparation of various products. A flow chart for the extraction of pulp is presented in Fig. 4.1.

Hand peeling of soursop fruit is a difficult process to develop commercially. Mororo *et al.* (1997) suggested the use of a processor. Attempts to remove the peel by the lye peeling method are not effective because of the development of reddish-brown discoloration, which affects pulp quality. Soursop pulp should be processed below 93°C and frozen to maintain its quality (Torres and Sánchez, 1992). The percentage of pulp recovery ranges from 62% to 85.5% (Pinto *et al.*, 2005). The variation is due to the type of equipment, extraction method, cultivar, cultural practices and number of seeds per fruit (Nakasone and Paull, 1998). Freezing is the most commonly used pulp conservation method, but it often involves inadequate control and poor raw-material handling and processing practices (da Costa *et al.*, 2014). Freezing is usually carried out using conventional freezers, which make it very slow, and leads to the production of large ice crystals that can harm the product's inner structures, especially the texture (Teixeira *et al.*, 2006). Pasteurization is effective to control enzyme activity but has negative effects on sensory properties, producing a bitter flavour. Adding antioxidants (0.15% ascorbic acid, 0.2% citric acid or 0.02% EDTA-$Na_2$) can maintain the sensory properties during storage (Olaeta and Abufom, 1986).

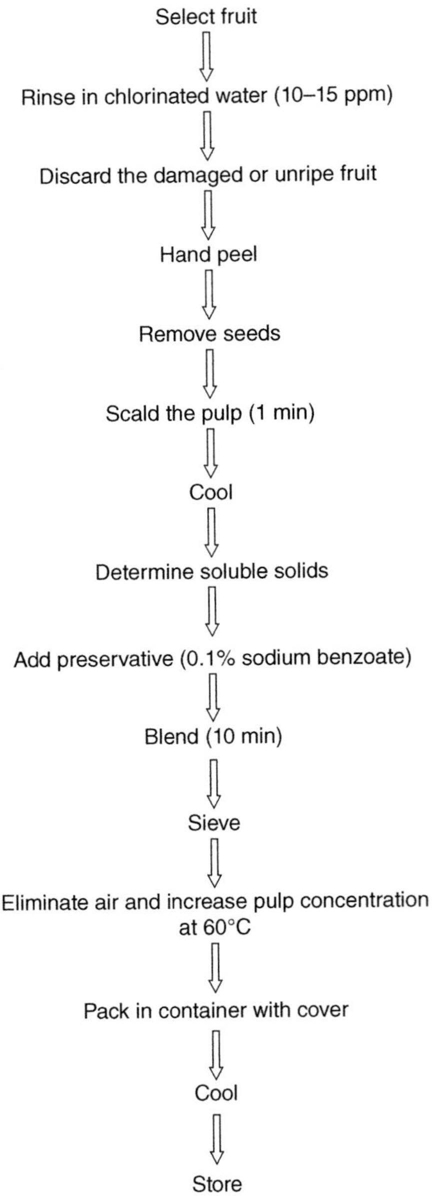


**Fig. 4.1.** Flowchart for the extraction of soursop pulp. Developed after Torres and Sánchez (1992) and Mororo *et al.* (1997).

#### *4.2.1.2 Dehydration*

Juice dehydration for use in instant beverages has become an alternative to artificial juice available on the market (de Oliveira *et al.*, 2007). Soursop powder obtained by freeze-drying (lyophilization) was found to have a water content 2.1 g 100 $g^{-1}$, pH of 4.7, total soluble solids (TSS) of 39.4°Brix, total titratable acidity of 0.85 g 100 $g^{-1}$ and vitamin C (ascorbic acid) content of 32.6 mg 100 $g^{-1}$ (Mata *et al.*, 2005). Pulps are initially concentrated in phases with different soluble solids content, and a constant fraction of maltodextrin is fixed in the final pulp samples (Telis-Romero *et al.*, 2007). Maltodextrin is added to the pulp to prevent stickiness between particles. Da Costa *et al.* (2014) standardize the spray-drying method for pulp.

Different concentrations of maltodextrin DE 20 (15%, 30% and 45%) were added to the commercial soursop pulp, which was then dehydrated. The results obtained for the three powder treatments of maltodextrin were, respectively: water activity of 0.19 ± 0.00, 0.20 ± 0.00 and 0.18 ± 0.01; moisture of 1.17 ± 0.12, 1.47 ± 0.05 and 1.82 ± 0.06 g 100 g$^{-1}$; pH of 3.75 ± 0.05, 3.73 ± 0.06 and 3.70 ± 0.03; soluble solids of 89.67 ± 0.00, 89.84 ± 0.00 and 90.00 ± 0.06°Brix; acidity of 3.01 ± 0.02, 1.91 ± 0.03 and 1.24 ± 0.03 g 100 g$^{-1}$; ascorbic acid levels of 18.90 ± 0.00, 14.48 ± 0.00 and 11.26 ± 0.78 mg 100 g$^{-1}$; hygroscopicity of 5.93 ± 0.40, 3.82 ± 0.16 and 3.28 ± 0.38 g 100 g$^{-1}$; degree of caking of 78.36 ± 2.86%; 35.38 ± 6.07% and 24.77 ± 4.89%; and rehydration time of 2.03 ± 0.46, 1.16 ± 0.50 and 0.59 ± 0.30 s. These results suggested that the atomization process significantly reduced the water content of pulp, which resulted in a product with moisture and water activity stability in terms of product preservation. Soursop powders with 30% and 45% maltodextrin showed few significant differences in the physico-chemical and hygroscopic analysis. It was thus suggested to use 30% maltodextrin for pulp atomization.

#### 4.2.2.3 Nectar

Nectar is the unfermented but fermentable product obtained by adding water with or without the addition of sugar honey and/or syrup and/or sweeteners as a food additive to a mixture of these products (CODEX, 2005) Sanchez-Nieva *et al.* (1953) prepared soursop nectar by dispersing the fruit pulp in water. They suggested that the TSS should be 6–8°Brix. To obtain the correct acid:sugar ratio, the pH of the nectar should be 3.7 (by addition of citric acid), corresponding to a total acidity of 0.4%; the TSS should be 11–15°Brix (by adding sugar). The nectar should be pasteurized at 90.6°C and canned in plain tin cans. De Oliveira *et al.* (1994) suggested adding ascorbic acid to the pasteurized puree at a rate of 0.5–1.5 g 450 g$^{-1}$, which improved retention of the nectar's flavour and served as an antioxidant to control polyphenol oxidase-mediated pulp darkening of the fruit. Peters *et al.* (2007) investigated soursop nectar storage by adjusting the pH to 3.6–3.7 with 0.1% xanthan gum, and 6°Brix or 8°Brix pulp adjusted to 13°Brix or 15°Brix with sucrose (4°C or 30°C, 12 weeks) using pasteurized pulp (fresh or frozen). Frozen pulp nectars had lower pH and higher acidity. Browning increased in most nectars, but one specific formulation (unstored 8°Brix pulp adjusted to 15°Brix) remained high quality. All nectars showed microbial growth after 8–12 weeks.

Pasteurization causes degradation of the bioactive compounds, loss of nutrients and unfavourable changes in physical properties such as stability, viscosity and colour due to the application of high temperatures (Cruz-Cansino *et al.*, 2021) Thermo-ultrasound (TUS), which combines the use of ultrasound with temperatures below 100°C (usually 50–60°C). Zafra-Rojas *et al.* (2023) compared thermosonicated and pasteurized nectar. The soursop nectar (25%) was processed and response surface methodology was used to determine the optimal conditions for TUS. TUS (75–90% amplitude, 3.15–15 min) was applied and 2% stevia and 6% agave inulin were added as sweeteners. The optimum TUS condition was found to be 82% amplitude for 9.15 min. The total phenolic content was expressed as gallic acid equivalents (GAE; 38.40 mg 100 ml$^{-1}$) and antioxidant activity (measured using the ABTS assay) was expressed as Torlex equivalents (TE) (31.28 μmol 100 ml$^{-1}$). Compared with pasteurized the sample, the TUS-treated sample had high values of L* (lightness), h° (hue angle), total phenolic content, ABTS and 2,2-diphenyl-1-picrylhydrazyl (DPPH) and low pectin methylesterase activity. TUS improved the physico-chemical properties and *in vitro* bioaccessibility of antioxidants in the soursop nectar. This beverage also offers low caloric and probiotic properties that benefit consumer health (Zafra-Rojas *et al.*, 2023).

Kumar *et al.* (2015) analysed samples every 30 days for 90 days. Over time, acidity, TSS, total sugar and reducing sugar were found to increase, while non-reducing sugar decreased. Nectar with 20% pulp, 0.3% acidity, and 17% TSS from *A. squamosa* genotype IGCA-38 had the highest acidity (0.40%), TSS (18.88%), total sugar (13.64%) and reducing sugar (4.58%), making it suitable for commercial production. Shrivastava *et al.* (2013) suggested the use of 20% pulp, 0.3% acidity and 25% TSS, which was deemed suitable for commercial production.

#### 4.2.1.4 Alcoholic beverages

Low-alcohol beverages with appreciable qualities and acceptability can be produced from ripe soursop fruit. Minh (2015) investigated the technical factors affecting soursop juice fermentation and suggested using 0.03% pectinase, incubation at 40°C for 2 h, an 8% yeast (*Saccharomyces cerevisiae*) ratio and maintaining a pH of 4.0. The fermentation temperature was suggested as 26°C for 60 h. An 8% concentrated syrup (65°Brix) at 30% was suggested to be added along with 1.5% caramel. The prepared beverage should be pasteurized at 80°C for 10 min. De Oliveira *et al.* (2016) investigated the effect of varying concentrations of pulp (300, 400 and 500 g) and sucrose syrup (50°Brix, 60°Brix and 70°Brix) on the physical and chemical characteristics of beverages. In the production process, the pulp was macerated with distilled alcohol for 20 days, after which it was filtered, added to the syrup, bottled in polyethylene bottles and subjected to accelerated ageing. All drinks resulted in alcohol levels (15.25–16.69%, v/v) and total sugar (12.63–17.97%) in accordance with the standards.

#### 4.2.1.5 Jam

Soursop jam is prepared by cooking the fruit purees at 100°C with sucrose and pectin until 65°Brix is reached. Fruit purees are first obtained by homogenization of the fresh pulp using a blender. The pectin (1.5 g) is dispersed in 45 g sucrose and the fruit purees cooked with the addition of sugar to about 65°Brix. The pH was adjusted at the end of cooking to about 3.2–3.4 by adding citric acid (10%, w/v). The prepared jams are poured into sterile glass jars, packed and immediately cooled to room temperature. Chemical analysis of the jam has shown that it has a moisture content of 21.52 g 100 $g^{-1}$, carbohydrates at 77.83 g 100 $g^{-1}$, pH 3.41 and 68.60°Brix (Orsi *et al.*, 2012).

#### 4.2.1.6 Marmalade

To prepare marmalade, 17.8% pulp, 10.7% sugar, 0.02% sodium benzoate, 0.02% sodium metabisulfide and water to 100% are treated at 100°C for 15 min (Villachica *et al.*, 1996). The pH of marmalade ranges from 3.1 to 3.3 and it contains 60% concentrated pulp (Villachica *et al.*, 1996).

#### 4.2.1.7 Encapsulation of fruit pulp

Soursop constitutes a rich source of bioactive compounds with great potential to develop functional foods. The fresh fruit is often processed as a pulp and then preserved by pasteurization or freezing for industrial processing and commercialization. However, several active compounds from soursop fruit can be degraded by the effect of applied treatments (López-Córdoba *et al.*, 2016). Ionic gelation is a simple and fast way of obtaining encapsulating systems, which allow protection of the encapsulated compounds from factors that may cause deterioration such as temperature, light, pH, moisture and oxygen (López-Córdoba *et al.*, 2014). López-Córdoba *et al.* (2016) added alginate hydrogels of different concentrations (5%, 10% and 20% by weight) of soursop pulp obtained by ionic gelation. The hydrogels had diameters of around 4 mm and high circularity. After drying, the systems exhibited water content values of 8–10% and water activity 0.3–0.4. Micrographs obtained by scanning electron microscopy showed that incorporation of soursop pulp at 20% led to better preservation of the morphology of the beads. The results suggest that the calcium alginate beads containing soursop pulp could have useful applications as healthy food ingredients.

Figure 4.2 shows some of the soursop (guanabana) products available in Mexican and Brazilian markets.

### 4.2.2 Cherimoya

#### 4.2.2.1 Fresh cut

Cherimoya fruit are suitable for fresh-cut processing, considering its high number of seeds, large size and inedible peel, resulting in a significant non-edible portion when consumed as a whole fresh fruit (Campos-Vargas *et al.*, 2008). Working with cherimoya cultivar 'Concha Lisa', Campos-Vargas *et al.* (2008) suggested using fully ripe fruit with firmness values close to 1 kgf. Before processing, the fruit were washed with a 100 mg $l^{-1}$ sodium hypochlorite solution at room temperature to eliminate surface contamination. The fruit was then hand peeled with a sharp knife, deseeded and the pieces immersed for 2 min in 0.5% L-cysteine and then drained for 1 min. The fresh-cut cherimoyas were stored in

**Fig. 4.2.** (A–C) Guanaba juice. (D) Ice cream mix. (E) Milk-like drink. (F) Frozen pulp. Photographs courtesy of Dr E.M. Gonzalez (A–C) and Dr Carlos Santos (D–F).

0.5 l plastic containers at 0°C and 90–95% relative humidity. L-cysteine was found to be effective in reducing browning development, without affecting other quality attributes after 6 or 12 days of storage.

##### *4.2.2.2 Flour*

Cherimoya flour can be prepared from pulp by dehydrating in a convective dryer at 50°C with an air velocity of 3 m $s^{-1}$ until constant weight is attained. Subsequently it is ground and sieved using an 80-mesh sieve to obtain the flour (Alvarez-Yanamango *et al.*, 2019).

### 4.2.3 Atemoya

#### *4.2.3.1 Liquor*

Fermented beverages from atemoya can be made using fruit pulp (5%), water (78%), acidulant (0.04% citric acid), pH regulator (0.04% diammonium phosphate) sugar (15%) and yeast (1.92% *S. cerevisiae*). This process of liquor production includes selection, cleaning and disinfection of the fruit, characterization of the pulp, pulping, mixing, fermentation, filtration, clarification and packaging. The results obtained show a good sensory acceptance in relation to the texture, colour, flavour and smell. A pH of 3.78, volatile acidity of 0.08%, density of 1.10 g $ml^{-1}$, reducing sugar of 10.56%, alcohol grade of 9.770 GL, methanol at 67 p.p.m., *Escherichia coli* absent, aerobic mesophiles count of 90 colony-forming units (CFU) $ml^{-1}$ and yeast and fungal counts of 50 CFU $ml^{-1}$ showed that the obtained liquor is a physico-chemically, microbiologically and sensorial stable drink (Martínez-Girón *et al.*, 2019).

### 4.2.4 Sugar apple

#### *4.2.4.1 Pulp extraction*

Sugar apple fruit pulp contain multiple seeds, along the inner side of the peel making pulp extraction a difficult task. Moreover, the peel disintegrates while cutting and handling. Because of this complexity, manual pulp extraction is mostly used (Fig. 4.3). Pulp recovery varies among the different extraction methods. Dhumal (1994) reported 46.4%, 46.8% and 46.7% recovery of pulp from sugar apple using cold hand press, hot hand press and enzymatic extraction methods, respectively. Rao (2009) developed a manual pulp extraction method using a round hair comb in a stainless-steel sieve, employing a rotary motion on the scooped fruit, at the Central Research Institute for Dryland Agriculture (CRIDA), Hyderabad, India. This pulp is used by the ice cream industry.

An automatic sugar apple pulper (Fig. 4.4) was developed by Nambi *et al.* (2011) and contains three mechanisms: fruit cutting, fruit scooping and a pulping mechanism. The machine is fully automatic, with pneumatic power and electronic controls. The fruit are fed into the fruit holding cups and then cut into two halves. The peel and pulp are separated with the help of a peel holding sieve and pressing and scooping mechanisms. Both pulp and peel are collected in separate outlets. The separated pulp along with the seeds are fed into the pulping section for separation of pulp and seed. After pulping, the seeds are collected in a separate outlet and the pulp is separated into two forms, fine and intact pulp.

#### *4.2.4.2 Pulp processing*

Fresh pulp is either directly placed into containers or processed by blending with potassium metabisulfite at a concentration of 0.4 g $kg^{-1}$ for freezing, or 0.8 g $kg^{-1}$ for storage at room temperature (Fig. 4.5).

George and Nissen (1993) reported two methods for removing sugar apple pulp, a screen processing process and a reaming process. Frozen pulp treated with preservatives such as ascorbic acid (1500–2000 mg $kg^{-1}$ pulp) or

**Fig. 4.3.** Manual extraction of sugar apple pulp.

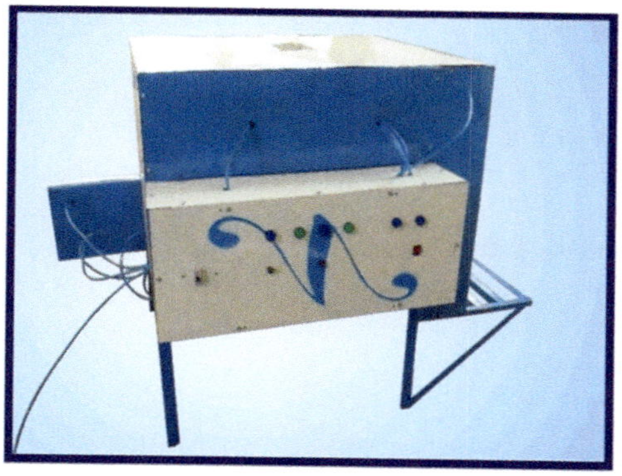

**Fig. 4.4.** Automatic sugar apple pulper.

Select fruit
⇩
Cut into halves
⇩
Extract pulp and remove seeds
⇩
Freshly extracted pulp
⇩
Add potassium metabisulfite at 0.4 g $kg^{-1}$
⇩
Transfer into jars/bottles heated at 70°C for 3 min
⇩ Freeze ⇩ Store for further usage
⇩ Final filling in jars/bottles
⇩
Boil/heat at 95°C for 15 min
⇩
Store for further usage

**Fig. 4.5.** Flowchart for processing and storage of sugar apple pulp. Developed after Sinthiya and Poornima (2017).

potassium metabisulfite can be stored for 120 days without losing its flavour or colour after thawing. Heating the pulp at 83°C for 2 min reduced polyphenol oxidase activity but resulted in lower sensory characteristics; this effect could be mitigated by adding 2000 p.p.m. of ascorbic acid (Prasanna *et al.*, 2000).

Minimal processing of sugar apple pulp by treating with 0.1–0.5% ascorbic acid and storing it in modified-atmosphere packages at 0°C for 4 weeks was studied by Gamage *et al.* (1997). Vacuum sealing with a 65 μm five-layer co-extruded bag containing polyethylene, nylon and linear low-density polyethylene maintained the pulp's creamy colour during storage and after exposure to ambient conditions for 3 h. Kolekar (2002) preserved sugar apple pulp with varying sugar content (1:1, 1:2 and 1:4) and concluded that the pulp can be stored for up to 90 days under frozen condition with minimal changes in physico-chemical properties, with the 1:4 ratio yielding particularly good results. Pulp mixed with saccharified starch and stored under frozen conditions remained better for up to 4 months than pulp packed with nitrogen gas (Geeta, 2000).

Mixing the pulp with 250 mg ascorbic acid and 500 g sugar $kg^{-1}$ and storing it at −40°C can maintain its quality for up to 90 days (Kolekar *et al.*, 2003).

Pulp treated with 100 p.p.m. ascorbic acid, packed in 200-gauge polyethylene and frozen at −25°C with subsequent storage at −18°C was found to maintain pulp quality (Chikhalikar *et al.*, 2000). These authors used a blend of 2% glycerol, 1% propylene glycol, 10% glucose syrup and 10% maltodextrin, which demonstrated an extended shelf life of 75.33 days for ascorbic acid degradation, compared with 34.82 days for the control.

Shashirekha *et al.* (2008) studied changes in volatile compounds in fruit pulp using different processing methods (Fig. 4.6). They observed that frozen pulp stored for 12 months maintained a flavour profile similar to that of fresh pulp. Heating the fresh pulp at 55°C and 85°C also enhanced its flavour.

Steam blanching of pulp and storage at −4°C and −18°C shows improved sensory attributes including appearance, colour, flavour, taste, texture and overall acceptability. Heat application can alter the pulp's molecular composition, potentially affecting viscosity, flavour, taste and colour due to processes such as protein denaturation, volatile component evaporation and starch gelatinization (Kamble and Soni, 2010). Polyphenol oxidase activity in sugar apple pulp can be effectively inhibited by heat treatment

Select ripe fruit

↓

Cut into halves

↓

Remove the seeded pulp manually and discard peel

↓

Collect seeded pulp

↓

Process the seeded pulp with a pulper

↓

Obtain pulp (can be used fresh, stored or heat-treated)

↓

Weigh out 100 g of pulp (fresh-weight basis)

↓

Perform extraction using a chilled mixture of dichloromethane and *n*-pentane (1:1 ratio), using 150 ml solvent for each extraction, and repeat three times

↓

Combine all collected extracts

↓

Pass the pooled extracts through anhydrous sodium sulfate to remove moisture

↓

Concentrate the filtrate using a Vigreux column at a temperature of <35°C

↓

Collecting the final flavour extract (~1 ml)

**Fig. 4.6.** Steps involved in extraction of flavour volatiles from sugar apple fruit. Developed after Shashirekha *et al.* (2008).

and the use of antioxidants. Steaming the pulp at 82°C for 5 min caused complete inhibition of polyphenol oxidase activity. Additionally, storing the pulp at −18°C resulted in a slower increase in total polyphenol content compared with storage at 5°C (Pawar *et al.*, 2011). Deepa and Mohapatra (2020) investigated foam-mat drying (50–70°C) of pulp using glycerol monostearate as the foaming agent at concentrations ranging from 0.5% to 4.5%, with 0.5% methylcellulose as the stabilizer, and whipping times of 2–10 min with foam thicknesses of 2, 4 and 6 mm. Optimal conditions were determined as 3.5% glycerol monostearate and 0.5% methylcellulose whipped for 6 min, which caused 61.11% foam expansion, 94% foam stability and a density of 0.60 g $cm^{-3}$. Tray drying of foamed pulp took 240–450 min, achieving significant time savings compared with non-foamed pulp. Shrivastava *et al.* (2020) investigated the storage of pulp at −20°C for 8 months. Despite the natural tendency of pulp to develop an unattractive colour and off-flavours during processing and rapid deterioration during storage, the pulp maintained above-average organoleptic scores throughout the period, making it suitable for consumption. The addition of potassium metabisulphite as a preservative further enhanced sensory stability, allowing the pulp to maintain its quality and sensory characteristics over the storage duration.

#### *4.2.4.3 Juice*

Fresh juice is highly perishable and spoils within a day of extraction. The processing involves extracting juice from pulp using methods such as a screw press or manual hand pressing, filtering through a muslin cloth, and clarifying with 1.4 g polyvinylpyrrolidone $l^{-1}$ of juice. Sugar is then added, and the juice is heated and then cooled, with 10 g sodium benzoate added as a preservative. The final juice is bottled in sterilized bottles, sealed with crown cork and stored in a cool, dry place (Fig. 4.7) (Sinthiya and Poornima, 2017).

#### *4.2.4.4 Fruit pulp concentrate*

Concentrates are derived from the pulp of sugar apples. These concentrates are valued for their rich flavour, nutritional benefits and versatility in food and beverage applications. The production process involves selecting ripe fruit, peeling, deseeding and processing the pulp to concentrate its natural sugars and flavours while preserving its nutritional content. The process typically includes heating the pulp to eliminate microbial contamination, adding preservatives such as sodium benzoate to extend shelf life, and concentrating the pulp by removing water through evaporation (Fig. 4.8). The resulting concentrate retains the fruit's natural taste and aroma, making it an excellent ingredient for juices, smoothies, desserts and confectionery products (Fig. 4.9).

Fruit pulp processed below 93°C and frozen in polythene bags retains its taste and aroma. To prevent oxidation, it has been suggested to heat the pulp to 70°C for 20 min and add 0.5% ascorbic acid. It can be stored for 1 month in polythene bags at 5°C. For a longer storage, the pulp needs to be frozen at −18°C (Sinthiya and Poornima, 2017).

#### *4.2.4.5 Jam*

Sugar apple jam is a delightful spread made from the pulp, known for its creamy texture and sweet flavour. This jam captures the essence of the fruit, making it a popular choice for breakfast spreads, desserts and baking. Sugar apple jam processing involves selecting ripe sugar apples, peeling, deseeding and extracting the pulp. Sugar is then added (1:1) and heated until it has dissolved, followed by the addition of citric acid

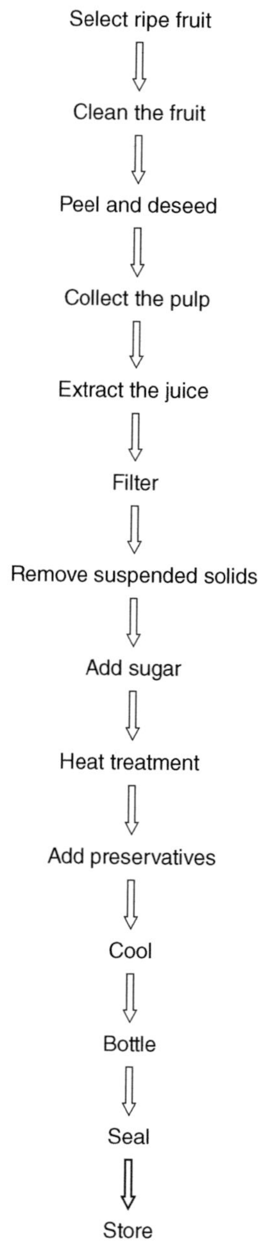


**Fig. 4.7.** Flowchart for processing of sugar apple juice. Developed after Sinthiya and Poornima (2017).

Select fresh, ripe fruit

⇓

Peel and remove seeds manually

⇓

Heat fruit pulp to 80°C

⇓

Add 0.1% sodium benzoate and mix well

⇓

Strain pulp to remove solids

⇓

Mix with sugar

⇓

Remove trapped air via compression

⇓

Concentrate pulp by heating at 100°C

⇓

Cool and store in sterilized containers

**Fig. 4.8.** Flowchart for processing of fruit pulp concentrate. Developed after Sinthiya and Poornima (2017).

**Fig. 4.9.** Sugar apple fruit pulp concentrate. Photograph courtesy of Sisir Mitra.

to improve the taste. The pulp is mixed with potassium metabisulphite (0.1 g kg$^{-1}$ in water) to increase the shelf life. After cooking, the hot jam is poured into sterilized jars, sealed and cooled (Fig. 4.10).

Sugar apple jam with 50% pulp showed high sensory value and was edible for 4 months. (Singh *et al.*, 2006). Saket *et al.* (2018b) investigated mixing papaya fruit pulp with sugar apple pulp in the preparation of jam using varying proportions of sugar apple pulp (40–70%), papaya pulp (60–30%) and sugar (30–60 g). Storing the prepared jam at room temperature for up to 100 days was found to affect the organoleptic scores for colour (8.59–7.53), flavour (8.73–8.26), texture (7.86–7.40), taste (8.60–7.93) and overall acceptability (8.80–8.40). The highest overall acceptability score (8.80) was achieved with the recipe comprising 60% custard apple pulp and 40% papaya pulp, with 40% sugar and 1.0 g citric acid. Over time, overall acceptability was highest on day 0 and slightly decreased with longer storage.

#### *4.2.4.6 Jelly*

The jelly captures the distinctive taste and fragrance of sugar apples, making it a versatile spread for toast, topping for desserts or flavourful ingredient in various culinary creations. Its natural sweetness and smooth texture make sugar apple jelly a favoured choice among fruit preserves. Shashirekha *et al.* (2003) developed a method for producing sugar apple jelly that is devoid of discoloration, bitterness and off-flavours. This method includes diluting pulp with water to achieve a mixture containing 60–80% weight by weight of pulp. The mixture is then filtered under a pressure of 1–3 bars to obtain a filtrate containing 45–60% weight by weight of pulp. Subsequently, sweeteners, preservatives, settling agents and other food additives are added. The mixture is boiled at a temperature of 90–100°C, followed by cooling to yield the final sugar apple jelly.

#### 4.2.4.7 *Syrup*

Syrup is a concentrated liquid extract made from the pulpy flesh of sugar apples. The process involves blending of pulp and straining to remove any fibres, yielding a smooth and flavoured liquid. Sugar along with citric acid is added to the strained pulp, and the mixture is heated to dissolve the sugar completely. This blend is then cooked down to a thick consistency, enhancing the natural sweetness and capturing the essence of sugar apples. The clear syrup is cooled and filled in bottles. The bottles are sealed using crown caps and sterilized for 15 min at 95°C (Sinthiya and Poornima, 2017) (Fig. 4.11).

The syrup can be used in a variety of ways, such as drizzling over pancakes, waffles or desserts, mixing into beverages, or as a flavouring agent in cocktails and mocktails. Its rich flavour and versatile applications make sugar apple syrup a popular choice for adding a sweet and fruity twist to dishes and drinks alike.

#### 4.2.4.8 *Toffee*

Toffee is a delectable confectionery made from sugar apple pulp with the addition of sugar, and often dairy ingredients such as butter or condensed milk. Toffee preparation consists of peeling the

Select ripe fruit
↓
Wash thoroughly
↓
Peel and remove seeds
↓
Extract pulp
↓
Mix pulp with sugar (1:1 ratio) and heat until sugar dissolves
↓
Add citric acid/lemon juice for flavour balance
↓
Continue heating until thickened
↓
Add preservative (e.g. potassium metabisulphite at 0.1 g $kg^{-1}$)
↓
Pour hot jam into sterilized jars
↓
Seal jars immediately
↓
Cool at room temperature
↓
Store in a cool, dry place

**Fig. 4.10.** Flowchart for processing of jam. Developed after Sinthiya and Poornima (2017).

Extract juice and remove astringency
↓
Add sugar (0.37 kg $l^{-1}$ of juice) and citric acid (1 g $l^{-1}$) and mix well
↓
Boil the mixture at 70°C for 3 min
↓
Cool the clear syrup
↓
Fill jars or bottles with syrup
↓
Seal bottles with crown caps
↓
Sterilize sealed bottles for 15 min at 95°C
↓
Cool and store

**Fig. 4.11.** Flowchart for processing of syrup. Developed after Sinthiya and Poornima (2017).

ripe fruit, deseeding it and extracting of pulp. This pulp is then cooked with sugar and butter or condensed milk over medium heat until it reaches a thick, caramel-like consistency. The mixture is stirred continuously to prevent burning and ensure even caramelization, followed by pouring into moulds or trays. It is left to cool and set, and then cut into bite-sized pieces. Sugar apple toffee delights with its creamy texture, rich caramel flavour and hint of sugar apple's unique sweetness, making it a favourite treat among confectionery enthusiasts. Toffees made with 55% pulp have the highest sensory score for appearance, colour, texture and overall acceptability (Mundhe *et al.*, 2008).

#### *4.2.4.9 Milkshake*

A milkshake is a creamy and refreshing beverage that blends the sweet and aromatic flavour of sugar apples with milk and sugar. The most preferred sugar apple milkshake is prepared using a ratio of 90:10 of buffalo milk:pulp, and ranks highly from 'like very much' to 'like extremely' in sensory evaluations (Poul *et al.*, 2009). Bakane *et al.* (2016), however, reported that blending 35% sugar apple pulp with 65% cows' milk resulted in the most preferred milkshake. They also reported that increasing the pulp percentage increased the TSS, acidity, total sugar and viscosity of the milk shake, while pH decreased. Figure 4.12 shows the process of making milkshakes.

Select ripe fruit

⇩

Wash and cut open fruit

⇩

Scoop out pulp

⇩

Separate seeds from pulp

⇩

Transferring pulp to blender

⇩

Add chilled milk

⇩

Add sweetener (sugar, honey or jaggery) as desired

⇩

Blend until milkshake is smooth

**Fig. 4.12.** Flowchart for preparation of milkshakes. Developed after Bakane *et al.* (2016) and Poul *et al.* (2009).

#### *4.2.4.10 Ice cream*

Sugar apple-flavoured ice cream offers a delightful twist on traditional flavours, blending the creamy texture of ice cream with the sweet and aromatic taste of sugar apples (Fig. 4.13). Sugar apple ice cream, incorporating 15% pulp and 15% sugar, has an energy value ranging from 61.42 to 97.27 kcal 100 $g^{-1}$. Alternatively, a low-fat version can be made with the same proportions of pulp and sugar but with 10% fat (Pawar *et al.*, 2011).

Markam *et al.* (2022) investigated the sensory properties of sugar apple pulp ice cream. The ice cream formulation with 100% pulp, 50% whipping cream, 20% condensed milk, 5% sugar and 5% milk powder received the highest scores for flavour (7.60), texture (8.47), taste (8.22) and overall acceptability (8.65) in organoleptic evaluation. Makwana *et al.* (2018) explored the physical, chemical and sensory attributes of sugar apple pulp ice cream containing 5–30% pulp. Stored at −35°C to −38°C, the ice cream samples were analysed over 0–60 days for physico-chemical properties and sensory qualities. Ice cream with 15% pulp scored highest for colour, texture, flavour, taste, overall acceptability and palatability in sensory evaluations, although sensory parameters deteriorated during storage.

#### *4.2.4.11 Dahi (yogurt)*

Sugar apple-flavoured dahi offers a refreshing and nutritious twist on traditional yogurt, combining fruity sweetness with probiotic benefits, making it a delightful treat or snack option. Dahi may be prepared following the flowchart (Fig. 4.14).

Probiotic dahi can be made using *Lactobacillus acidophilus* starter culture and enriched cows' milk with 1–4% soursop powder to increase

Select ripe fruit

↓

Pulp the fruit mechanically

↓

Add food-grade preservative (potassium metabisulphite or nitrite at 0.5 g $kg^{-1}$ of pulp)

↓

Store the preserved pulp at −20°C

↓

Incorporate whipping cream and blend thoroughly for 2–3 min

↓

Add sweetener

↓

Add condensed milk

↓

Add milk powder to enhance texture and richness

↓

Blending the mixture until smooth and uniform

↓

Freeze the mixture at −18°C for 7–8 h

↓

Pack and store the final product at −5°C

**Fig. 4.13.** Flowchart for preparation of ice cream. Developed after Pawar *et al.* (2011).

the total solids content. Chemical analyses have revealed that fat and protein content ranges from 3.35% to 3.51% and from 3.23% to 3.31%, respectively, while lactic acidity and total solids vary from 0.65% to 0.83% and from 12.00% to 15.03%. *L. acidophilus* counts were $22.5 \times 10^6$–$26.25 \times 10^6$ CFU $g^{-1}$, with minimal coliform, yeast and mould counts. Sugar apple powder significantly ($P < 0.05$) increased total solids and enhanced sensory quality. Optimal probiotic custard apple dahi formulation included 12% total solids, 2% sugar apple powder and 1% *L. acidophilus* starter culture (Patil *et al.*, 2015).

#### *4.2.4.12 Alcoholic beverages*

Alcoholic beverages involve fermenting the fruit's pulp to convert sugars into alcohol through yeast fermentation (Fig. 4.15). Minh (2014) examined the factors affecting sugar apple wine fermentation, favouring grinding over osmotic extraction for fruit pulp. *S. cerevisiae* yeast was used with a 10% supplement to achieve a density of $5 \times 10^6$ CFU $ml^{-1}$. Initial soluble dry matter was 20%, with primary fermentation completed in 4 days, resulting in 8.2% ethanol and 7.25°Brix soluble dry matter. Jagtap and Bapat (2015) studied sugar apple fruit wine fermented

Milk

⇩

Warm by heating to 45°C

⇩

Filter and clarify

⇩

Standardize fat to 3.5%

⇩

Pasteurize at 85°C for 15 min

⇩

Cool to 30–35°C

⇩

Add sugar apple powder (0%, 1%, 2% or 3%)

⇩

Add *Lactobacillus acidophilus* starter culture (21%) and mix thoroughly

⇩

Fill cups

⇩

Incubate at 35–37°C for 6–8 h until dahi (yoghurt) forms

⇩

Cool and store at 5°C

**Fig. 4.14.** Flowchart for preparation of custard apple dahi (yogurt). Developed after Patil *et al.* (2015).

Prepare fruit pulp

⇩

Add potassium metabisulfite

⇩

Add sugar

⇩

Dilute with water for preparation of must

⇩

Prepare mother culture/yeast activation

⇩

Pasteurize must

⇩

Prepare substrate and inoculate (7 days)

⇩

Carry out fermentation

⇩

Filter the mixture

⇩

Perform distillation at 60°C

⇩

Allow wine to mature

**Fig. 4.15.** Flowchart for preparation of sugar apple fruit wine. Developed after Kumar (2010).

using *S. cerevisiae* (NCIM 3282). Evaluating the antioxidant capacity, total phenolic content and DNA damage protection, it was observed that the wine effectively scavenged DPPH and DMPD free radicals and showed robust ferric-reducing antioxidant capacity. It contained gallic acid, protocatechuic acid, gentisic acid, caffeic acid and *p*-coumaric acid.

Kumar (2011) prepared sugar apple wine by diluting pulp with water (at ratios of 1:2, 1:3 or 1:4) with or without 0.1% diammonium hydrogen phosphate (DAHP). Potassium metabisulfite was added at 125 p.p.m. and fermented with *S. cerevisiae* var. *ellipsoideus*. During fermentation, TSS, sugars and pH decreased, while alcohol and acidity increased. Fermentation was finished by day 12 for the 1:2 and 1:3 dilutions with DAHP. The 1:4 dilution with DAHP had higher alcohol (8.14%, v/v) and lower sugars (1.91% reducing sugars, 5.53% total sugars), acidity (0.56%), phenols (231.66 $\mu g\ ml^{-1}$) and pH (3.72). It also showed the fastest fermentation rate (1.32°Brix in 24 h), while the 1:3 dilution without DAHP had the slowest fermentation rate (1.25°Brix in 24 h). The fermentation efficiency was highest (88.05%) with the 1:3 dilution with DAHP and lowest (79.98%) with the 1:2 dilution without DAHP.

#### 4.2.4.13 Pulp powder

Sugar apple powder is a versatile product derived from the fruit through dehydration and pulverization, yielding a fine powder that preserves the fruit's nutritional and flavour characteristics (Fig. 4.16). A flowchart for the preparation of sugar apple pulp powder is shown in Fig. 4.17. Packed with vitamins, minerals and dietary fibre, it serves as a beneficial ingredient in a wide range of foods and drinks. This powder enhances the flavour of desserts, smoothies, ice creams and yogurt, and can be included in baking mixes and nutritional supplements. Valued for its extended shelf life and convenience, sugar apple pulp powder provides an easy way to savour the distinctive taste and health advantages of sugar apple throughout the year.

The pulp powder can also be prepared by freeze-drying fresh pulp treated with 5% maltodextrin at −40°C for 41 h, ensuring high retention of vitamins and carbohydrates. Its shelf life can be extended to 90 days at room temperature by vacuum sealing in 50 μm polyethylene bags. The pulp powder can be used in acetaminophen tablets and affects their properties; tablets with polyvinylpyrrolidone as a binder showed doubled disintegration time, while replacing the polyvinylpyrrolidone with pulp powder and using starch as a disintegrant significantly reduced disintegration time (Thube *et al.*, 2011).

Convective tray drying can also be used to dehydrate sugar apple pulp (Soni *et al.*, 2021). Convective tray-dried pulp was shown to have a moisture content of 7.44%, solubility index of 8.00 ml, bulk density of 359.97 kg $m^{-3}$, tapped density of 423.84 kg $m^{-3}$, hygroscopicity of 0.83 g kg $solids^{-1}$ $min^{-1}$, TSS of 53.86°Brix, water absorption capacity of 0.66%, degree of caking of 50.97% and colour lightness of 77.69. The powder showed a favourable compressibility (15.06%) and Hausner ratio (1.18), indicating good flow behaviour and desirable physical properties.

#### 4.2.4.14 Vinegar

Vinegar production from sugar apple pulp requires a 30-day fermentation period. The resulting vinegar was shown to have the following characteristics: specific gravity of 1.019 g $ml^{-1}$, alcohol concentration of 1%, acetic acid level of 5.39% (v/v), 2.0°Brix reading and pH of 2.8. Consumer taste tests using a nine-point scale indicated a 'like very much' rating for the sugar apple vinegar, while commercial vinegar samples received

**Fig. 4.16.** Sugar apple fruit pulp powder. Photograph courtesy of Sisir Mitra.

Select mature fruit
↓
Separate pulp and discard seeds
↓
Add ascorbic acid (100 ppm) and potassium metabisulfite (350 ppm)
↓
Liquefy with pectinolytic enzymes
↓
Strain liquid through a fine cloth
↓
Add maltodextrin (15%, 17% or 20%)
↓
Add tricalcium phosphate (1%)
↓
Mix until consistent
↓
Dry using a spray dryer
↓
Powder

**Fig. 4.17.** Flowchart for preparation of sugar apple pulp powder. Developed after Patil (2011).

Select ripe fruit and extract pulp
↓
Prepare must (adjust sugar and pH and pasteurize)
↓
Prepare yeast starter and inoculate must
↓
Primary fermentation (7 days)
↓
Filter and clarify the fermented must
↓
Add alcohol wort for vinegar production
↓
Inoculate with acetic acid bacteria
↓
Ferment with aeration in dark
↓
Age the vinegar for flavour development
↓
Pasteurize and bottle
↓
Store for further use

**Fig. 4.18.** Flowchart for preparation of sugar apple vinegar. Developed after Raichurkar and Dadagkhair (2017).

a 'like extremely' rating (Raichurkar and Dadagkhair, 2017). Figure 4.18 shows the process for vinegar production.

#### *4.2.4.15 Non-alcoholic beverages*

Non-alcoholic sugar apple beverages or squashes have been developed with various formulations to enhance taste and shelf life. A blend of sugar apple and lime juice (3:2) with 15% TSS and 0.2% acidity was found to be optimal for colour, taste and overall acceptance (Pilania *et al.*, 2010). Potassium metabisulfite as a preservative was effective in maintaining microbial and sensory quality (Markam and Singh, 2012). The squash showed increased TSS, acidity, ascorbic acid and sugar content during storage, while tannin content and organoleptic score decreased. The best recipe for nectar has been suggested as 20% pulp, 0.3% acidity and 25% TSS (Shrivastava *et al.*, 2013).

Carbonated non-alcoholic beverages are refreshing drinks made by blending pulp or juice with carbonated water, often enhanced with sweeteners and flavourings (Fig. 4.19). Patil *et al.* (2011) developed a carbonated beverage using 100 g sugar apple pulp, 100 g sugar, 90 ml water and 3 g citric acid. They assessed its physico-chemical characteristics and organoleptic qualities, and found it to be a good source of carbohydrates (12.9%), protein (0.7%) and fat (0.18%).

Prepare fruit
↓
Peel and deseed
↓
Extract juice using pectinase enzymes
↓
Filter the extracted juice
↓
Combine juice with filtered syrup
↓
Beverage production (ready-to-serve)
↓
Refine mixture
↓
Fill bottles
↓
Add carbon dioxide
↓
Cap with crown caps
↓
Store

**Fig. 4.19.** Flowchart for preparation of carbonated beverages. Developed after Patil *et al.* (2011).

Select ripe, fresh fruits
↓
Wash, peel and deseed
↓
Extract fruit pulp
↓
Homogenize pulp until texture is uniform
↓
Add sweeteners and food-grade preservatives
↓
Spread mixture on to lined drying trays
↓
Dry
↓
Cool and cut into bar shapes
↓
Wrap with food-grade packaging material
↓
Store in cool, dry conditions

**Fig. 4.20.** Flowchart for making leather from custard apple. Developed after Saniya *et al.* (2023).

#### *4.2.4.16 Leather*

Leather is a dehydrated fruit product made from pulp. The pulp is blended with sugar and sometimes other flavourings, then spread thinly and dried, often using solar or oven drying methods, until it forms a leathery texture (Fig. 4.20). Sugar apple leather offers a convenient, portable and shelf-stable snack option.

Saniya *et al.* (2023) made leather using varying pulp (65–100%) and sugar (0–35%) levels. Extracted pulp was defrosted, homogenized and mixed with sugar, citric acid (0.5%), pectin (1%) and potassium metabisulphite (0.01%). The mixture was spread on stainless steel trays and dried in an oven at 55°C until the moisture content reached 15–20%. The dried sheets were then packed in aluminium foil and stored for up to 150 days to assess the organoleptic qualities. Leather prepared with 65% pulp, 35% sugar and 0.5% citric acid had the highest score for texture, taste and overall acceptability.

Saket *et al.* (2018a) evaluated the chemical characteristics of mixed fruit bars made with sugar apple (60–90%) and papaya pulp (10–40%) with 10–15% sugar. Over 100 days of storage at room temperature, they observed changes in TSS (41.70–43.20°Brix), acidity (1.35–1.31%), pH (3.92–3.32), ascorbic acid (106.0–93.00 mg 100 $g^{-1}$), moisture (18.48–16.12%), reducing sugar (49.78–53.90%) and total sugar (77.24–79.30%).

#### *4.2.4.17 Basundi*

Basundi is a well-loved Indian dessert, especially in Maharashtra, Gujarat and Karnataka. It is a creamy, dense, milk-based sweet created by slowly reducing milk over heat. Common flavour additions include spices such as cardamom and saffron, and it may contain nuts and other ingredients. Gite *et al.* (2017) prepared basundi using different combinations of basundi and sugar apple pulp. The basundi prepared with 30% pulp had good sensory scores for all attributes, indicating superior taste, flavour, increased nutritional value and overall acceptability compared with plain basundi. Figure 4.21 shows a flowchart for preparation of basundi.

Pulp levels of 5–15% in basundi increased the sucrose content while altering the fat, protein, carbohydrate and ash levels, ultimately affecting total solids content, with the highest acceptability score at 10% pulp concentration (Bawale *et al.*, 2019).

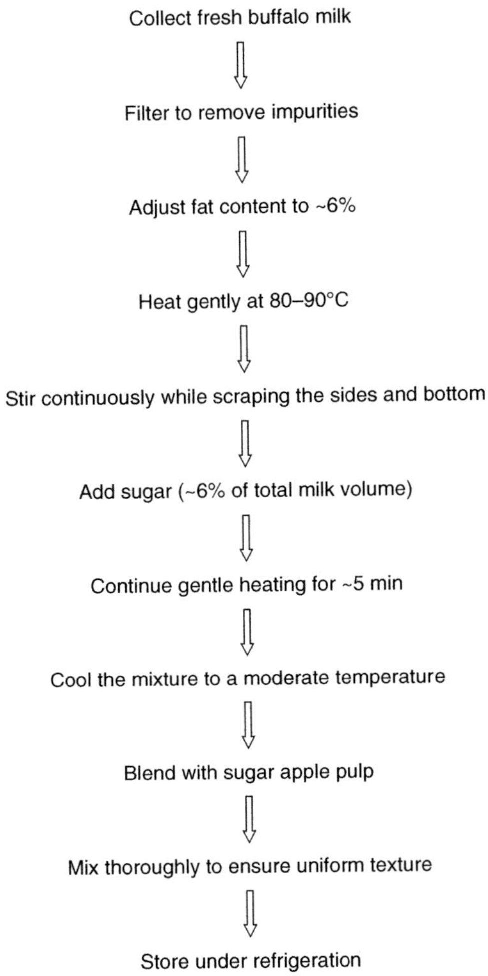


**Fig. 4.21.** Flowchart for preparation of basundi. Developed after Gite *et al.* (2017).

#### *4.2.4.18 Microencapsulated sugar apple powder*

Microencapsulation of sugar apple powder using the spray-drying method was explored by Sangamesh *et al.* (2023). They used pulp (100 g) diluted in water (1:5 ratio) with 20% maltodextrin. The mixture was homogenized for 8–10 min and then filtered. Spray-drying parameters varied with inlet air temperatures (175°C, 180°C and 185°C) and outlet air temperatures (70°C, 80°C and 90°C), at a feed flow rate of 2 ml min$^{-1}$ The resulting powder was packed in 50 μm aluminium pouches and showed optimal characteristics under conditions of inlet temperature of 185°C and outlet temperature of 90°C, with high recovery (20.42%), low moisture content (4.92%) and a water activity of 0.27.

## 4.3 Waste Utilization and Sustainability

Waste utilization and sustainability in annona fruit processing involve efficient management of by-products such as seeds, peels and pulp residues to minimize environmental impact and maximize economic value. Sugar apple seeds, often discarded as waste, contain bioactive compounds that can be extracted for various purposes, including pharmaceuticals and cosmetics. Industrial processing generates significant waste in the form of seeds, peel and seedcoats, often overlooked and termed seed waste. These by-products contain a wealth of bioactive compounds, suggesting potential for extraction and utilization in the food processing sector. Sugar apple seeds have been used in traditional hair tonics for lice removal and other applications such as insecticides and abortifacients. Recent research has highlighted their rich phytochemical composition, with over 400 active compounds identified in *A. squamosa*.

Of particular importance, annonaceous acetogenins and cyclopeptides found in the seeds exhibit antibacterial, anti-inflammatory and antitumor properties. Studies have shown their efficacy against cancers such as liver, prostate, cervical and pancreatic cancers, attributing these effects to various compounds including phenolics, alkaloids and flavonoids. Sugar apple seed extracts also demonstrate hepatoprotective and cytotoxic effects, making them promising for pharmaceutical and food innovation (Kumari *et al.*, 2022).

The peel and pulp residues can be utilized for composting or as feed for livestock, reducing landfill waste and greenhouse-gas emissions. By implementing sustainable practices such as composting, recycling and energy recovery from processing residues, sugar apple processors can contribute to environmental conservation and resource efficiency while enhancing economic viability within the industry. These efforts align with broader sustainability goals by promoting circular economy principles and reducing the ecological footprint of custard apple production.

Dhivyadharshini *et al.* (2023) investigated biodiesel production from sugar apple seeds using Soxhlet extraction. They utilized a catalyst derived from banana leaf, obtained through calcination, which exhibited a blue colour indicative of high potassium content. The biodiesel was synthesized via transesterification, yielding crude biodiesel subsequently purified using various techniques. Gas chromatography–mass spectrometry analysis revealed peaks at retention times of 19 and 20.7 min, corresponding to methyl hexadecanoate and 9-hexadecenoic acid methyl ester. Optimal conditions included a catalyst weight percentage of 1 g, a reaction time of 60 min, and an oil:methanol ratio of 1:3, achieving a biodiesel yield of 85%.

Sugar apple bagasse flour was used in cookie formulations at varying levels (5–50%) to assess its acceptability (Souza *et al.*, 2018). The flour underwent analysis for physico-chemical properties, proximate composition, minerals, phenolic content and antioxidant capacity. The results showed that the flour met the specifications for food use, with mineral content contributing over 20% of daily nutrient intake, including copper, iron, manganese, zinc, calcium and magnesium, from 100 g of flour. The phenolic levels in flour and cookies ranged from 200 to 658 mg GAE 100 $g^{-1}$, comparable to other tropical fruit residues. Inhibition ranged from 9.68% to 10.75%. Cookies made with the flour were highly accepted, indicating its potential for nutritional enhancement in food products (Souza *et al.*, 2018).

## References


Alvarez-Yanamango, E., Chuqulin-Goicochea, R., Huayta, F. and Arias, G. (2019) Functionality and physicochemical properties of the cherimoya flour (*Annona cherimola* Mill.) cv. Cumbe. *Italian Journal of Food Science* 31, 39–49.

Bakane, P.H., Khakare, M.M., Gajabe, M.H. and Khedkar, M.B. (2016) Standardization of process for custard apple milk shake. *International Journal of Environment, Agriculture and Biotechnology* 1, 708–712. DOI: 10.22161/ijeab/1.4.12

Bawale, S.M., Narwade, S.G. and Kamble, N.S. (2019) Studies on preparation of basundi blended with custard apple pulp. *Pharma Innovation Journal* 8, 307–309.

Campos-Vargas, R., Defilippi, B.G., Romero, Q.P., Valdes, G.H. and Robeldo, M.P. (2008) Effect of harvest time and l-cysteine as an antioxidant on flesh browning of fresh-cut cherimoya (*Annona cherimola* Mill). *Chilean Journal of Agricultural Research* 68, 217–227. DOI: 10.4067/S0718-58392008000300001

Chikhalikar, N.V., Sahoo, A.K., Singhal, R.S. and Kulkarni, P.R. (2000) Studies on frozen pourable custard apple (*Annona squamosa* L) pulp using cryoprotectants. *Journal of the Science of Food and Agriculture* 80, 1339–1342. DOI: 10.1002/1097-0010(200007)80:9<1339::AID-JSFA629>3.0.CO;2-C

CODEX STAN 247 (2005) *Codex Alimentarius. General Standard for Fruit Juices and Nectars*. Food and Agriculture Organization of the United Nations and World Health Organization (FAO/WHO), Geneva, Switzerland.

Cruz-Cansino, N.D.S., Ariza-Ortega, J.A., Alanis-Garcia, E., Ramirez-Mrreno, E., Velazquez-Estrada, R.M. *et al.* (2021) Optimal thermo ultrasound processing of jackfruit (*Artocarpus heterophyllus* Lam.) nectar: physicochemical characteristics antioxidant properties, microbial quality and fatty acid profile comparison with pasteurized nectar. *Journal of Food Processing and Preservation* 45: e15029. DOI: 10.1111/jfpp.15029

da Costa, J. de P., Rocha, É.M. de F.F. and da Costa, J.M.C. (2014) Study of the physicochemical characteristics of soursop powder obtained by spray-drying. *Food Science and Technology (Campinas)* 34, 663–666. DOI: 10.1590/1678-457X.6380

de Oliveira, A.R.G., Borges, S.V., Faria, R.K., Endo, E. and Gregorio, S.R. (2007) Influencia das condicoes de secagem por atomlzacao sobre as caracteristicas sensoriais de sucos marcuja (*Passiflora edulis*) e abacaxi (*Ananas comosus*) desidratados. *Revista de Ciencias Agronomica* 38, 251–256.

de Oliveira, E.N.A., da Costa Santos, D., dos Santos, Y.M.G., Buchweitz, P.R. and Gomes, J.P. (2016) Soursop liquor processing: influence of the process variables on the physical and chemical characteristics. *Revista Caatinga* 29, 246–256.

de Oliveira, S.L., Guerra, N.B., Maciel, M.I.S. and Livera, A.V.S. (1994) Polyphenoloxidase activity, polyphenols concentration and browning intensity during soursop (*Annona muricata* L.) maturation. *Journal of Food Science* 59, 1050–1052. DOI: 10.1111/j.1365-2621.1994.tb08187.x

Deepa, K. and Mohapatra, M. (2020) Foam-mat drying characteristics of custard apple pulp. *Journal of Food and Nutrition Sciences* 8: 89. DOI: 10.11648/j.jfns.20200804.13

Dhivyadharshini, M.V., Shobika, K. and Sanjana, L. (2023) Biodiesel production from custard apple seeds. *International Research Journal of Modernization in Engineering Technology and Science* 5, 1668–1681.

Dhumal, N.S. (1994) Studies on processing of custard apple (*Annona squamosa* L.). MSc thesis, Mahatma Phule Krishi Vidyapeeth, Rahuri, India.

Gamage, T.V., Yuen, C.M.C. and Wills, R.B.H. (1997) Minimal processing of custard apple (*Annona atemoya*) pulp. *Journal of Food Processing and Preservation* 21, 289–301. DOI: 10.1111/j.1745-4549.1997.tb00784.x

Geeta, G. (2000) Studies on preservation of custard apple pulp for value addition. MSc thesis, Dr Panjabrao Deshmukh Krishi Vidyappeth, Akola, India.

George, A.P. and Nissen, R.J. (1993) Annonaceous fruits. In: Macrae, R., Robinson, R.K. and Sadler, M.J. (eds) (1993) Annonaceous fruits. In: *Encyclopedia of Food Science, Food Technology and Nutrition*. Academic Press, London, pp. 195–199.

Gite, A.S., More, D.R. and Satwadhar, P.N. (2017) Development and standardization of custard apple basundi. *Journal of Pharmacognosy and Phytochemistry* 6, 1170–1172.

Jagtap, U.B. and Bapat, V.A. (2015) Phenolic composition and antioxidant capacity of wine prepared from custard apple (*Annona squamosa* L.) fruits. *Journal of Food Processing and Preservation* 39, 175–182. DOI: 10.1111/jfpp.12219

Kamble, K.J. and Soni, S.B. (2010) Effect of steam blanching on quality of custard apple pulp. *Journal of Horticulture and Forestry* 2, 104–107. DOI: 10.5897/JHF.9000041

Kolekar, T.T. (2002) Extraction and preservation of custard apple (*Annona squamosa* L.) pulp. MSc thesis, Mahatma Phule Krishi Vidyapeeth, Rahuri, India.

Kolekar, T.T., Masalkar, S.D., Chavan, J.K. and Kadam, S.S. (2003) Preservation of custard apple pulp. *Beverage and Food World* 20–21.

Kumar, K., Dikshit, S.N., Chandel, N. and Kesharwani, A. (2015) Effect of recipe treatment and storage period on biochemical composition of nectar prepared from custard apple genotypes. *Annals of Plant and Soil Research* 17, 399–403.

Kumar, V. (2010) Studies on fermentation of custard apple pulp with *Saccharomyces cerevisiae* var. *ellipsoideus* at different dilutions for wine preparation. MSc thesis, Andhra Pradesh Horticultural University Hyderabad, India.

Kumar, V. (2011) Preparation and evaluation of custard apple wine: effect of dilution of pulp on physico-chemical and sensory quality characteristics. *International Journal of Food and Fermentation Technology* 1, 247–253.

Kumari, N., Prakash, S., Kumar, M., Radha, Zhang, B. *et al.* (2022) Seed waste from custard apple (*Annona squamosa* l.): a comprehensive insight on bioactive compounds, health promoting activity and safety profile. *Processes* 10: 2119. DOI: 10.3390/pr10102119

López-Córdoba, A., Deladino, L. and Martino, M. (2014) Release of yerba mate antioxidants from corn starch-alginate capsules as affected by structure. *Carbohydrate Polymers* 99, 150–157. DOI: 10.1016/j.carbpol.2013.08.026

López-Córdoba, A., Morales, V.E.M., Pucheco, V.M.M. and Navarro, A.S. (2016) Encapsulation of soursop (*Annona muricata*) fruit pulp in calcium alginate hydrogels. *Agronomia Columbiana* 34 (Suppl.), 1315–1318.

Makwana, A., Varu, D.K., Malam, V.R., Bhad, M. and Maheta, P. (2018) Effect of different concentrations of custard apple pulp on the properties of ice cream. *HortFlora Research Spectrum* 7, 126–132.

Markam, P.S., Shankar, D., Ganjeer, P.K., Siwana, H. and Kujur, A. (2022) Study on standardization of recipe for custard apple ice-cream. *Pharma Innovation Journal* 11, 1708–1710.

Markam, R. and Singh, V. (2012) Studies on microbial and sensory quality of custard apple RTS beverage. *Asian Journal of Horticulture* 7, 460–464.

Martínez-Girón, J., Arias-Rodas, A.L., Hernández-Montaña, J.C., Agudelo-Burbano, J.H. and Reina-Hernandez, B. (2019) Physicochemical, microbiological and sensory characterization of liquor obtained from atemoya fruit (*Annona cherimola* × *Annona squamosa*). *Scientia et Technica* 24, 636–642.

Mata, M.E.R.M.C., Duarte, M.E.M., Alsemo, G.C.S., Rodrigues, E., Guedes, M.A. *et al.* (2005) Obtenção de Graviola em po pelo processo de Liofilização. *Revista Brasileira de Produtos Agroindustriais* 7, 165–172.

Minh, N.P. (2014) Different factors affecting to custard apple (*Annona squamosa*) wine fermentation. *International Journal of Multidisciplinary Research and Development* 1, 165–167.

Minh, N.P. (2015) Production of fermented beverage from soursop fruit. *International Journal of Pure and Applied Bioscience* 3, 231–236.

Mororo, R.C., Freire, E.S. and Sacramento, C.K. (1997) Processumento da graviola para Obtencao de Polpa. In: Sao Jose Boas, A.R., Morais, M. and Reboucas, T.N.H. (eds) *Anonaceas: Poduocao e Mercado (Pinha, Graviola, Atemoia e Cherimolia)*. Universidade Estadula do Sudoeste da Bahia, Brazil, pp. 263–274.

Mundhe, S.A., Kshirsagar, R.B., Kulkarni, D.N. and Patil, B.M. (2008) Studies on utilization of custard apple pulp in toffee. *Indian Journal of Nutrition and Dietetics* 45, 472–478.

Nakasone, H.Y. and Paull, R.E. (1998) Annonas. In: Nakasone, H.Y. and Paull, R.E. (eds) *Tropical Fruits*. CAB International, Wallingford, UK, pp. 45–75.

Nambi, V.E., Gupta, R.K. and Vishwakarma, R.K. (2011) Automatic machine for scooping out the pulp from custard apple fruits. Indian Patent No. 334054.

Olaeta, J.A. and Abufom, J. (1986) Technical factibility of freezing cherimoya (*Annona cherimola* Mill) fruit pulp cv. Concha Lisa and Bronceada. *Acta Horticulturae* 194, 273–278. DOI: 10.17660/ActaHortic.1986.194.29

Orsi, D.C., Carvalho, V.S., Nishi, A.C.F., Damiani, C. and Asquieri, E.R. (2012) Use of sugar apple, atemoya and soursop for technological development of jams: chemical and sensorial composition. *Ciência e Agrotecnologia* Lavara 36, 560–566. DOI: 10.1590/S1413-70542012000500009

Patil, M.S., Dhole, P.T. and Chavan, K.D. (2015) Studies on quality evaluation of probiotic custard apple (*Annona reticulata*) dahi. *Indian Journal of Nutrition* 2, 1–6.

Patil, S.R., Kurhekar, S.P. and Patil, R.R. (2011) Study on development of custard apple carbonated beverage. *International Journal of Processing and Post Harvest Technology* 2, 56–58.

Patil, T.B. (2011) Standardization of a pre-treatments for liquefaction of custard apple pulp for spray drying. MSc thesis, Mahatma Phule Krishi Vidyapeeth, Rahuri, India.

Pawar, N., Kardile, W.G. and Hashmi, S.I. (2011) Effect of heat processing on discoloration of custard apple (*Annona squamosa* L.) fruit pulp and changes in quality characteristics during storage. *Electronic Journal of Environmental, Agricultural and Food Chemistry* 10, 2098–2113.

Peters, M., Badrie, N. and Comissiong, E. (2007) Processing and quality evaluation of soursop (*Annona muricata* L.) nectar. *Journal of Food Quality* 24, 361–374. DOI: 10.1111/j.1745-4557.2001.tb00616.x

Pilania, S., Dashora, L.K. and Singh, V. (2010) Standardization of recipe and juice extraction method for preparation of ready-to-serve beverage from custard apple (*Annona squamosa* L.). *International Journal of Processing and Post-Harvest Technology* 1, 65–72.

Pinto, A.C. de Q., Corderio, M.C.R., de Andrade, S.R.M., Ferreira, F.R., Filgueiras, H.A. de C. *et al.* (2005) *Annona Species.* International Centre for Underutilized Crops. University of Southampton, UK.

Poul, S.P., Sontakke, A.T., Munde, S.S., Adangale, A.B. and Jadhav, P.B. (2009) Process standardisation for custard apple milk shake. *Journal of Dairying, Foods and Home Sciences* 28, 202–205.

Prasanna, K.N.V., Sudhakar Rao, D.V. and Krishnamurthy, S. (2000) Effect of storage temperature on ripening and quality of custard apple (*Annona squamosa* L.) fruits. *Journal of Horticultural Science and Biotechnology* 75, 546–550. DOI: 10.1080/14620316.2000.11511283

Raichurkar, S.J. and Dadagkhair, R.A. (2017) Studies on preparation of custard apple vinegar. *International Journal of Advanced Engineering Research and Science* 4, 101–106. DOI: 10.22161/ijaers.4.7.15

Rao, V.S. (2009) *Post Harvest Technology and Value Addition in Fruits*. Central Research Institute for Dryland Agriculture, Hyderabad, India.

Saket, R.K., Bisen, B., Pandey, S.K., Singh, G., Mishra, P. *et al.* (2018a) Effect of various recipes on chemical characteristics of mixed fruit bar. *Multilogic in Science* 8 (Special Issue D), 247–251.

Saket, R.K., Bisen, B.P., Pandey, S.K., Pandey, S. and Sharma, H.L. (2018b) Effect of various recipes on organoleptic evaluation of custard apple jam. *Journal of Pharmacognosy and Phytochemistry* 7 (Special Issue 2), 300–304.

Sanchez-Nieva, F., Igaravidez, L. and Lapez-Ramos, B. (1953) The preparation of soursop nectar. *University of Puerto Rico, Agriculture Experimental Station Technical Paper* 11, 5–19.

Sangamesh, M., Karadiguddi, M., Gorabal, K., Patil, R.T., Nadaf, A.M. *et al.* (2023) Microencapsulation of custard apple powder through spray drying. *Biological Forum* 15, 609–611.

Saniya, D., Kaushik, R.A., Pilania, S. and Rajpurohit, D. (2023) Standardization and organoleptic evaluation of custard apple leather and ice cream. *Pharma Innovation Journal* 12, 746–751.

Shashirekha, M.N., Rajarathnam, S., Vijayalakshmi, M.R. and Revathy, B. (2003) A process for the preparation of jelly from custard apple. PCT patent application no. PCT 0434.

Shashirekha, M.N., Baskaran, R., Rao, L.J.M., Vijayalakshmi, M.R. and Rajarathnam, S. (2008) Influence of processing conditions on flavour compounds of custard apple (*Annona squamosa* L.). *LWT – Food Science and Technology* 41, 236–243. DOI: 10.1016/j.lwt.2007.03.005

Shrivastava, A., Bharad, S.G., Panchbhai, D.M., Patil, S.R. and Dalal, S.R. (2020) Sensory stability of *Annona* species pulp: influence of storage conditions and duration. *International Journal of Chemical Studies* 8, 2826–2828. DOI: 10.22271/chemi.2020.v8.i6an.11210

Shrivastava, R., Dubey, S., Dwivedi, A.P., Pandey, C.S. and Banafar, R.N.S. (2013) Effect of recipe treatment and storage period on biochemical composition of custard apple (*Annona squamosa* L.) nectar. *Progressive Horticulture* 45, 110–114.

Singh, P., Patel, D., Agrawal, S. and Panigrahi, H. (2006) Value addition of custard apple through jam for year round availability. In: *Proceedings of the National Symposium on Production, Utilization and Export of Underutilized Fruits with Commercial Potentialities*, Mohanpur, Bidhan Chandra Krishi Viswavidyalaya, India, pp. 262–265.

Sinthiya, R. and Poornima, K. (2017) Value added products from annona fruit. *IOSR Journal of Environmental Science, Toxicology and Food Technology* 11, 1–5. DOI: 10.9790/2402-1108010105

Soni, N., Sharma, M., Mudgal, V.D., Champawat, P.S. and Kohli, D. (2021) Physical characteristics and flow behaviour of custard apple powder. *Octa Journal of Biosciences* 9, 119–122.

Souza, F.T.C., Santos, E.R., Silva, J. da C., Valentim, L.B., Rabelo, T.C.B. *et al.* (2018) Production of nutritious flour from residue of custard apple (*Annona squamosa* L.) for the development of new products. *Journal of Food Quality* 2018: 5281035. DOI: 10.1155/2018/5281035

Teixeira, C.K.B., Neves, E.C.A. and Pena, R.S. (2006) Study of the pasteurization process of graviola's pulp. *Alimentos e Nutricao Araraquara* 17, 251–257.

Telis-Romero, J., Beristain, C.I., Gabas, A.L. and Telis, V.R.N. (2007) Effect of apparent viscosity on the pressure drop during fluidized bed drying of soursop pulp. *Chemical Engineering and Processing* 46, 684–694. DOI: 10.1016/j.cep.2006.09.001

Thube, R., Purohit, S. and Gothoskar, A. (2011) Study of effect of custard apple pulp powder as an excipient on the properties of acetaminophen tablet. *World Applied Sciences Journal* 12, 364–371.

Torres, W.E. and Sánchez, L.A. (1992) *Fruticultura Colombiana Guanabano*. Manual de Asistencia Tecnica no. 57. Instituto Colombiano Agropecuario, ICA, Bogota.

Villachica, H., Carvalho, J.E.U., Muller, C.H., Diaz, S.C. and Almanza, M. (1996) Guanabana: *Annona muricata* L. In; *Frutales y Hortalizas Promisorios de la Amazonia*. Trata Do de Cooperation Amazonica, Secretaria Pro-Tempore, Lima, Peru, pp. 130–136.

Zafra-Rojas, Q.Y., Jiménez-Hernández, J.L., Olloqui, E.J., Cruz-Cansino, N. del S., Alanís-García, E. *et al.* (2023) Optimization of thermoultrasound process of soursop (*Annona muricata*) nectar and comparison of its physicochemical properties and *in vitro* bioaccessibility of antioxidants with pasteurized sample. *Food Technology and Biotechnology* 61, 536–548. DOI: 10.17113/ftb.61.04.23.8180

# 5 Propagation and Nursery Management

**Gisela Ferreira[1], Carolina Ovile Mimi[1*], Samuel Bassani Lins de Miranda[1], Stephane Catarine Rosa Kim[1] and Ivan De-La-Cruz-Chacón[2]**
*[1]Department of Biodiversity and Biostatistics, Institute of Biosciences, São Paulo State University (UNESP), Botucatu, São Paulo, Brazil; [2]Laboratorio de Fisiología y Química Vegetal, Instituto de Ciencias Biológicas, Universidad de Ciencias y Artes de Chiapas, Tuxtla Gutiérrez, CHIS, Mexico*

## 5.1 Introduction

The family Annonaceae, which belongs to the order Magnoliales, is composed of approximately 108 genera and 2400 species (Rainer and Chatrou, 2006; Chatrou *et al.*, 2012). In the Neotropics, it is represented by approximately 40 genera and 950 species, with 31 genera and 387 species occurring in Brazil, which stands out for its great diversity in the Amazon Rainforest and Atlantic Forest (Mendes-Silva *et al.*, 2024a).

Preservation of biodiversity is necessary both from an evolutionary point of view and to support the technological advances needed to implement more productive cultivation areas. The perpetuation of species is a noble task, as is the production of seedlings that will initially be responsible for high cultivation productivity. Thus, this chapter aims to present the main characteristics of sexual propagation, with all the peculiarities involving seeds, as well as some responses of members of the Annonaceae to the ability to regenerate new individuals from cuttings, grafts or micropropagation.

## 5.2 Seeds

The formation of seeds in species of the genus *Annona* occurs from hermaphrodite flowers that present protogynous dichogamy, a mechanism that ensures cross-fertilization. After pollination, seed development begins inside syncarpous fruit, which are formed by the union of several carpels (Vidal-Lezama *et al.*, 2023).

Annonaceae seeds vary in size and morphological characteristics, which influence their dispersal and survival strategies. In *Annona purpurea*, for example, seeds measure approximately 2.64 cm in length and 1.35 cm in width, dimensions similar to those found in *Annona cherimola*, *A. reticulata* and *A. squamosa* (1.5–2 cm seed length) (Manica *et al.*, 2023; Vidal-Lezama *et al.*, 2023). Seeds are composed of ruminated endosperm, which surrounds a relatively small embryo. The embryo of *A. purpurea* is approximately 3.4 mm in length and 0.58 cm in width, located close to the hilum, which indicates a small embryo immersed in endosperm, a common characteristic in basal species of angiosperms (Chatrou, 1999; Forbis *et al.*, 2002; Vidal-Lezama *et al.*, 2023). The embryo size, although small, is generally differentiated at the time of dispersal, as observed in studies with *Annona emarginata* (Corsato *et al.*, 2012).

The seedcoat of *Annona* is usually lignified and may contain phenolic compounds, such as tannins, which limit oxygen absorption and therefore contribute to seed dormancy. This characteristic is well documented in species such

* Email: carolina.mimi@unesp.br


DOI: 10.1079/9781800627628.0005

as *Annona cacans*, whose seeds demonstrate morphophysiological dormancy, hindering immediate germination after dispersal (Dalanhol *et al.*, 2013). Imbibition studies carried out by Dalanhol *et al.* (2013) in *A. cacans* seeds indicated that dormancy is not caused by barriers to water entry but by the presence of tannins, which inhibit embryo respiration.

The presence of specialized metabolites is not limited to defence against predators. During germination, stored metabolites in the seed play an essential role, ensuring that seeds are able to establish themselves in the soil. For example, germinating *A. purpurea* seeds produce anthocyanins as a protection strategy against intense sunlight. In contrast, *A. emarginata* seeds begin to produce alkaloids 36 h after the beginning of imbibition, demonstrating their defence capacity shortly after the beginning of metabolism reactivation for germination (da Silva *et al.*, 2022; Ferreira *et al.*, 2006; Oliveira *et al.*, 2010).

### 5.2.1 Desiccation tolerance

In the final stage of seed formation, a reduction in water content occurs, known as desiccation, characterizing orthodox seeds (Roberts, 1973). In contrast, some seeds experience minimal or no water loss, and these are classified as intermediate or recalcitrant, depending on their ability to tolerate desiccation (Corsato *et al.*, 2012).

For seeds to tolerate significant water loss (90–95%), soluble sugars, such as sucrose, play a crucial role in protecting cells during desiccation. Furthermore, accumulation of late embryogenesis abundant (LEA) proteins at the end of seed formation and an active antioxidant system during imbibition are characteristics to be considered in determining the ability of seeds to tolerate desiccation (Koster and Leopold, 1988; Pammenter and Berjak, 2000; Corsato *et al.*, 2012). Sucrose, together with raffinose and stachyose, are sugars that contribute to desiccation tolerance (Peterbauer and Richter, 2001; Corsato *et al.*, 2012).

*Annona* spp. seeds vary in terms of desiccation tolerance, with several species considered orthodox, such as *A. cherimola* Mill., *A. glabra* L., *A. hayesii* Saff., *A. muricata* L., *A. reticulata* L., *A. spraguei* Saff., *A. squamosa* L. (Royal Botanic Gardens, Kew, 2016), *A. macroprophyllata* Donn Sm. (González-Esquinca *et al.*, 2015), *A. mucosa* (Jacq.) Baill. H. Rainer (Ferreira *et al.*, 2009), *A. purpurea* Moc. & Sessé ex Dunal (González-Esquinca *et al.*, 2016; Ferreira *et al.*, 2016) and *A. crassiflora* Mart. (da Silva, 2007).

There are also intermediate species, such as *A. emarginata* (Schldtl.) H. Rainer, whose seeds tolerate drying up to 5% moisture content, have LEA proteins, an active antioxidant system and the presence of sugars but do not tolerate storage for more than 3 months (Corsato *et al.*, 2012).

An interesting fact is the behaviour of *A. glabra*, a species originally from mangrove regions whose seeds have a greater capacity to tolerate desiccation than *A. crassiflora* seeds, a species from the Cerrado, Brazil (De-Pieri-Oliveira, 2019).

Therefore, knowing the ability of seeds to tolerate drying is important because it directly impacts the seed storage time (Corsato *et al.*, 2012), which in the case of Annonaceae can vary from 1 to 10 months (da Silva *et al.*, 2007; Ba *et al.*, 2012; Ferreira *et al.*, 2019).

The formation time of an *Annona* seed is approximately 49–116 days after flowering for *A. squamosa*, *A. emarginata* and *A. macroprophyllata* (Ferreira *et al.*, 2019), and at the end of the formation time, they can be dispersed dormant or quiescent. In this context, *Annona* seeds generally present some dormancy mechanism.

### 5.2.2 Dormancy

When seeds receive all the environmental conditions to germinate but do not germinate, they may be dead or present some dormancy mechanism. The dormancy classification has undergone several changes over time, but in general, dormancy can be divided into primary (acquired during formation) and secondary (acquired after dispersal), and exogenous (related to the seedcoat) and endogenous (related to the embryo and its coats) (Ferreira *et al.*, 2022).

There are several types of dormancy reported for seeds of *Annona* spp. from those determined by the seedcoat impermeability, which would be classified as exogenous dormancy, to endogenous dormancy of morphological, physiological and morphophysiological types

(Ferreira *et al.*, 2022). However, regarding the seedcoat impermeability to water, Ferreira *et al.* (2014a) found that *Annona* spp. seeds are not impermeable to water but present slow imbibition, ranging from 27–47 h in atemoya (*Annona atemoya* Mabb., *A. cherimola* × *A. squamosa*) to 109 h in *A. emarginata*. In contrast, *A. cacans* warm seeds are considered impermeable to gases due to the presence of phenolic compounds that prevent the passage of oxygen to the embryo (Dalanhol *et al.*, 2013).

In this context, the main dormancy mechanisms in *Annona* spp. are physiological and morphophysiological, because although the embryos are small, measuring millimetres (Vidal-Lezama *et al.*, 2023), they are differentiated and complete at the time of dispersal, and therefore this is a characteristic of members of the genus *Annona*, and cannot be confused with morphological dormancy (Ferreira *et al.*, 2016, 2020).

Physiological dormancy can occur in mature embryos, which have all the basic differentiated structures but present metabolic blocks, such as the presence of germination inhibitors like abscisic acid (ABA). This dormancy is frequently overcome through treatments with gibberellins, which alter the hormonal balance in relation to ABA and act on the degradation of reserves, providing energy for the resumption of embryo development. In *Annona* spp., dormancy-overcoming responses with gibberellins have been observed in species such as *A. crassiflora*, *A. squamosa*, *A. muricata*, *A. macroprophyllata* Donn Sm. and *A. purpurea* Moc. & Sessé ex Dunal (Bernardes *et al.*, 2007; Cavalcante *et al.*, 2007; Ferreira *et al.*, 2016; Barros *et al.*, 2019). These treatments are effective in overcoming dormancy and in accelerating germination and seedling development and consequently improving plant uniformity in the nursery (Doran, 1997; Alouani and Bani-Aameur, 2004; Pompelli *et al.*, 2023).

Working with *A. macroprophyllata* Donn Sm. and *A. purpurea*, Ferreira *et al.* (2020) found that the seeds presented physiological dormancy due to high ABA concentrations at the time of seed dispersal, in relation to those of gibberellins. The authors found that with exogenous application of a mixture of gibberellins and cytokinins (gibberellic acid 4 + 7 ($GA_{4+7}$) and 6-benzylaminopurine (BAP)), dormancy was overcome, with a reduction in ABA concentrations and increase in gibberellins over time, allowing visible germination in approximately 14 days.

*A. crassiflora* Mart. (da Silva, 2007), *A. macroprophyllata* Donn Sm., *A. purpurea* Moc. & Sessé ex Dunal (Ferreira *et al.*, 2016; González-Esquinca *et al.*, 2016) and atemoya (Gimenez *et al.*, 2014a) present physiological dormancy, while *A. coriacea* Mart. (Dresch *et al.*, 2014), *A. spraguei* Saff. (RBG Kew, 2016) and *A. sylvatica* A. St.-Hil. (Mayer *et al.*, 2008) seeds present morphophysiological dormancy.

One of the most widely methods used to overcome *Annona* seed dormancy involves the use of plant growth regulators. An example is the combination of $GA_{4+7}$ with benzyladenine (BA), applied to species such as *A. macroprophyllata* and *A. purpurea* (Ferreira *et al.*, 2016). For atemoya, gibberellic acid ($GA_3$) has been shown to be effective in the germination process (Stenzel *et al.*, 2003).

In addition to plant regulators, other methods are also used, such as mechanical scarification. Scarification is used to accelerate water entry, allow oxygen transit or reduce inhibitory substances present in the seedcoat, as reported in *A. glabra* (Zamora-Cornelio *et al.*, 2010) and *A. mucosa* (Ferreira *et al.*, 2009). There are also reports of the use of acid scarification in *A. muricata*, *A. senegalensis* and *A. squamosa* seeds (Ba *et al.*, 2012). In *A. cacans*, scarification is used to facilitate oxygen entry into the embryo, which is prevented by the presence of compounds such as tannins in the seedcoat (Dalanhol *et al.*, 2013).

Environmental variations in water availability, temperature and light are closely linked to the adaptation and maintenance of species, and can affect biochemical reactions (de Menezes *et al.*, 2009) and can act to overcome or maintain species dormancy (Honório *et al.*, 2022).

### 5.2.3 Germination

Seed germination is a complex mechanism that involves reactivation of the embryo's metabolism so that it resumes its development, which is reduced at the end of maturation, and transforms into a seedling and later an adult plant (Bewley *et al.*, 2013).

#### 5.2.3.1 Germination phases

Imbibition is the first phase of germination. This process is dependent on the variation of water potential between seeds and medium, and therefore is a physical phase, depending mainly on the variation of the seed's matrix potential. At the end of imbibition, seeds reach an average of 31% water, and water input stabilizes momentarily (Ferreira *et al.*, 2014a). This is the fastest phase of the germination process; however, speed is relative. While common bean seeds (*Phaseolus vulgaris*) take 2–3 h to imbibe, *Annona* spp. seeds can take 27–47 h (atemoya), 50 h (*A. macroprophyllata*), 60–84 h (*A. purpurea*) (Ferreira *et al.*, 2014a; Vidal-Lezama *et al.*, 2024), 72 h (*A. squamosa*) (Pompelli *et al.*, 2023) and up to 109 h (*A. emarginata*) (Gimenez *et al.*, 2014b).

The study of this phase allows evaluation of the permeability of seeds to water, and is also an important tool for determining the time required for seeds to remain in treatments with regulators. It is noteworthy that when treating *Annona* seeds under immersion, it is necessary to oxygenate the immersion solution in order to avoid hypoxia, due to the length of time required for *Annona* seeds to complete the imbibition period. In a study with *A. emarginata* seeds, Gimenez *et al.* (2017) studied the effects of immersion for long periods in aqueous solutions. The authors found that after 122 h of immersion without proper oxygenation, seeds begin to produce ABA, which can lead to dormancy during the imbibition process. Thus, immersion treatments, especially with germination regulators, become dormancy inducers if the medium is not properly oxygenated (Gimenez *et al.*, 2017).

During imbibition, there is the first increase in the respiratory activity of seeds, and this is also when the degradation of seed reserves begins (Mimi *et al.*, 2023). Atemoya 'Gefner' seeds begin to degrade after 24 h of imbibition, especially when treated with gibberellins and cytokinins ($GA_{4+7}$ + phenylmethyl aminopurine), as also observed in *A. macroprophyllata* and *A. purpurea* seeds (Ferreira *et al.*, 2022).

The second germination phase is when there is the greatest mobilization of reserves. In *A. emarginata* seeds, this has an average duration of 60–72 h, and at the end of this phase, the primary root is emitted, completing the germination process, which occurs approximately 768 h after the beginning of imbibition in this species (Costa *et al.*, 2011).

#### 5.2.3.2 Factors that affect seed germination

Water is the main factor that affects germination, without which the metabolism cannot be reactivated. As discussed, it is acquired very slowly during the germination process in *Annona* spp. Another relevant factor is temperature. *A. montana* seeds present higher germination rates when germinated at 30°C (Oliveira *et al.*, 2005). However, *A. emarginata* and atemoya seeds germinate faster and with higher germination rates when submitted to temperatures ranging from 20°C to 30°C, without the use of regulators (Costa *et al.*, 2011; da Silva *et al.*, 2022).

## 5.3 Asexual Propagation

### 5.3.1 Grafting

Plant propagation by grafting is the technique most widely used for annonaceous (Almeida *et al.*, 2010; Baron *et al.*, 2016). The technique consists of joining (morphologically and physiologically) two parts of a living plant, namely a root system (rootstock) and a shoot system (scion) (Martínez-Ballesta *et al.*, 2010; Cookson and Ollat, 2013). The new plant formed by grafting presents the union of a rootstock tolerant to biotic and abiotic stresses with a commercial scion, selected based on both fruit quality and high production regularity (Ferreira *et al.*, 2019).

The use of grafting requires union techniques that assist in the formation of meristematic tissues in the graft region and their differentiation into a functional vascular system. Different methods of grafting such as soft wood (Dharmik *et al.*, 2022; Kamble *et al.*, 2024), cleft (da Silva *et al.*, 2023), whip-and-tongue (Baron *et al.*, 2016; Almeida *et al.*, 2010), wedge (Panchal *et al.*, 2022), side (Panchal *et al.*, 2022), veneer (Ogden *et al.*, 1981) and bench (Kumar and Shukla, 2008) grafting have been tried in different *Annona* spp. with varying degree of success. In Annonaceae, the most commonly used union techniques are cleft grafting or budding (Baron *et al.*, 2016).

However, different types can be used depending on the grafter's experience, time of year, material availability and the regeneration capacity of the new plant, which is called compatibility.

Cleft grafts are commonly employed in nurseries and widely used for their simplicity, which ensures good contact between the vascular cambium of the scion and the rootstock. This contact is crucial for nutrient exchange and successful graft union. Proper wrapping techniques, such as using parafilm or vinyl tape, help maintain humidity and prevent desiccation during the initial healing phase (Brokaw, 1998). In cherimoya (*A. cherimola*) propagation, for example, this type of grafting is highly preferred, especially for commercial growers in California, due to its ability to ensure consistent fruit quality and uniform orchard trees (Brokaw, 1998).

In Brazil, the cleft graft and whip-and-tongue graft methods are commonly used for atemoya (Fig. 5.1), particularly the 'Gefner' and 'Thompson' cultivars, with both methods demonstrating successful scion adherence to rootstocks such as *A. emarginata* ('terra-fria' and 'mirim' morphotypes) and *A. mucosa* (Baron *et al.*, 2016).

The cleft grafting carried out on atemoya 'Gefner' grafted on to *A. glabra* and *A. squamosa* was shown to be efficient in a study carried out by Leite *et al.* (2013), with emphasis on the full cleft, whip graft and tongue graft methods presenting the highest grafting success rates of, respectively, 93.33%, 60% and 76.67% for *A. glabra* and 75%, 81.67% and 83.33% for *A. squamosa*, while the

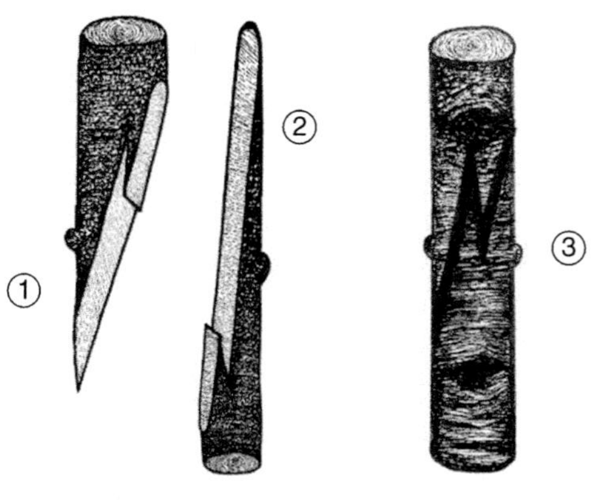


**Fig. 5.1.** Propagation by whip-and-tongue graft. 1. Scion. 2. Stock. 3. Grafting in position with stock and scion matched. Illustrations courtesy of Luiza Pângaro Pedrosa Barros.

budding method was not efficient for the grafting of atemoya 'Gefner' in the two species studied.

For the grafting of atemoya 'Thompson' on to *A. emarginata* ('terra-fria' morphotype), Almeida *et al.* (2010) used different grafting methods, and observed that using the so-called splice grafting method, the grafting success rate was 90%, while with the lateral cleft method, the grafting success rate was 75%. When evaluating the grafting success rate of atemoya 'African Pride' into *A. sylvatica*, da Silva *et al.* (2023) observed a 100% success rate with the cleft and whip-and-tongue graft methods. Softwood grafting was reported to be a successful method of propagation for sugar apple (*A. squamosa*). In this method, 10 days before grafting, the selected scion shoots were cured by removing the leaves. The rootstock comprised 1-year-old seedlings. Working with three sugar apple cultivars ('Arka Sahan', 'Island Gem' and 'Balanagar'), Kamble *et al.* (2024) observed 97.77% graft success after 30 days in 'Arka Sahn' when grafting was done in February and the plants were kept in a mist house. Keeping the grafted plants in the mist house also caused maximum survivability of 90.74% after 90 days of grafting. Soft-wood grafting was performed on 1-year-old rootstock of uniform size. For this purpose, rootstocks were beheaded at 15–20 cm above ground level. A vertical slit of 3–5 cm length was made on the rootstock using a budding knife. A scion shoot of 8–10 cm in length with about five buds and a pencil-sized girth thickness was selected, and a similar matching cut was prepared in a slanting manner on both the surfaces in the lower portion of scion shoot. The prepared scion was then inserted in the incision on the rootstock. The grafted portion was wrapped using transparent polythene strip (1.5 cm wide and 200 gauge), and immediately after completion of the grafting operation, the scion and graft union was covered with polycap to avoid desiccation of the scion and to create a good microclimate (Fig. 5.2). Dharmik *et al.* (2022) stated that rootstock age and grafting time significantly influence grafting success. They suggested the use of 7-month-old seedlings as the rootstock and softwood grafting during the third week of January, which showed 95.11% success in sugar apple.

In semi-arid regions of India where the temperature during summer is very high, bench grafting using the cleft grafting method was

**Fig. 5.2.** Softwood grafting of *Annona squamosa*. (A) Scion sticks of cultivar 'Arka Sahan'. (B) scion is cut into a wedge shape downwards. (C) Rootstock for grafting. (D) Insertion of scion into the rootstock. (E) Grafted portion tied with polythene strip. (F) Sprouting of grafts. Photographs courtesy of Dr A.K. Kamble.

found to be successful (Kumar and Shukla, 2008). Bare-rooted stocks of sugar apple 'Balanagar' of 0.6 cm diameter were bench grafted with a dormant scion of 0.6 cm thickness and showed 96.4% graft success when performed on 29 May. The bare-rooted grafts were planted in polythene bags (22 × 15 cm) that were then filled with soil, farm yard manure and leaf mould (1:1:1, v/v/v). Immediately after placing the grafts in the polythene bags, they were covered with a polycap of size 20 × 2.5 cm, and the lower open top end of the cap was buried in the filling mixture to check evapotranspiration loss and desiccation of the grafts. Thereafter, they were kept under partial shade and watered properly. The cap was removed after a week of scion sprouting (Kumar and Shukla, 2008).

In southern Florida, spring is considered the best time of year for grafting annonas (Ogden *et al.*, 1981), after the leaf drop in winter and just before bud break in spring. The buds should be collected just as they are beginning to swell. Ogden *et al.* (1981) prepared scions to use for veneer grafts in the standard manner, but instead of cutting the bud sticks in half longitudinally, only a superficial cut drawn to the cambium was made, removing the bark and some cortex, leaving three-quarters to seven-eighths of the diameter of the bud sticks intact. An oblique cut was made in the front surface of the scion to match a corresponding notch on the prepared stock. Rollinia, cherimoya, soursop (*A. muricata*), sugar apple and atemoya were veneer grafted in August 1979 and May 1980. Bud wood that had been prepared during May and August had 90–100% grafting success, depending on the species. Panchal *et al.* (2022) compared the time and methods of grafting of sugar apple 'Sindhan' in South Gujarat, India. The results suggested that wedge grafting was superior to side grafting, and the best time was 15 March or 15 April. However, higher graft survival (82% at 90 days after grafting) was recorded when grafting was done on 15 April.

### 5.3.2 Budding

Dos Santos *et al.* (2005) budded sugar apple in biribá (*A. mucosa*) using the inverted-T budding method and observed a success rate of 4%, indicating that this method is not effective in the production of sugar apple plants. In contrast, the success rate for sugar apple was 95.8% with the plate budding method (Dantas *et al.*, 1993).

Using 1-year-old sugar apple seedlings as rootstock and scion wood of the past season's growth of cultivar 'Balanagar' for budding and grafting purposes, Joshi *et al.* (2011) reported 70% bud take and 43.33% final survivability when shield budding was performed in August, and 66.66% and 36.66% respectively, when the patch budding method was used.

The search for rootstocks that are tolerant to diseases and pests and that are compatible with the commercial scion is an important task in the production process of *Annona* spp. seedlings, as the quality of fruit produced begins with the quality of seedlings used in the orchard establishment. Ideally, seedlings should present phytosanitary quality that genetically guarantees the characteristics of selected matrices.

### 5.3.3 Compatibility and rootstocks

The first aspect to be explored in order to define which species to use as rootstock is its compatibility with the scion. Compatibility is a complex term to define, because in order to state that one plant is compatible with another, a series of factors and evaluation levels must be analysed. Among the factors to be considered, genetic, enzymatic, hormonal, phytopathological, nutritional and environmental characteristics stand out, and one must also consider whether the compatibility is early (observed days after grafting) or late (after years in the field), in addition to the grafter's experience.

There are several species that are candidates for rootstocks of commercial species, such as *A. emarginata* ('terra-fria' and 'mirim' morphotypes), *A. mucosa*, *A. muricata*, *A. squamosa* and *A. crassiflora*, which have already been reported as compatible with different commercial species (Table 5.1) and are discussed in more detail below. There are other species that are still little known, such as *A. neoinsignis*, which shows great potential as a rootstock for atemoya in hot climates, but field tests are still in the initial phase and require further compatibility studies. *A. cherimola* has also been grafted but requires further studies regarding the best rootstocks to be used.

#### 5.3.3.1 *A. emarginata* (Schltdl.) H. Rainer (syn. *Rollinia emarginata* Schltdl.)

*A. emarginata* has two morphotypes known as 'terra-fria' and 'mirim' (Mimi *et al.*, 2021). This species is adapted to different environmental conditions, provides greater longevity to the graft and does not cause hypertrophy of the grafted tissue or dwarfism in plants (Tokunaga, 2005). Furthermore, the species is tolerant to pathogens that attack roots and stems (Bettiol Neto *et al.*, 2006), which indicates that it is an important rootstock for atemoya (Mantoan *et al.*, 2016; Baron *et al.*, 2018a,b).

In this context, and with the aim of studying the compatibility between *A. emarginata* and atemoya, Baron *et al.* (2018a) evaluated rootstocks of *A. emarginata* 'terra-fria' and 'mirim', biribá (*A. mucosa*) and atemoya 'Thompson' for the atemoya 'Thompson' scion in the mid-western region of São Paulo, Brazil. After grafting compatibility was observed in the first few days after its completion, with no mortality. From anatomical analyses, it was possible to detect over time that regeneration of the grafted region occurred more quickly with plants grafted on to *A. emarginata*, and regeneration occurred in all combinations used.

Subsequently, and considering this variation in tissue formation time, several analyses were performed to understand metabolically how this regeneration occurred. One of the analyses was to study expression of the UGP gene and the activity of the UDP-glucose pyrophosphorylase enzyme (UGPase), important in the production and metabolism of UDP-glucose, a key metabolite for sucrose and cell-wall biosynthesis (Lerouxel *et al.*, 2006). In this context, Baron *et al.* (2016) demonstrated that UGP expression levels were higher in atemoya grafts on *A. emarginata* 'terra-fria' than in non-grafted

**Table 5.1.** Grafting compatibility between scion and rootstock in some *Annona* spp. Adapted from Sanewski (1991), Pinto *et al.* (2005) and Ferreira *et al.* (2014b).

| | Scion species | | | | |
|---|---|---|---|---|---|
| Rootstock species | *A. cherimola* | *A. muricata* | *A. reticulata* | *A. squamosa* | *A. atemoya* |
| *A. cherimola* | C | NC | C | C | C |
| *A. glabra* | C | LC | C | C | – |
| *A. atemoya* | C | NC | NC | C | C |
| *A. montana* | – | C | – | – | – |
| *A. muricata* | C | C | C | NC | – |
| *A. reticulata* | C | C | C | C | – |
| *A. squamosa* | NC | LC | C | C | C |
| *A. emarginata* | C | C | – | NC | C |
| *A. crassiflora* | – | NC | – | – | – |
| *A. mucosa* | – | C | C | LC | C |

C, compatible; LC, low compatibility; NC, not compatible; –, unknown compatibility.

atemoya species, indicating a rapid union of plant tissues after grafting.

In atemoya 'Thompson', Baron *et al.* (2018a) observed changes in photosynthetic rate and ionic changes in atemoya plants grafted on to *A. emarginata* 'terra-fria' that resulted directly in vegetative growth of the graft, with a significant increase in plant height and number of leaves being observed 90 days after grafting. Regarding hormonal balance, Baron *et al.* (2018b) reported that *A. emarginata* 'terra-fria' and 'mirim' had lower ABA concentrations in the regions above and below the graft, but showed no differences in the graft region when compared with atemoya grafted on to atemoya.

Other studies with *A. emarginata* have been carried out to evaluate its ability to tolerate dry or flooded soils. The literature and some personal reports have reported this species as being able to tolerate both situations; however, Honório *et al.* (2024) demonstrated that the species presents greater damage to the photosynthetic apparatus in flooded conditions, causing greater difficulty in plant recovery when compared with plants in drought conditions.

In addition to physiological characteristics, studies have investigated the ability of *Annona* spp. to produce defence substances, such as alkaloids. In this context, it is considered that species with greater capacity to produce defence substances may be potentially more tolerant. Furthermore, such species can be used in the bioprospecting of molecules of interest, both to combat phytopathogens and pests, and also to supply molecules for the pharmaceutical and cosmetic industries. *A. emarginata* produces the following alkaloids: liriodenine, norpredicentrine, *N*-methyllaurotetanine, norglaucine, discretin, xylopinine, xylopine, asimilobine, laurotetanine, oxoglaucine, lanuginosine (Mimi *et al.*, 2021), anonaine, nornuciferine, lysicamine and oxoxylopine (da Silva *et al.*, 2024). In a study conducted by da Silva *et al.* (2024), the alkaloidal extract of this species showed potential antifungal activity against *Colletotrichum fructicola* and *C. theobromicola*, which cause anthracnose, a disease that causes damage and reduces commercial production in *Annona* spp. such as atemoya.

Among the species used as rootstock, *A. emarginata* has been the most studied since the mid-2000s, and this is undoubtedly the most widely used rootstock for atemoya. Araticum mirim, considered an *Annona emarginata* morphotype, is also an alternative for grafting atemoya. However, it is most commonly used when seedlings will be used to implement orchards in regions with warmer climates and the scion used is atemoya 'Gefner'.

#### 5.3.3.2 *A. mucosa* Jacq. (syn. *Rollinia mucosa* (Jacq.) Baill.)

This is popularly known as biribá and is native to Brazil with occurrences in the phytogeographical domains of Amazonia, Cerrado and Atlantic Forest (Mendes-Silva *et al.*, 2024b).

Some studies using biribá as rootstock for commercial Annonaceae have been conducted, among them the study carried out by Baron *et al.* (2016), where it was observed that atemoya grafted on to biribá produces, 90 days after grafting, an increase in the graft plant height and in the number of leaves when compared with other combinations (i.e. atemoya on to *A. emarginata* 'mirim' and on to atemoya itself). Almeida *et al.* (2010) reported that the grafting of atemoya 'Thompson' on to biribá showed a plant survival rate of 45.8% for the splice graft method and 24.2% for the lateral cleft method. In other studies, it was reported that sugar apple (*A. squamosa*) grafted on to biribá had a grafting success rate of 19.2% with the splice graft method (dos Santos *et al.*, 2005), and soursop (*A. muricata*) cultivar 'BRS Cerradina' grafted on to biribá had the highest productivity and plant survival index values during 14 years of evaluation (Zacaroni *et al.*, 2014).

#### 5.3.3.3 *A. muricata* L.

Soursop is another species intended for fruit production, whose orchards are not always developed from grafted seedlings but rather from seeds, as observed in Mexico, which is the world's largest soursop producer (Queiroga *et al.*, 2024). When grafting is performed, the soursop itself is used as the rootstock (Ferreira *et al.*, 2014b; Queiroga *et al.*, 2024), and rootstocks aged 8–12 months are normally used (Queiroga *et al.*, 2024).

#### 5.3.3.4 *A. squamosa* L.

The production of sugar apple seedlings still occurs, in most cases from seeds in orchards in north-eastern Brazil and Mexico. When the grafting method is used, it is recommended to use the species itself as the rootstock, but other species have also been tested, such as atemoya, *A. reticulata* and *A. glabra* (Ferreira *et al.*, 2014b, 2019). The main budding techniques recommended for the species are: lateral and top cleft budding, green strip budding, inverted-T budding and plate budding, and the efficiency of each technique depends on factors such as environmental conditions, graft and rootstock size, and the grafter's experience (Cristina *et al.*, 2000).

A recent study on *Annona* spp. grafting, specifically with *A. squamosa* and *A. cherimola*, focused on cross- and self-cleft grafting methods to enhance growth and adaptation in diverse soil types (Hammoud *et al.*, 2022). This research evaluated scion/rootstock combinations of *A. squamosa* on to *A. cherimola* (Sq/Ch), *A. squamosa* on to *A. squamosa* (Sq/Sq), and *A. cherimola* on to *A. cherimola* (Ch/Ch) under two soil types: calcareous 'white soil' and clay-loamy 'red soil, with and without iron fertilization. The results highlighted that plants grafted in red soil with iron fertilization, particularly the Sq/Ch combination, displayed superior performance, including higher plant height and increased trunk diameter and yield, as well as an increased number of primary roots and fruit. In red soil, iron fertilization in the Sq/Ch graft combination also improved root length, chlorophyll index and iron content in the leaves, significantly boosting the plant's resilience and productivity (Hammoud *et al.*, 2022).

#### 5.3.3.5 *A. crassiflora* Mart.

Popularly known as marolo, this is widely found in the Brazilian Cerrado, has good capacity to recover from drought (Honório *et al.*, 2021), and has records of the presence of alkaloids with potential phytopathogenic activity (Egydio *et al.*, 2013; da Silva *et al.*, 2017; Honório *et al.*, 2021). However, seedlings have slow development and require a great amount of skill from nurseries in order to avoid a very long time to obtain seedlings at the grafting point. There are few reports on its use as a rootstock for custard apple. In a study conducted by Zacaroni *et al.* (2014), grafts of soursop 'BRS Cerradina' on to *A. crassiflora* rootstock resulted in a good grafting success rate of around 60%, but after 2 years in the field, all seedlings died, which was possibly caused by physiological incompatibility.

As reported, it has been observed that the great challenge is not the grafting technique itself, but the search for compatible and tolerant rootstocks for the various scion species. In this context, one way to overcome the issue of possible incompatibilities is the possibility of producing seedlings from cuttings of highly selected matrices.

### 5.3.4 Cuttings

The use of cuttings is a vegetative propagation method that uses plant segments such as stems, roots or leaves, in which the formation of roots is induced in order to re-establish a new individual genetically identical to the plant that gave rise to it. This detached part of the plant that will be rooted is called a cutting (Hartmann *et al.*, 2011). The method is fast, low-cost and allows the formation of uniform, high-density orchards, with production beginning a few years after planting (Sanewski, 1991).

The main application of cuttings in *Annona* spp. is to prepare seedlings of commercial species and cultivars intended for the production of high-quality fruits, such as atemoya and cherimoya. Cuttings can also be used to produce rootstocks on to which commercial scion trees will later be grafted (Scaloppi Júnior and Martins, 2014). In addition, cuttings can assist in the production of *Annona* spp. seedlings intended for the recovery of degraded areas, timber production, extraction of specialized bioactive metabolites, and preservation of species at risk of extinction or conservation. Therefore, it is a method that should be used when large-scale multiplication of plant material is required, while maintaining genetic characteristics and in a short period of time.

When the objective is to produce seedlings of commercial species by cuttings, the soil of orchards in which seedlings will be planted should be evaluated, as commercial species such as atemoya and cherimoya are susceptible to soil pathogens, which can be a problem for this propagation method (Sanewski, 1991). Therefore, seedlings from cuttings of commercial species should be sent to locations in which the soil where seedlings will be planted does not present phytopathological or pest problems, or when it is not well established which rootstock species will be used, in the case of grafting.

The rooting capacity of cuttings of a species depends on a series of factors related to its genetic characteristics and environmental conditions. Thus, the development stage of the mother plant, time of year, type of cutting, hormonal balance and environmental variations are factors that must be analysed for each *Annona* spp. to be propagated by cuttings (Figueirêdo *et al.*, 2013).

Plants can be divided into three classes according to rooting difficulties: (i) easy-to-root plants, which have endogenous substances essential for root initiation, such as auxins; (ii) intermediate-rooting plants, which respond to the application of exogenous auxins; and (iii) difficult-to-root plants, which lack the activity of one or more cofactors and under these conditions, do not respond or respond very little to the application of auxins (Hartmann *et al.*, 2011). Species of the genus *Annona* are generally considered difficult to root (Casas *et al.*, 1984). However, there are variations depending on the cultivar; for example, approximately 60% rooting is obtained with atemoya 'African Pride', while 'Pink's Mammoth' cuttings present rooting difficulties (Sanewski, 1991). Scaloppi Júnior and Martins (2003) also observed variations in responses among species (*A. glabra*, *A. montana*, *A. emarginata* and *A. mucosa*) and in relation to the season of the year. Therefore, there is a need for different strategies according to the peculiarities of each species and cultivar, as can be seen in the advances presented below. Semi-hardwood cuttings (15–20 cm in length) of custard apple (*A. reticulata*) rooting occurred at 36.60–49.75% when propagated between February and August, and the highest rooting (49.75%) was recorded in April (Hazarika *et al.*, 2022).

#### 5.3.4.1 *A. atemoya* Mabb.

Atemoya is one of the species intended for the production of high-quality fruits but is susceptible to pathogens; for example, the cultivar 'African Pride' is susceptible to bacteria (Sanewski, 1991). Therefore, it is common to graft on pathogen-tolerant rootstocks such as *A. emarginata* 'terra-fria' morphotype (da Silva *et al.*, 2024). However, while other species such as *A. mucosa* and *A. emarginata* 'mirim' morphotype have been evaluated as potential rootstocks for atemoya (Baron *et al.*, 2018a,b), seedlings from cuttings are an alternative for the establishment of orchards. Furthermore, it is noteworthy that in some regions of the State of São Paulo, Brazil, there are orchards with atemoya seedlings grafted on to atemoya itself, and in the State of Paraná, Brazil, there are reports of seedlings originating from seeds (Stenzel *et al.*, 2003), due to the difficulties in finding rootstock compatible with atemoya.

In Brazil, the first advances in research on the rooting of atemoya cuttings was carried out in the 1990s by Professor Ede Cereda, at UNESP in Botucatu, São Paulo. In initial tests with atemoya 'Gefner' cuttings, 28.6% rooting was obtained in median cuttings and 10.25% in apical cuttings, but there was no effect of indole-3-butyric acid (IBA) and naphthaleneacetic acid (NAA) at different concentrations (0, 1000, 3000, 5000 mg $l^{-1}$) in quick-dip treatments for 5 s (Ferreira and Cereda, 1999).

Subsequently, in 'Gefner' cuttings collected in summer, kept with a pair of leaves cut in half under nebulization and treated with 0.5% NAA and 0.5% IBA in the form of talc, Ferreira *et al.* (2008) observed, from surviving cuttings, 80% rooting in apical cuttings treated with NAA, 96.67% in those treated with IBA, 80% in untreated cuttings (control), and also 68.33% in median cuttings treated with NAA. However, in another experiment using IBA in a prolonged soaking treatment with median and apical cuttings, Scaloppi Júnior and Martins (2014) reported that apical cuttings of 'Gefner' atemoya did not react to IBA treatment, while in the median cuttings, 36% rooting was obtained by the use of 3000 IBA mg $l^{-1}$.

Comparing the application of auxin in quick-dip (5 s) and prolonged soaking (24 h) mode in medium-sized cuttings of 'Gefner', collected in summer, Ferreira and Ferrari (2010) concluded that the quick-dip treatment, with 200 mg NAA $l^{-1}$ increased the process of rooting. A similar effect was also observed with IBA, irrespective the of concentration used (500, 1000, 2000, 3000, 4000 or 5000 mg $l^{-1}$). Savazaki (2013) conducted research to compare the rooting potential of medium-sized cuttings of three atemoya cultivars collected in summer and treated with 0–3000 mg $l^{-1}$ of IBA for 5 s and kept under nebulization conditions. The author found that cuttings of the cultivar 'Bradley' had higher rooting potential (60.42%) compared with 'Gefner' (35.77%), while 'PR-3' showed an intermediate response (50.35%). Summer has proven to be the ideal period for the collection of 'Gefner' cuttings, as rooting rates of 32.65% were obtained using 0.5% IBA and 35.12% with 0.75% IBA compared with 20% in control cuttings, while in winter, there was a reduction in these rooting rates to 17.32% (with 0.5% IBA), 18.25% (with 0.75% IBA) and 10.25% with the control. In this experiment, auxin was used in the form of talc (da Silva *et al.*, 2019).

In summary, the most important considerations for rooting atemoya cuttings are to maintain part of the leaves in median or apical cuttings from branches from the last growing season (Figs 5.3 and 5.4), with collection preferably carried out in summer, and optional use of auxins (IBA or NAA). The use of a nebulization system is necessary to increase the relative humidity of the propagation house and reduce transpiration, preventing leaf fall, until rooting is obtained. The system should be kept on for 10 s at 4 min intervals until rooting is observed, which requires approximately 90 (Savazaki, 2013) to 180 (Ferreira *et al.*, 2008) days.

#### 5.3.4.2 *A. cherimola* Mill. *(cherimoya)*

In cherimoya, treatment with NAA at 5000 mg $l^{-1}$ caused 20–25% rooting of hardwood cuttings from 1-year-old plants (Duarte *et al.*, 1974). Tazzari *et al.* (1990) stated that cherimoya is a species with low rooting capacity. In addition to the variable rooting capacity among cultivars, there was also a seasonal effect that limits rooting. The cultivar 'Tardiva' showed 39% rooting when cuttings were collected in June while all cuttings of 'Precoce' failed to root. The authors considered that the phenological phase seems to be the determining factor for rooting and that the best rooting period was immediately after the first seasonal growth.

#### 5.3.4.3 *A. crassiflora* Mart.

This species is native to the Brazilian Cerrado and produces fruits used in the production of beverages, sweets and jellies; however, its production is still restricted mainly to small areas in the south-eastern region of Brazil. Propagation by cuttings of this species remains a challenge. The difficulty in rooting has been attributed to the low availability of reserves in cuttings of this species (Scaloppi Júnior and Martins, 2014). Pimenta *et al.* (2017) stated that *A. crassiflora* cuttings do not present a continuous fibre barrier,

**Fig. 5.3.** Growth stage branch intended for the preparation of apical cuttings of *Annona atemoya* with two leaves cut in half and two buds at the base. Photographs courtesy of Edson Tadashi Savazaki.

**Fig. 5.4.** Growth stage branch intended for the preparation of median cuttings of *Annona atemoya* cultivar 'Gefner', with the leaves removed. Photographs courtesy of Edson Tadashi Savazaki (2005).

which can prevent the emergence of the root primordium. However, even with the application of IBA (1500, 3000, 4500 or 6000 mg $l^{-1}$) for 10 s, rooting did not occur and cuttings presented total mortality. The authors concluded that this is a difficult-to-root species, with high degree of phenolic exudation oxidation and a low amount of starch in stem tissues. These difficulties have also been reported by producers in the region of Paraguaçu, Minas Gerais, Brazil, which leads to the production of seedlings from seeds for cultivation in the field or for the formation of rootstocks for commercial species (Carvalho *et al.*, 2017).

#### *5.3.4.4 A. emarginata* Rainer (syn. *Rollinia emarginata*)

This species is recognized for having two morphotypes, 'terra-fria' and 'mirim' (Mimi *et al.*, 2021).

Bettiol Neto *et al.* (2006) obtained 56% rooting in the 'terra-fria' morphotype using 3000 mg IBA $l^{-1}$. In a study comparing the rooting of cuttings from adult and young plants (2 years old), cuttings were treated for 12 h with a combination of IBA (0, 100, 200 or 400 mg $l^{-1}$) and boron (0, 50, 100 or 150 mg $l^{-1}$). Twenty-five per cent rooting was observed in cuttings from young mother plants, while cuttings from adult plants did not root (Scaloppi Júnior and Martins, 2014). Cuttings of 'mirim' with one pair of leaves and 15 cm in length treated with 0, 1000, 2000 or 3000 mg IBA $l^{-1}$ also failed to root (Bettiol Neto *et al.*, 2006).

However, Scaloppi Júnior and Martins (2003) recorded 19.4% rooting of apical cuttings collected in summer, 15.5% from autumn cuttings and 6% from winter cuttings, regardless of the IBA concentrations used (1000–7000 mg $l^{-1}$).

#### *5.3.4.5 A. glabra* L.

The rooting of *A. glabra* (pond apple) apical cuttings is influenced by the season. Scaloppi Júnior and Martins (2003) obtained 94% rooting in summer, 19.8% in autumn and 3.8% in winter, but there was no effect of treatments with IBA concentrations ranging from 1000 to 7000 mg $l^{-1}$ as a quick-dip for 5 s.

#### *5.3.4.6 A. montana* Macfad. *(mountain soursop, wild soursop)*

*A. montana* is commonly known as mountain soursop or wild soursop. For propagation, apical cuttings should be collected in summer, as these show the highest rooting rates (48%) and survival (76.3%) The lowest rooting rate was observed in autumn (8.1%) and winter (14.7%). Use of IBA for promotion of rooting showed no benefit at concentrations between 1000 and 7000 mg $l^{-1}$ (Scaloppi Júnior and Martins, 2003).

#### *5.3.4.7 A. muricata* L.

Casas *et al.* (1984) were unable to promote rooting in *A. muricata* (soursop) cuttings with or without leaves obtained from girdled and non-girdled branches, even with the use of plant growth regulators. Later, Santos *et al.* (2011) obtained rooting of approximately 70% using IBA (2000 mg $kg^{-1}$ in talc) applied to the base of herbaceous cuttings (6–12 cm in length) with two leaves, originating from young branches of 8-year-old 'Giant of Alagoas' trees during spring. In another study with mini cuttings measuring

7 cm and containing three to four buds, de Figueirêdo *et al.* (2013) achieved 88%, 97% and 88% rooting using subapical cuttings from field and greenhouse plants, and apical cuttings from greenhouse plants, respectively, 80 days after the experiment implementation, which occurred between spring and summer.

Soursop cuttings collected in summer and treated with 0.5% IBA resulted in 25% rooting compared with 14% rooting in untreated cuttings. In contrast, cuttings collected in winter resulted in 18% rooting (with 0.5% IBA) or 17% rooting (with 0.75% IBA), compared with 10% in the control (da Silva *et al.*, 2019). Although the rooting rate was not altered, treatment with NAA (1000 p.p.m.) and IBA (1000 p.p.m.) reduced the days required for sprouting, and increased the number and length of roots, number of shoots, number of leaves and length of leaves (Maheswari *et al.*, 2021).

#### *5.3.4.8* *A. mucosa* Jacq. Baill. (syn. *Rollinia mucosa* (Jacq.) Baill.)

Other synonyms are *A. reticulata* var. *mucosa* (Jacq.) Willd.; *Rollinia biflora* Ruiz & Pav. ex G. Don; *R. orthopetala* A. DC.; *R. deliciosa* Saff.; *R. jimenezii*; *R. mucosa* var. *macropoda* (Mendes-Silva *et al.*, 2024a). In a study with apical cuttings and quick-dip treatment (5 s) with IBA (1000, 3000, 5000 or 7000 mg $l^{-1}$), Scaloppi Júnior and Martins (2003) obtained 7.7% rooting in cuttings collected in summer, 2.2% in autumn and 2.5% in winter. They evaluated the effect of juvenility on the rooting of *A. mucosa*, with cuttings from 1-year-old, 2-year-old and adult mother plants. Forty per cent rooting was obtained with cuttings from 1-year-old plants.

#### *5.3.4.9* *Annona neosalicifolia* H. Rainer

Two cutting lengths (15 and 30 cm) collected in summer and three substrates (carbonized rice husk, coarse sand and commercial substrate Plantmax®) were tested by Silveira (2020). At 60 days from the date of planting, the results demonstrated that *A. neosalicifolia* has low potential for vegetative propagation via cuttings, with approximately 3.5–5.5% rooting with no difference between treatments. The length of cuttings and type of substrate did not influence the rooting, survival or sprouting responses of the cuttings (Silveira, 2020).

#### *5.3.4.10* *A. rugulosa* Schl. (syn. *Rollinia rugulosa* Schl.)

Semi-hardwood cuttings from branches of mother trees collected in summer, autumn and winter did not root, while cuttings collected in spring showed 4% rooting following treatment with 6000 mg IBA $l^{-1}$ and 2000 mg IBA $l^{-1}$ + 2000 mg NAA $l^{-1}$ (Pinto *et al.*, 2003). Similarly, Santos *et al.* (2011) observed low rooting capacity of stem cuttings with IBA treatment; however, the use of root cuttings proved to be more promising, resulting in approximately 10% rooting in cuttings from the control group.

#### *5.3.4.11* *A. squamosa* L.

In *A. squamosa* (sugar apple), Bourke (1976) obtained rooting rates of less than 5%, while Bankar (1989) reported 26% rooting with the use of IBA (2500–3000 mg $l^{-1}$), compared with 4% rooting in the control.

The time of year is an important factor in the rooting of sugar apple cuttings, with summer being considered more suitable for rooting than winter. This factor, associated with cutting treatments with 0.5% and 0.75% IBA in the form of talc and maintenance of three pairs of leaves (reduced by half), resulted in the highest rooting rates (28.55% and 28.96%, respectively), compared with the control (15.91%). However, NAA treatment proved ineffective for rooting of cuttings of this species (da Silva *et al.*, 2019). When evaluating the effects of IBA treatment (0, 1000, 2000, 3000 or 4000 mg $l^{-1}$) in powder and liquid form with semi-hardwood cuttings, Salvador *et al.* (2014) obtained higher rooting rates (70%), regardless of treatment. Treatment with IBA in powder form was found to be significantly superior to the liquid method, with 86% (powder) and 66% (liquid) rooting, regardless of IBA concentration used.

#### *5.3.4.12* *A. sylvatica* (A. St-Hil.) Mart. (syn. *Rollinia sylvatica* (A. St-Hil.) Mart.)

*A. sylvatica* is another species with rooting difficulty. Cuttings treated with IBA (1000, 3000, 5000 or 7000 mg $l^{-1}$) using the quick-dip method for 5 s resulted in only 6% rooting with no visible effect of IBA treatments. Using a prolonged soaking treatment (12 h) with IBA

(0, 50, 100, 200 or 400 mg $l^{-1}$), the rooting rate was even lower (1%) (Scaloppi Júnior and Martins, 2014).

### 5.3.5 Marcotting (air layering)

Marcotting or air layering can also be used to propagate cultivars that grow from cuttings. The marcots should be ready to remove from the tree in about 3 months if started in summer/monsoon season.

Cherimoya cannot easily be propagated by cuttings or marcotting (Sanewski 1991). Soursop (*A. muricata*) and custard apple (*A. reticulata*) can be successfully propagated by air layering. Hazarika *et al.* (2021) obtained the highest success (71.55%) in air layering of custard apple by etiolating the shoots 30 days prior to air layering in August without the use any auxin. Addition of IBA at 4000 p.p.m. and use of sphagnum moss as the rooting medium improved rooting, resulting in 100% success. Treatment also improved the number of roots and root length of layers (Chacko *et al.*, 2023).

### 5.3.6 Stooling

Sugar apple has also been successfully propagated by stooling or mound layering. Rathore (1976) obtained 93.8% rooting in the stools of sugar apple after treating with 20,000 p.p.m. IBA. The rooted stools had 80% survival on planting. Three-year-old seedlings were headed back at ground level in April. After about 3 months when several new shoots were developed, a 3 cm-wide ring of bark was taken from the base of each emerged shoot bearing one bud to serve as a nurse shoot. The auxin IBA prepared as a lanolin paste was smeared above the ringed portion followed by mounding of all the shoots. The rooted shoots were detached from the mother plant in September and planted in the nursery under shade.

### 5.3.7 Micropropagation

In addition to propagation by seeds, cuttings and grafts as described, *in vitro* micropropagation can also be used, allowing the production of large quantities of healthy seedlings in a short period of time (Grattapaglia and Machado, 1998; Santana *et al.*, 2008). This technique not only accelerates production but also plays a crucial role in the conservation of endangered species, such as *A. senegalensis* and *A. deceptrix*, which are under severe pressure due to human exploitation and environmental degradation (Ferreira *et al.*, 2019; Ba *et al.*, 2021; Nevárez Loor *et al.*, 2024). Micropropagation generally refers to the aseptic culture of cells, tissues, organs and their components under defined chemical and physical conditions *in vitro* (Gupta *et al.*, 2020).

#### *5.3.7.1 Organogenesis and callogenesis*

Adventitious organogenesis, a technique that induces organ formation directly from explants, has been successful in several *Annona* spp. In *A. cherimola*, for example, the application of plant growth regulators and optimized culture media was able to improve the shoot regeneration rate (George and Nissen, 1993; Padilla and Encina, 2003; Encina *et al.*, 2014). In *A. deceptrix*, the use of WPM medium supplemented with BAP (1 mg $l^{-1}$) and $GA_3$ (0.25 mg $l^{-1}$) promoted healthy development of shoots and stems (Nevárez Loor *et al.*, 2024). Combination of these techniques with callogenesis also increased the regeneration efficiency from juvenile and adult tissues (Encina *et al.*, 2014).

Experimental results showed that, for *A. muricata*, MS medium (Murashige and Skoog, 1962) supplemented with BAP (2 mg $l^{-1}$) produced an average of 2.87 shoots per cotyledonary node, while for *A. squamosa*, MMS medium (modified Murashige and Skoog medium) with 2 mg BAP $l^{-1}$ was the most effective, generating an average of 3.12 shoots at terminal apices (Ba *et al.*, 2021). The addition of 0.1 and 0.2 mg NAA $l^{-1}$ provided better results in terms of elongation and number of nodes in *A. muricata*, with shoots reaching 9.11 cm in length and an average of 5.62 nodes. For *A. squamosa*, MM7 medium (supplemented with BAP and kinetin (6-furfurylaminopurine)) generated shoots of 9.05 cm and with an average 5.62 nodes (Ba *et al.*, 2021).

Plant bioregulators such as auxins and cytokinins are essential to promote shoot multiplication and rooting. In *A. muricata*, the use of BAP was particularly effective for the induction of new shoots and elongation, while in *A. squamosa*, the combination of BAP and kinetin had a positive impact on elongation and the number of nodes (Ba *et al.*, 2021).

#### 5.3.7.2 Rooting and acclimatization

Root induction, or rhizogenesis, is a crucial step to ensure micropropagation success. In *A. squamosa*, rhizogenesis induction with IBA (50 mg $l^{-1}$) and activated charcoal (2 g $l^{-1}$) for 3 days in the dark resulted in a rooting rate of 66.67%. In *A. muricata*, induction treatment for 5 days led to a rooting rate of 83.33%. After acclimatization, 83.33% of *A. muricata* plants originating from terminal apices and *A. squamosa* plants derived from cotyledonary nodes survived (Ba *et al.*, 2021).

#### 5.3.7.3 Clonal propagation and conservation

Clonal propagation through micropropagation is a viable solution for conserving endangered species, such as *A. senegalensis* and *A. deceptrix*. This technique allows the large-scale production of genetically identical clones, preserving the genetic diversity of these species, which is essential for the restoration of degraded areas (Ferreira *et al.*, 2019; Ba *et al.*, 2021; Nevárez Loor *et al.*, 2024). The acclimatization of *in vitro*-propagated plants resulted in a survival rate of 83.33%, indicating efficiency in the process for the conservation of valuable species (Ba *et al.*, 2021).

## 5.4 General Aspects of Nursery Management

Seedling production nurseries are areas where all activities involving seed management, cutting preparation and grafting or budding are carried out. Therefore, special care is required for each stage of the seedling formation process in order to guarantee the quality of the material that will be used in the development of orchards and other production areas.

Seedlings must comply with strict production and management criteria established by the control agencies of each country. However, there are some general characteristics in the management of *Annona* spp. that can guide producers of any part of the world. The main points to be considered are water availability, plant nutrition and soil health. The plants to be used as the matrix must be of known origin, meeting health and genetic characteristics criteria. When the aim is to produce seedlings from seeds, it should be kept in mind that seeds of some *Annona* spp. do not withstand storage for longer than 3–4 months (*A. atemoya*, *A. emarginata*), while *A. crassiflora* seeds require storage for about 6 months to overcome dormancy. Once it has been ensured that seeds are within their viability period, one of the methods for overcoming dormancy should be used, if this is a characteristic of the seeds. *Annona* spp. seeds can be treated with $GA_3$ or with mixtures of $GA_{4+7}$ + cytokinin in order to improve the germination rate and speed. Sowing can be performed in sand boxes or beds, or in Styrofoam trays filled with sand or substrate made from rice husks, plant waste and vermiculite. At this stage, irrigation is usually done by sprinkling, and the bed should be sprayed with fungicide as a preventative measure, as recommended by the agronomist responsible for the nursery. As soon as seedlings have two to three pairs of leaves, they are transplanted into bags containing substrate. The plants should be kept in the nursery for 1 year before they are planted in the orchard or used as rootstock for grafting or budding.

For grafting, tools such as grafting knives and pruning shears must be disinfected and disinfested. The grafting method to be used depends on the time of year, type of material to be used (budding or grafting) and the grafter's experience. After grafting, the plants are observed daily until buds appear, indicating the success of the process. When grafting, for example, atemoya on *A. emarginata*, the first signs of grafting success are observed in the first few weeks, and after buds begin to appear, seedlings will be ready to go to the field within another 3–4 months. During the period of rootstock formation and after grafting, care must be taken to prevent attacks by pests and diseases, and an irrigation system must be maintained, which is usually by sprinkler.

When rootstocks are developed from cuttings, the process is quite complex and requires a great amount of care. In general, cuttings should be collected from branches at a particular time of the year, keeping about two leaves cut in half and then, just below leaves. Two to three buds should be kept. However, attention should be paid to the specific characteristics of each species and cultivar to be propagated. For rooting to occur, the base of the cutting should be treated with auxins and then the cuttings should be inserted into propagation benches or Styrofoam trays containing commercial substrate, vermiculite or sand.

Cuttings from species such as atemoya require the leaves to be maintained until rooting is observed. To do so, it is necessary to increase the relative humidity of the environment so that the leaves do not lose water to the atmosphere. If cuttings lose their leaves, they will probably not root. Therefore, maintaining leaves is an important step, which is achieved by implementing a nebulization system. The nebulization system should be the type that produces a mist, increasing the relative humidity without forming water droplets on leaves.

After rooting, the seedlings should be transferred to a bag and gradually acclimatized until they have adapted to the field conditions where the orchard will be installed.

## 5.5 Conclusion

The propagation of *Annona* spp. has been the focus of numerous studies, with significant advances from both scientific and technological perspectives. Regarding seeds, the main progress has been in understanding tolerance to desiccation and dormancy mechanisms, particularly related to physiological dormancy, involving a balance between ABA and gibberellins.

For grafting, several studies support the use of *A. emarginata* as a rootstock for atemoya. These studies range from the effect of this species on altering the photosynthesis of the atemoya canopy to the rootstock's ability to tolerate abiotic stresses (dry or waterlogged soils) and biotic stresses such as fungi like *Colletotrichum* spp. attacking the atemoya. Additionally, it has been possible to select candidate trees (accessions) within a select group of plants to provide seeds for rootstock production. In this context, *A. emarginata* has been one of the most studied species as a rootstock for atemoya. Advances have also been made in understanding the molecular mechanisms involving compatibility between the canopy and the rootstock, confirming the use of *A. emarginata* as a rootstock for atemoya and indicating new compatible alternatives, such as *A. montana* and *A. emarginata* 'mirim' morphotype.

Regarding cuttings, the main advancements have been in rooting of atemoya cuttings; however, *Annona* spp. still represent a challenge for propagation by cuttings, often being considered difficult to root. These difficulties suggest the need for further studies, including those on micropropagation methods, which could be an important tool to explore.

## 5.6 Funding

This research was funded by Coordenação de Aperfeiçoamento de Pessoal de Nível Superior - Brasil (CAPES) – Finance Code 001 and National Council for Scientific and Technological Development (CNPq), grant numbers 140073/2022-1 and 308728/2022-0.

## References

Almeida, L.F.P.D., Alencar, C.M.D. and Yamanishi, O.K. (2010) 'Thompson' atemoya grafting onto Rollinia rootstocks. *Revista Brasileira de Fruticultura* 32, 653–656. DOI: 10.1590/S0100-29452010005000058

Alouani, M. and Bani-Aameur, F. (2004) Argan (*Argania spinosa* (L.) Skeels) seed germination under nursery conditions: Effect of cold storage, gibberellic acid and mother-tree genotype. *Annals of Forest Science* 61, 191–194. DOI: 10.1051/forest:2004011

Ba, O., Sagna, M. and Sy, M.O. (2012) Germination capacity of *Annonaceae* seeds (*Annona muricata* L., *A. squamosa* L. and *A. senegalensis* Pers.) cultivated under axenic conditions. *International Journal of Science and Advanced Technology* 2, 21–34.

Ba, O., Dieme, A., Ndoye, A.L. and Sy, M.O. (2021) *In vitro* clonal propagation from juvenile and different explant types of two edible *Annonaceae* species: *Annona muricata* L. and *Annona squamosa* L. *Advances in Bioscience and Biotechnology* 12, 458–480. DOI: 10.4236/abb.2021.1212029

Bankar, G.J. (1989) Vegetative propagation in annonas (*Annona squamosa* L). *Haryana Journal of Horticultural Sciences* 18, 10–13.

Baron, D., Bravo, J.P., Maia, I.G., Pina, A. and Ferreira, G. (2016) UGP gene expression and UDP-glucose pyrophosphorylase enzymatic activity in grafting annonaceous plants. *Acta Physiologiae Plantarum* 38: 79. DOI: 10.1007/s11738-016-2097-7

Baron, D., Amaro, A.C.E., Macedo, A.C., Boaro, C.S.F. and Ferreira, G. (2018a) Physiological changes modulated by rootstocks in atemoya (*Annona* × *atemoya* Mabb.): gas exchange, growth and ion concentration. *Brazilian Journal of Botany* 41, 219–225. DOI: 10.1007/s40415-017-0421-0

Baron, D., Gimenez, J.I. and Ferreira, G. (2018b) Abscysic acid and compatibility of atemoya (*Annona* × *atemoya* Mabb.) grafted onto native species. *Revista Brasileira de Fruticultura* 40: e954. DOI: 10.1590/0100-29452018954

Barros, A.P.G., dos Santos, L.A., Ferreira dos Santos, A.K., Sedrez da Rosa e Silva, P., Araújo, C. and Pimenta, A.C. (2019) Caracterização de frutos e sementes de araticum (*Annona crassiflora* Mart.) nativos do Cerrado *matogrossense*. *Agropecuária Científica no Semiárido* 14, 280–286. DOI: 10.30969/acsa.v14i4.1001

Bernardes, T.G., Estrêla, C.T., Naves, R.V., Rezende, C.F.A., Mesquita, M.A.M. and Pires, L.L. (2007) Efeito do armazenamento e de fitohormônios na qualidade fisiológica de sementes de araticum (*Annona crassiflora* Mart.). *Pesquisa Agropecuária Tropical* 37, 163–168.

Bettiol Neto, J.E., Pio, R., Bueno, S.C.S., Bastos, D.C. and Scarpare Filho, J.A. (2006) Enraizamento de estacas dos porta-enxertos Araticum-de-Terra-Fria (*Rollinia* sp.) e Araticum-Mirim (*Rollinia emarginata* Schltdl.) para anonáceas. *Ciência e Agrotecnologia* 30, 1077–1082. DOI: 10.1590/S1413-70542006000600005

Bewley, J.D., Bradford, K.J., Hilhorst, H.W.M. and Nonogaki, H. (2013) *Seeds: Physiology of Development, Germination and Dormancy*, 3rd edn. Springer, New York.

Bourke, D.O.D. (1976) *Annona* spp. In: Garner R.J. and Chaudhri, S.A. (eds) *The Propagation of Tropical Fruit Trees*. Horticultural Review No. 4, Commonwealth Bureau of Horticulture and Plantation Crops, Kent, UL, pp. 223–248.

Brokaw, R. (1998) Propagation. In: Faber, B. (ed.) *The CCA Cherimoya Growers Handbook*. California Cherimoya Association, Saticoy, California, USA. Available at: https://ucanr.edu/sites/default/files/2024-06/398848.pdf (accessed 18 August 2025)

Carvalho, G.A., Carvalho, L.M. and Palhão, S.A. (2017) *Marolo: Um fruto, várias ideias*. Gráfica e Editora Gilcav, Machado, Minas Gerais, Brazil.

Casas, M.H., Victória, S.M.A. and Zarate, R.R.D. (1984) Preliminary trials on sexual and asexual propagation of soursop (*Annona muricata* L.). *Acta Agronomica* 4, 66–81.

Cavalcante, T.R.M., Naves, R.V., Braga Filho, J.R. and Silva, L.B. (2007) Influência de substratos e do armazenamento de sementes sobre a emergência e crescimento de plântulas de araticum (*Annonaceae*). *Bioscience Journal* 23, 11–20.

Chacko, I., Ranchana, P., Gopi, V., Srinivasan, J., Dinesh Kumar, P. *et al.* (2023) Effect of time and IBA concentrations on soursop (*Annona muricata*): air layering. *Pharma Innovation Journal* 12, 815–818.

Chatrou, L.W. (1999) The *Annonaceae* and the *Annonaceae* project: a brief overview of the state of affairs. *Acta Horticulturae* 497, 43–58. DOI: 10.17660/ActaHortic.1999.497.2

Chatrou, L.W., Erkens, R.H.J., Richardson, J.E., Saunders, R.M.K. and Fay, M.F. (2012) The natural history of *Annonaceae*. *Botanical Journal of the Linnean Society* 169, 1–4. DOI: 10.1111/j.1095-8339.2012.01242.x

Cookson, S.J. and Ollat, N. (2013) Grafting with rootstocks induces extensive transcriptional re-programming in the shoot apical meristem of grapevine. *BMC Plant Biology* 13: 147. DOI: 10.1186/1471-2229-13-147

Corsato, J.M., Ferreira, G. and Barbedo, C.J. (2012) Desiccation tolerance in seeds of *Annona emarginata* (Schldtl.) H. Rainer and action of plant growth regulators on germination. *Brazilian Journal of Plant Physiology* 24, 253–260. DOI: 10.1590/S1677-04202012000400004

Costa, P.N., Bueno, S.S.C. and Ferreira, G. (2011) Germination phases of *Annona emarginata* (Schltdl.) H. Rainer seeds under different temperatures. *Revista Brasileira de Fruticultura* 33, 253–260. DOI: 10.1590/S0100-29452011005000025

Cristina, M., Cordeiro, R., Carlos de Queiroz, A., Víctor, P. and Ramos, H.V. (2000) *O Cultivo da Pinha, Fruta-do-Conde ou Ata no Brasil*. Embrapa Cerrados, Planaltina, Brazil.

Dalanhol, S.J., Mombach, T.C., Toderke, M.L., Nogueira, A.V. and Bortolini, M.F. (2013) Dormência em sementes de *Annona Cacans* Warm. (Annonaceae). *Revista Acadêmica Ciência Animal* 11: 183. DOI: 10.7213/academica.10.s01.ao21

Dantas, A.P., Pedrosa, A.C., Lederman, I.E., Bezerra, I.E.F. and Melo Neto, M.L. (1993) Técnicas de enxertia na propagação da pinheira (*Annona squamosa* L.) em viveiro. *Revista Brasileira de Fruticultura* 15, 235–238.

da Silva, C.P.D., Ono, E.O., Rodrigues, J.D., Corrêa, L.D.S. and Boliani, A.C. (2019) Enraizamento de estacas apicais de pinheira, gravioleira e atemoeira tratadas com auxinas. *Revista Agraria Academica* 2, 143–156. DOI: 10.32406/v2n22019/143-156/agrariacad

da Silva, E.A.A, de Melo, D.L.B., Davide, A.C., de Bode, N., Abreu, G.B. *et al.* (2007) Germination ecophysiology of *Annona crassiflora* seeds. *Annals of Botany* 99, 823–830. DOI: 10.1093/aob/mcm016

da Silva, E.C., Villa, F., da Silva, D.F.D., Possenti, J.C., da Silva, A.D.C.C. and Masiero, M.A. (2023) Potential potential of a native araticum species as the rootstock of atemoya cultivars propagated by grafting methods. *Acta Scientiarum. Agronomy* 45: e60964. DOI: 10.4025/actasciagron.v45i1.60964

da Silva, G.C., De-la-Cruz-Chacón, I., Honório, A.B.M., Martin, B.C., Sousa, M.C. *et al.* (2022) Temperature and $GA_3$ as modulating factors in the biosynthesis of alkaloids during imbibition and early development of *Annona* x *atemoya* Mabb. cv. 'Gefner' seedlings. *Horticulturae* 8: 766. DOI: 10.3390/horticulturae8090766

da Silva, G.C., Dutra, L.M., da Silva Almeida, J.R.G., da Silva, F.M.A., Harakava, R. *et al.* (2024) Alkaloid screening of *Annona emarginata* (Schltdl.) H. Rainer rootstocks to increase fungal tolerance in *Annona atemoya* Mabb. crops revealed by MS and NMR chemical profiling. *Industrial Crops and Products* 212: 118335. DOI: 10.1016/j.indcrop.2024.118335

da Silva, M.A., Rosa, C.P., Bastos, R.G., da Silva, G.A., da Marques, M.J. and Espuri, P.F. (2017) Triagem fitoquímica, atividade antioxidante e leishmanicida do extrato hidroetanólico 70% (v/v) e das frações obtidas de (*Annona crassiflora* Mart.). *Revista Fitos* 10. DOI: 10.5935/2446-4775.20160036

de Figueirêdo, G.R.G., Vilasboas, F.S., de Oliveira, S.J.R., Sodré, G.A. and do Sacramento, C.H. (2013) Propagation of soursop tree through minicuttings. *Revista Brasileira de Fruticultura* 35, 860–865. DOI: 10.1590/S0100-29452013000300024

de Menezes, N.L., Franzin, S.M. and Bortolotto, R.P. (2009) Dormência em sementes de arroz: causas e métodos de superação. *Revista de Ciências Agro-Ambientais* 7, 35–44.

De-Pieri-Oliveira, M. (2019) Mecanismos de tolerância à dessecação em sementes de *Annona crassiflora* Mart. e *Annona glabra* L. MSc thesis, Universidade Estadual Paulista 'Júlio de Mesquita Filho', Instituto de Biociências de Botucatu, Botucatu, São Paulo, Bazil. Available at: https://repositorio.unesp.br/handle/11449/190934 (accessed 18 August 2025).

Dharmik, Y.B., Bagde, E.D., Goramnagar, H.B., Dawaneand, P.N. and Patinge, S.P. (2022) Standardization of time of grafting and age of rootstock in custard apple (*Annona squamosa* L.). *Biological Forum* 14, 325–328.

Doran, J.C. (1997) Seed, nursery practice and establishment. In: Waterhouse, D.F. (ed.) *The Major Invertebrate Pests and Weeds of Agriculture and Plantation Forestry in the Southern and Western Pacific*. Australian Centre for International Agricultural Research, Canberra, pp. 1–29.

dos Santos, C.E., Roberto, S.R. and Martins, A.B.G. (2005) Propagação do biribá (*Rollinia mucosa*) e sua utilização como porta-enxerto de pinha (*Annona squamosa*). *Acta Scientiarum. Agronomy* 27, 433–436. DOI: 10.4025/actasciagron.v27i3.1404

Dresch, D.M., Scalon, S.P.Q. and Masetto, T.E. (2014) Effect of storage in overcoming seed dormancy of *Annona coriacea* Mart. seeds. *Anais Da Academia Brasileira de Ciências* 86, 2077–2085. DOI: 10.1590/0001-3765201420130276

Duarte, O., Villagarcia, J. and Franciosi, R. (1974) Efecto de algunos tratamientos en la propagación del chirimoyo por semillas, estacas e injertos. *Proceedings of the Tropical Region of American Society Horticulture Science* 18, 41–48.

Egydio, A.P.M., Valvassoura, T.A. and Santos, D.Y.A.C. (2013) Geographical variation of isoquinoline alkaloids of *Annona crassiflora* Mart. from Cerrado, Brazil. *Biochemical Systematics and Ecology* 46, 145–151. DOI: 10.1016/j.bse.2012.08.006

Encina, C.L., Martin, E.C., Lopez, A.A. and Padilla, I.M.G. (2014) Biotechnology applied to *Annona* species: a review. *Revista Brasileira de Fruticultura* 36, 17–21. DOI: 10.1590/S0100-29452014000500002

Ferreira, G. and Cereda, E. (1999) Efeito da interação entre fitorreguladores, substratos e tipos de estacas no enraizamento de atemóia (*Annona cherimola* Mill x A. *squamosa* L.). *Revista Brasileira de Fruticultura* 21, 79–83.

Ferreira, G. and Ferrari, T.B. (2010) Rooting of atemoya (*Annona cherimola* Mill. × *A. squamosa* L.) cv. Gefner cuttings subjected to slow and fast treatment with auxins. *Ciência e Agrotecnologia* 34, 329–336. DOI: 10.1590/S1413-70542010000200009

Ferreira, G., Guimarães, V.F., Pinho, S.Z.D., Oliveira, M.C.D., Richart, A. *et al.* (2006) Water absorption curve in seeds of atemoya (*Annona cherimola* Mill. × *Annona squamosa* L.) cv. Gefner. *Revista Brasileira de Fruticultura* 28, 121–124. DOI: 10.1590/S0100-29452006000100033

Ferreira, G., Ferrari, T.B., Pinho, S.Z.D. and Savazaki, E.T. (2008) Rooting of atemoya 'Gefner' cuttings treated with auxins. *Revista Brasileira de Fruticultura* 30, 1083–1088. DOI: 10.1590/S0100-29452008000400039

Ferreira, M.G.R., Santos, M.R.A., Silva, E.O., Gonçalves, E.P., Alves, E.U. and Bruno, R. de L.A. (2009) Superação de dormência em sementes de biribá (*Rollinia mucosa* (Jacq.) Baill). *Revista Brasileira de Sementes* 31, 95–99. DOI: 10.1590/S0101-31222009000400011

Ferreira, G., Esquinca, A.R.G. and De-la-Cruz-Chacón, I. (2014a) Water uptake by *Annona diversifolia* Saff. and *A. purpurea* Moc. & Sessé ex Dunal seeds (Annonaceae). *Revista Brasileira de Fruticultura* 36, 288–295. DOI: 10.1590/S0100-29452014000500034

Ferreira, G., Kavati, R., Boaro, C.S.F., Ferrari, T.B. and Lenoel, S. (2014b) *Anonáceas: Propagação e Produção de Mudas*. FEPAF, Botucatu, São Paulo, Brazil.

Ferreira, G., De-la-Cruz-Chacón, I. and González-Esquinca, A.R. (2016) Overcoming seed dormancy in *Annona macroprophyllata* and *Annona purpurea* using plant growth regulators. *Revista Brasileira de Fruticultura* 38: e234. DOI: 10.1590/0100-29452016234

Ferreira, G., De-la-Cruz-Chacón, I., Boaro, C.S.F., Baron, D. and de Lemos, E.E.P. (2019) Propagation of Annonaceous plants. *Revista Brasileira de Fruticultura* 41: e500. DOI: 10.1590/0100-29452019500

Ferreira, G., De-la-Cruz-Chacón, I. and González-Esquinca, A.R. (2020) Changes in hormonal balance as key to reserve degradation after dormancy overcoming in *Annona macroprophyllata* and *Annona purpurea* seeds. *Revista Brasileira de Fruticultura* 42: e664. DOI: 10.1590/0100-29452020664

Ferreira, G., Pegorin, P., Seraphim, R.G., Torres, A.M., Ferreira, J.J. da S. *et al.* (2022) *Dormência de Sementes: Provocações e Reflexões*. Universidade Estadual Paulista Júlio de Mesquita Filho, Campus de Botucatu, Instituto de Biociências, Botucatu, São Paulo, Brazil.

Forbis, T.A., Floyd, S.K. and Queiroz, A.D. (2002) The evolution of embryo size in angiosperms and other seed plants: implications for the evolution of seed dormancy. *Evolution*, 56, 2112–2125. DOI: 10.1111/j.0014-3820.2002.tb00137.x

George, A.P. and Nissen, R.J. (1993) Annonaceous fruits. In: *Encyclopedia of Food Science, Food Technology and Nutrition*. Academic Press, London, pp. 195–199.

Gimenez, J.I., Ferreira, G. and Cavariani, C. (2014a) Tetrazolium test for assessment of seed viability of atemoya (*Annona cherimola* Mill. × *A. squamosa* L.). *Journal of Seed Science* 36, 357–361. DOI: 10.1590/2317-1545v36n31004

Gimenez, J.I., Ferreira, G. and Corsato, J.M. (2014b) Soluble sugars and germination of *Annona emarginata* (Schltdl.) H. Rainer seeds submitted to immersion in $GA_3$ up to different water contents. *Revista Brasileira de Fruticultura* 36, 281–287. DOI: 10.1590/S0100-29452014000500033

Gimenez, J.I., Esteves Amaro, A.C., Machado, S.R. and Ferreira, G. (2017) Slow imbibition of *Annona emarginata* (*Annonaceae*) seeds: metabolic and ultrastructural evaluations. *Botany* 95, 1033–1040. DOI: 10.1139/cjb-2017-0110

González-Esquinca, A.R., De-la-Cruz-Chacón, I. and Domínguez-Gutú, L.M. (2015) Dormancy and germination of *Annona macroprophyllata* (*Annonaceae*): the importance of the micropylar plug and seed position in the fruits. *Botanical Sciences* 93, 509–515. DOI: 10.17129/botsci.166

González-Esquinca, A.R., De-la-Cruz-Chacón, I., Castro-Moreno, M. and Riley-Saldaña, C.A. (2016) Phenological strategies of *Annona* from deciduous forest of Chiapas, Mexico. *Botanical Sciences* 94, 531–541. DOI: 10.17129/botsci.645

Grattapaglia, D. and Machado, M.A. (1998) Micropropagação. In: Torres, A.C., Caldas, L.S. and Buso, J.A. (eds) *Cultura de Tecidos e Transformação Genética de Plantas*. Embrapa-SPI/Embrapa-CNPH, Brasília, pp. 43–76.

Gupta, N., Jain, V., Joseph, M.R. and Devi, S. (2020) A review on micropropagation culture method. *Asian Journal of Pharmaceutical Research and Development* 8, 86–93. DOI: 10.22270/ajprd.v8i1.653

Hammoud, M., El Sebaaly, Z., Alturki, S.M., Kattar, S. and Sassine, Y.N. (2022) Performance of *Annona* plants subjected to different graft combinations, soil types and iron fertilization. *Agrivita Journal of Agricultural Science* 44, 235–247. DOI: 10.17503/agrivita.v44i2.2626

Hartmann, H.T., Kester, D.E., Davies, F.T. and Geneve, R.L. (2011) *Hartmann and Kester's Plant Propagation: Principles and Practices*, 8th edn. Prentice Hall, Boston. Massachusetts.

Hazarika, J., Hazarika, D.N. and Langthasa, S. (2021) Standardization of propagation method of custard apple by air layering. *Journal of Agrisearch* 8, 222–228. DOI: 10.21921/jas.v8i03.1666

Hazarika, J., Hazarika, D.N., Boruah, P., Roy, M. and Ahmed, S. (2022) Standardization of propagation method of custard apple (*Annona reticulata*) by stem cuttings. *Pharma Innovation Journal* 11, 1742–1746.

Honório, A.B.M., De-la-Cruz-Chacón, I., Martínez-Vázquez, M., da Silva, M.R., Campos, F.G. *et al.* (2021) Impact of drought and flooding on alkaloid production in *Annona crassiflora* Mart. *Horticulturae* 7: 414. DOI: 10.3390/horticulturae7100414

Honório, A.B.M., Corrêa, P.L.C. and Ferreira, G. (2022) As relações da luz e temperatura com a dormência de sementes. In: Ferreira, G. (ed.) *Dormência de Sementes: Provocações e Reflexões*. Universidade Estadual Paulista "Júlio de Mesquita Filho", Instituto de Biociências de Botucatu, Botucatu, Brazil, pp. 144–157.

Honório, A.B.M., De-la-Cruz-Chacón, I., da Silva, G.C., Mimi, C.O., Campos, F.G. *et al.* (2024) Differential tolerance of primary metabolism of *Annona emarginata* (Schltdl.) H. Rainer to water stress modulates alkaloid production. *Horticulturae* 10: 220. DOI: 10.3390/horticulturae10030220

Joshi, P.S., Jadhao, B.J. and Chaudhari, G.V. (2011) Studies on vegetative propagation in custard apple. *Asian Journal of Horticulture* 6, 261–263.

Kamble, A.K., Lakshmi, K.V., Pavan, P.R., Jholgiker, P., Nanjappanavar, A.G. and Chittapur, R.B. (2024) Standardization of soft wood grafting for custard apple cultivars under various environmental condition. *Plant Archives* 24, 577–582. DOI: 10.51470/PLANTARCHIVES.2024.v24.no.2.081

Koster, K.L. and Leopold, A.C. (1988) Sugars and desiccation tolerance in seeds. *Plant Physiology* 88, 829–832. DOI: 10.1104/pp.88.3.829

Kumar, S. and Shukla, A.K. (2008) Bench grafting in custard apple (*Annona squamosa* L.) as influenced by scion thickness and time of grafting. *Indian Journal of Agroforestry* 10, 62–65.

Leite, G.A. Mendonça, V., Mendonça, L.F. de M., Dantas, L.L. de G.R. and Cunha, P.S. de C.F. (2013) Rootstocks and grafting methods for atemoya (*Annona squamosa* L. × *Annona cherimoya* Mill.) plants. *Semina: Ciências Agrárias* 34, 2117–2128. DOI: 10.5433/1679-0359.2013v34n5p2117

Lerouxel, O., Cavalier, D.M., Liepman, A.H. and Keegstra, K. (2006) Biosynthesis of plant cell wall polysaccharides – a complex process. *Current Opinion in Plant Biology* 9, 621–630. DOI: 10.1016/j.pbi.2006.09.009

Maheswari, T.U., Sinduja, S., Rajavel, A., Padmanaban, J. and Sudhagar, R. (2021) Effect of plant growth regulators in adventitious rooting of cutting in soursop (*Annona muricata* L.). *Plant Archives* 21 (Suppl. 1), 2013–2015. DOI: 10.51470/PLANTARCHIVES.2021.v21.S1.327

Manica, I., Midori, I.I., Pereira, J.K., Mancuso, da C.M., Santos, O.M.A. *et al.* (2003) *Frutas Anonáceas: Ata ou Pinha, Atemólia, Cherimólia e Graviola: Tecnologia de Produção, Pós-colheita e Mercado*. Cinco Continentes, Porto Alegre, Brazil.

Mantoan, L.P.B., Rolim de Almeida, L.F., Macedo, A.C., Ferreira, G. and Boaro, C.S.F. (2016) Photosynthetic adjustment after rehydration in *Annona emarginata*. *Acta Physiologiae Plantarum* 38: 157. DOI: 10.1007/s11738-016-2171-1

Martínez-Ballesta, M.C., Alcaraz-López, C., Muries, B., Mota-Cadenas, C. and Carvajal, M. (2010) Physiological aspects of rootstock–scion interactions. *Scientia Horticulturae* 127, 112–118. DOI: 10.1016/j.scienta.2010.08.002

Mayer, J.L.S., Alves, A.C.A., de Souza, L.A. and Bona, C. (2008) Morfoanatomia da plântula e tirodendro de *Rollinia sylvatica* (A. St.-Hil.) Mart. (Annonaceae). *Revista Brasileira de Botânica* 31, 689–700. DOI: 10.1590/s0100-84042008000400014

Mendes-Silva, I., Lopes, J.C., Silva, L.V. and Bazante, M.L. (2024a) *Annona* L. Flora e Funga do Brasil. Available at: https://floradobrasil.jbrj.gov.br/consulta/ficha.html?idDadosListaBrasil=110235 (accessed 18 August 2025).

Mendes-Silva, I., Lopes, J.C., Silva, L.V. and Bazante, M.L. (2024b) Annona mucosa Jacq. Flora e Funga do Brasil. Available at: https://floradobrasil.jbrj.gov.br/consulta/ficha.html?idDadosListaBrasil=110252 (accessed 18 August 2025).

Mimi, C.O., De-La-Cruz-Chacón, I., Sousa, M.C., Vieira, M.A.R., Marques, M.O.M. *et al.* (2021) Chemophenetics as a tool for distinguishing morphotypes of *Annona emarginata* (Schltdl.) H. Rainer. *Chemistry & Biodiversity* 18. DOI: 10.1002/cbdv.202100544

Mimi, C.O., Sousa, M.C., Corrêa, P.L.C., De-la-Cruz-Chacón, I., Boaro, C.S.F. *et al.* (2023) Impact of $GA_3$ on sugar and lipid degradation during *Annona* × *atemoya* Mabb. seed germination. *Horticulturae* 9: 388. DOI: 10.3390/horticulturae9030388

Murashige, T. and Skoog, F. (1962) A revised medium for rapid growth and bio assays with tobacco tissue cultures. *Physiologia Plantarum* 15, 473–497. DOI: 10.1111/j.1399-3054.1962.tb08052.x

Nevárez Loor, M.A., Macías Ponce, F.C. and Pico-Mendoza, J. (2024) Propagación *in vitro* de *Annona deceptrix* (Westra) H. Rainer (Annonaceae) una especie en peligro de extinción en Ecuador. *Bionatura* 1 1–12. https://doi.org/10.21931/bj/2024.01.01.3

Ogden, M.A.H., Campbell, C.W. and Lara, S.P. (1981) Grafting *Annonas* in Sothern Florida. *Proceedings of the Florida State Horticultural Society* 94, 355–358.

Oliveira, I.V. de M., de Andrade, R.A. and Martins, A.B.G. (2005) Influence of the temperature on seed germination of *Annona montana*. *Revista Brasileira de Fruticultura* 27, 344–345. DOI: 10.1590/S0100-29452005000200041

Oliveira, M.C.D., Ferreira, G., Guimarães, V.F. and Dias, G.B. (2010) Germination of atemoya seeds (*Annona cherimola* Mill. × *A. squamosa* L.) cv 'Gefner' subjected to treatments with gibberellic acid ($GA_3$) and ethephon. *Revista Brasileira de Fruticultura* 32, 544–554. DOI: 10.1590/S0100-29452010005000062

Padilla, I.M.G. and Encina, C.L. (2003) *In vitro* germination of cherimoya (*Annona cherimola* Mill.) seeds. *Scientia Horticulturae* 97, 219–227. DOI: 10.1016/S0304-4238(02)00160-7

Pammenter, N.W. and Berjak, P. (2000) Evolutionary and ecological aspects of recalcitrant seed biology. *Seed Science Research* 10, 301–306. DOI: 10.1017/S0960258500000349

Panchal, S.B., Patel, C.R. and Chaudhary, H.L. (2022) Effect of grafting time and method in custard apple (*Annona squamosa* L.) cv. Sindhan under South Gujarat condition. *Pharma Innovation Journal* 11, 1011–1016.

Peterbauer, T. and Richter, A. (2001) Biochemistry and physiology of raffinose family oligosaccharides and galactosyl cyclitols in seeds. *Seed Science Research* 11, 185–197. DOI: 10.1079/SSR200175

Pimenta, A.C., Amano, E. and Zuffellato-Ribas, K.C. (2017) Estaquia e anatomia caulinar de *Annona crassiflora* Mart. *Caderno de Ciências Agrárias* 9, 1–7.

Pinto, A.C. de Q., Cordiero, M.C.R., de Andrade, S.R.M., Ferriera, F.R., Filguieras, H.A. de C. *et al.* (2005) *Annona species*. International Centre for Underutilised Crops, University of Southampton, UK.

Pinto, L.S., Zuffellato-Ribas, K.C., Carpanezzi, A.A., Tavares, F.R. and Koehler, H.S. (2003) Indução do enraizamento de estacas de araticum-de-porco pela aplicação de fitorreguladores. *Scientia Agraria* 4: 41. DOI: 10.5380/rsa.v4i1.1064

Pompelli, M.F., Jarma-Orozco, A. and Rodriguez-Páez, L.A. (2023) Imbibition and germination of seeds with economic and ecological interest: physical and biochemical factors involved. *Sustainability* 15: 5394. DOI: 10.3390/su15065394

Queiroga, V. de P., Gomes, J.P., Mendes, N.V.B., Queiroz, A.J. de M., Lima, D. de C. *et al.* (2024) Sistema produtivo da graviola (*Annona muricata* L.). In: Queiroga, V. de P., Gomes, J.P., Mendes, N.V.B., Queiroz, A.J. de M., Lima, D. de C. *et al.* (eds) *Graviola (*Annona muricata*, L.) Tecnologias de Plantio e Utilização*. AREPB, pp. 10–147.

Rainer, H. and Chatrou, L.W. (2006) AnnonBase: world species list of Annonaceae – version 1.1, 12 Oct 2006. Available at: www.annonaceae.org (accessed 27 August 2025).

Rathore, D.S. (1976) Preliminary studies on the stooling of custard apple (*Annona squamosa* L.). *Indian Journal of Horticulture* 33, 244–245.

Royal Botanic Gardens, Kew (2016) *Annona* L. Plants of the World Online. Available at: https://powo.science.kew.org/taxon/urn:lsid:ipni.org:names:1966-1 (accessed 11 September 2025).

Roberts, E.H. (1973) Predicting the storage life of seeds. *Seed Science and Technology* 1, 499–514.

Salvador, T. de L., Salvador, T. de L., de Lemos, E.E.P., Barros, P.G. and Campos, R. da S. (2014) Rooting cuttings of sugar apple (*Annona squamosa* L.) with indolebutyric acid. *Revista Brasileira de Fruticultura* 36, 310–314. DOI: 10.1590/S0100-29452014000500037

Sanewski, G.M. (1991) Classification and cultivars. In: Sanewski, G.M. (ed.) *Custard Apples: Cultivation and Crop Protection*. Queensland Department of Primary Industries, Brisbane, Australia, pp. 2–11.

Santana, J.R.F.D., Oliveira, L.M.D., Paiva, R., Resende, R.K.S., Castro, E.M. and Pereira F.D. (2008) Anatomia foliar de seis espécies de anonáceas cultivadas in vitro e em casa de vegetação. *Ciência Rural* 38, 2362–2365. DOI: 10.1590/S0103-84782008000800043

Santos, M.Q.C., Lemos, E.E.P., Salvador, T.L., Rezende, L.P., Lima Salvador, T. *et al.* (2011) Rooting soft cuttings of soursop (*Annona muricata*) "Giant of Alagoas". *Acta Horticulturae* 923, 241–245. DOI: 10.17660/ActaHortic.2011.923.36

Savazaki, M. (2013) Efeito do ácido indolbutírico no enraizamento de estacas de três cultivares de atemoia e formação da muda. In: Ferreira, G., Kavati, R., Boaro, C.S.F., Ferrari, T.B. and Leonel, S. (eds) *Annonacaeas: Propagação e Produção de Mudas*. FEPAF, Brasil, pp. 75–109.

Scaloppi Júnior, E.J. and Martins, A.B.G. (2003 Vegetative propagation of four Annonaceae rootstocks. *Revista Brasileira de Fruticultura* 25, 286–289. DOI: 10.1590/S0100-29452003000200026

Scaloppi Júnior, E.J. and Martins, A.B.G. (2014) Anonas cutting. *Revista Brasileira de Fruticultura* 36, 147–156. DOI: 10.1590/S0100-29452014000500018

Silveira, R.S. (2020) Vegetative propagation of araticum (*Annona neosalicifolia* H. Rainer) using different types of stakes and substrates. *Brazilian Journal of Animal and Environmental Research* 3, 4257–4262. DOI: 10.34188/bjaerv3n4-121

Stenzel, N.M.C., Murata, I.M. and Neves, C.S.V.J. (2003) Overcoming atemoya and custard apple seed dormancy. *Revista Brasileira de Fruticultura* 25, 305–308. DOI: 10.1590/S0100-29452003000200031

Tazzari, L., Pestelli, P., Fiorino, P. and Parri, G. (1990) Propagation techniques for *Annona cherimola* Mill. *Acta Horticulturae* 275, 315–322. DOI: 10.17660/ActaHortic.1990.275.38

Tokunaga, T. (2005) *A Cultura da Atemóia*. Boletim Técnico No.2. CATI, Campinas, São Paulo, Brazil.

Vidal-Lezama, E., Villegas-Monter, Á., Vaquera-Huerta, H., Robledo-Paz, A., Martínez-Palacios, A. and Ferreira, G. (2023) Morphometry of chincuya seeds (*Annona purpurea* Moc. & Sessé ex *Dunal*) and embryonic growth under dry warm storage. *Revista Brasileira de Fruticultura* 45: e042. DOI: 10.1590/0100-29452023042

Vidal-Lezama, E., Villegas-Monter, Á., Vaquera-Huerta, H., Robledo-Paz, A., Martínez-Palacios, A. and Ferreira, G. (2024) Imbibition and germination of fresh chincuya (*Annona purpurea* Moc. & Sessé ex Dunal) seeds, by effect of gibberelic acid and pH of the soaking water. *Revista Brasileira de Fruticultura* 46: e604. DOI: 10.1590/0100-29452024604

Zacaroni, A.B., Junqueira, N.T.V., Sussel, A.A.B., Freitas, I.S., Braga, M.F. and Junqueira, K.P. (2014) Desempenho agronômico de gravioleira (*Annona muricata* L.) sobre diferentes espécies de porta-enxertos. *Cadernos de Agroecologia* 9, 1–5.

Zamora-Cornelio, L.F., Ochoa-Gaona, S., Vargas Simón, G., Castellanos Albores, J.H.J. and de Jong, B. (2010) Germinación de semillas y clave para la identificación de plántulas de seis especies arbóreas nativas de humedales del sureste de México. *Revista de Biología Tropical* 58, 717–732. DOI: 10.15517/rbt.v58i2.5241

# 6 Breeding and Cultivars

**Sisir Mitra***
*Former Professor, Faculty of Horticulture, Bidhan Chandra Krisi Viswavidyalaya (State Agricultural University), Mohanpur, West Bengal, India*

## 6.1 Introduction

The Annonaceae is the largest family in the early-divergent Magnoliidae clade of angiosperms with approximately 108 genera and 2400 species with a mainly pantropical distribution. However, only limited number of species belonging to two genera, *Annona* and *Asimina* (as the genus *Rollinia* is now included in the genus *Annona* (Rainer, 2007)), produce edible fruits.

Most *Annona* spp. and cultivars differ in environmental adaptation, productivity and fruit quality. Therefore, different conventional methods can be used in their breeding. Quantitative characteristics such as yield, controlled by numerous major and minor genes, are more difficult to manipulate than qualitative ones. In addition, traits whose expression is influenced by the environment are also more difficult to select than those unaffected by environmental factors. Introduction of superior genotypes and/or cultivars to establish a germplasm collection is, basically, the first requirement for any breeding programme. This can be complemented by the introduction in to the collection of some wild *Annona* spp. with useful genes, mainly for resistance to diseases. All accessions require comprehensive characterization and documentation, followed by evaluation and selection. Hybridization has tremendous scope in annonas because of the available genetic variability and comparatively easy intraspecific and interspecific hybridization. A large number of available *Annona* spp. may be intercrossed to evolve suitable cultivars and to develop rootstocks adaptable to various soil and climatic conditions.

## 6.2 Breeding Objectives

The first objective is determination of the specific characteristics that are important for the new cultivar – the ideotype. These characteristics may help a breeder to select parental groups to be used in an annona breeding programme in order to obtain desirable progeny. These progenies may not have all desirable characteristics but will at least have those most important to growers, retailers and consumers.

Yield is always an important trait in cultivar development, but genetic improvement for yield is the most expensive of all breeding objectives due to the complex nature of its inheritance and the numerous environmental factors influencing the trait. The different yield responses of the various *Annona* spp. and cultivars such as soursop (*Annona muricata*) (Pinto and Silva, 1996), sugar apple (*Annona squamosa*) (Senthilkumari

* Email: sisirm55@gmail.com


DOI: 10.1079/9781800627628.0006

*et al.*, 2024) and cherimoya (*Annona cherimola*) (Gardiazabal and Cano, 1999), in the same region prove the complexity of the genotype–environment interaction.

Quality is another important characteristic for the improvement of annonas and is as important or sometimes more so than yield, because market value is based on the fruit's appearance and its organoleptic characteristics (Pinto *et al.*, 2005). In cherimoya, a skin without protuberances above the carpel walls may be preferable, as this diminishes the susceptibility of the fruit to mechanical damage (Gardiazabal and Cano, 1999). Small soursop fruit are recommended for the fresh market, while large ones with a high acid content are more suitable for the processing industry (Pinto and Silva, 1996). The initial objectives of the atemoya (*A. cherimola* × *A. squamosa*) breeding programme in Australia was to select new cultivars with smooth skin, a symmetrical shape, a low seed number (fewer than ten seeds 100 $g^{-1}$ of pulp), low susceptibility to blemishes and high levels of self-pollination, eliminating the need for labour-intensive and expensive hand pollination (George *et al.*, 1999). Later, the hybrid progenies were evaluated with the objectives of red skin clones, yield, precocity, fruit quality and disease susceptibility (Bignell, 2022). Pinto *et al.* (2005) compiled the breeding objectives in different *Annona* spp. (Table 6.1).

The long reproductive cycles, higher levels of heterozygosity and the cost associated with evaluating large populations of crosses limits

**Table 6.1.** Main characteristics of cherimoya, soursop and sugar apple ideotypes (adult plants). Data from Anderson and Richardson (1990); George and Nissen (1987); Gardiazabal and Cano (1999); Higuchi *et al.* (1998); Paull and Durate (2011); Pinto *et al.* (2001); Scheldeman and van Damme (1999) and Viteri *et al.* (1999).

| Characteristic | Cherimoya | Soursop | Sugar apple |
|---|---|---|---|
| Plant | | | |
| Vegetative vigour | Medium to low | Medium to low | Medium to low |
| Reproductive vigour (yield) | High (>30 kg per tree year$^{-1}$) | High (>60 kg per tree year$^{-1}$) | High (>20 kg$^{-1}$ per tree year$^{-1}$) |
| Bearing | Regular | Regular | Regular |
| Rootstock/scion compatibility | High | High | High |
| Resistance to climate (temperature) | High temperature (>22°C) | Low temperature (>18°C) | Low temperature (>18°C) |
| Pest and disease | High | High | High |
| Flower number | Abundant (>150 flowers per tree) | Abundant (>100 flowers per tree) | Abundant (>180 flowers per tree) |
| Pollen fertility (20°C) | High (>76%) | | High (>76%) |
| Harvest | Out of season | Throughout year | Out of season |
| Fruit | | | |
| Size/weight | | | |
| For industry | Large, >650 g | Large, >2.5 Kg | Large, >400 g |
| For fresh consumption | Medium, 300–600 g | Small, 0.8–2.5 kg | Medium, 300–400g |
| Shape | Heart | Conical | Rounded to heart |
| Fruit set (naturally) | High (>27%) | High (>26%) | High (>25%) |
| Skin | Smooth (impressa) | Short protuberances | Tuberculate |
| Pulp | Sweet, fibreless | Subacid, low fibre | Sweet, low fibre |
| Flavour | Delicate | Subacid | Pleasant acidulous |
| Seed number/100 $g^{-1}$ of pulp | Low or absent, <6 seeds | Low, 10–30 seeds | Low or absent, <10 seeds |
| Shelf life (15–30°C) | Long (>10 days) | Long (>5 days) | Long (>5 days) |
| Transport resistance | High (pulp pressure >1.5 kg $cm^{-2}$) | High (not found) | High (not found) |

breeding of annonas. Early maturity, better fruit appearance and, in the subtropics, greater cold tolerance are the most frequent objectives (Paull and Durate, 2011).

## 6.3 Genetic Diversity

Initiatives to prioritize research on global plant genetic resources such as those led by the Food and Agriculture Organization of the United Nations (FAO) include calls for more inventories and surveys to increase our understanding of genetic diversity, reduction of which is referred to as genetic erosion, due to increased understanding of variation in plant genetic resources, explicitly referring to the application of molecular tools in such assessments (FAO, 2010, 2011).

There is a growing call for inventories that evaluate geographical patterns in the diversity of plant genetic resources maintained on farms and in natural populations of species in order to enhance their use and conservation. Such evaluations are relevant for useful tropical and subtropical species, as many of these species are still undomesticated or in the early stages of domestication, and local populations can offer as-yet-unknown traits to further increase domestication. For many outcrossing species, such as most trees, inbreeding depression can be an issue, and genetic diversity is important to sustain local production (van Zonneveld *et al.*, 2012).

Genetic improvement of cherimoya is made difficult by its long generation intervals, its highly heterozygous nature and its self-incompatibility (Rasai *et al.*, 1995). Breeding programmes aim to identify clones or develop hybrids that show good fruit shape, smooth skin, attractive skin colour, low seed content, good flavour and extended postharvest life (Mahdeem, 1990). Spanish researchers have concentrated on germplasm collection and *ex situ* evaluation and have collected around 280 accessions from diverse origins with a particular interest in accessions from Peru and Ecuador (Hermoso Gonzalez, 1999). In Mexico, germplasm collection characterization and subsequent selection led to the establishment of a commercial plantation with a superior accession, 'Cortés-II-31', showing promising results (Agustín, 1999). While studying characteristics such as fruit weight, percentage of pulp and seed, fruit soundness, type of skin and earliness, Agustín (1999) found great genetic variability among native fruits and proposed this as the basis for developing germplasm for commercial plantation. A very similar strategy has been used in Italy and Madeira (Portugal), with selection and evaluation of promising local types and introduced cultivars (Monastra, 1997; Nunes, 1997). The New Zealand cherimoya industry was estimated at 15,000 trees in the 1990s, mainly in Northland (Anderson and Richardson, 1992). The late cultivars 'Reretai' and 'Bronceada' make up approximately 23% and 24% of plantings respectively. 'Burtons Favourite', also a late maturity cultivar, comprises 44% of plantings while 'White', which matures mid-season, accounts for 10% of trees (Richardson and Anderson, 1993). The major producing countries have developed clonal selection well adapted to their climatic conditions, such as 'Fino de Jete' and 'Campas' in Spain, 'White' and 'Bays' in California, 'Concha Lisa' and 'Bronceada' in Chile, and 'Cumbe' in Peru (Escribano *et al.*, 2007). Considerable work has been done in Peru on the development of cultivars, but they are not widely known outside Peru; 'Cumbe' is considered exceptional because of its size, surface, smoothness, flavour and very few seeds (Paull and Durate, 2011). Several field collections of cherimoya have been established in Spain, Peru and Ecuador, preserving over 500 different accessions (Escribano *et al.*, 2007; CHERLA, 2008). In Ecuador, genetic variability was evaluated in 126 accessions of cherimoya conserved at Tumbaco Experimental Farm of the National Agricultural Research Institute (INIAP) and morpho-agronomic characterization identified promising materials within the cherimoya collection. The accessions were divided into three groups. The first group consisted of 27 accessions characterized by the cordate fruit shape in most accessions and resistance to abrasion. The second group had 40 entries. The texture of the pulp in some accessions in this group was creamy, while in others it was granular. In the third group of 30 accessions, difference in branches and leaves were recorded (Andrade, 2009).

In Mexico, Andrés-Agustín *et al.* (2006) characterized three commercially available cherimoya cultivars ('Campas', 'Buttons' and

'White'), one registered cultivar of Mexico ('Cortés-II-31') and seven Mexican germplasm selections (S-196, S-256, S-260, S-266, S-9651, S-Selene and S-Carapan) using morphometric traits of various organs of the adult plant, together with several agronomical and chemical characteristics. Principal component analysis showed that the traits that yielded the maximum separation between accessions were: leaf blade form factor, angle of the fifth vein of leaves, upper angle of leaves, area of the cross-section of the petals, sepal maximum projected area, weight of fruit, total soluble solids (TSS) in fruit, resistance of the skin of fruit to a penetrometer, and width:length ratio of the maximum projected area of seeds. Consequently, all of these characteristics may be of interest as descriptors for cherimoya cultivars. Four consistent groups of accessions were defined by the three-dimensional plot obtained by projecting the accessions on to the first three principal components and the tree-diagram or dendrogram obtained through cluster analysis. Two of these groups were made up of accessions with a well-defined common origin. The first consisted of selections S-196, S-256, S-260 and S-266 obtained from Coatepec Harinas in Mexico State, while the second comprised the selections S-Selene and S-Carpan from Purépecha native Indian Communities in Michoacán State.

The objectives of the improvement of sugar apple (*A. squamosa*) are the quality, appearance, consistency of the fruit, productivity and viability of pollen (Mahdeem, 1990). The Annonaceae Germplasm Bank of the State University of Montes Claros, Brazil, is working to improve this crop. The 'Brazilian Seedless' cultivar, originating from a natural change, was developed as a seedless sugar apple cultivar in Brazil, but the fruit are irregular in shape and have low yield (Pereira *et al.*, 2014). The seedlessness was due to the formation of stenospermal fruit with partially formed seeds due to its abortion after fertilization (Santos *et al.*, 2014). Some other selections (IPA-18-2, IPA-17-2 and IPA-17-3) developed by de Carvalho *et al.* (2000) have also been recommended for plantings. These selections yielded 10.6–11.7 kg fruit per plant year$^{-1}$. The cultivar 'Creole' was introduced in the 17th century in Bahia by the Count of Miranda and has since been domesticated and cultivated (Pinto *et al.*, 2005). 'Creole' has undergone several changes over the years. An interesting mutant of 'Creole' was observed with different morphological characteristics. In early 2000s, the cloned mutant of 'Creole' was named 'Green' because of its intense green coloration of leaves and fruit and the absence of the typical waxy cover on the leaves and fruit of the traditional 'Creole' cultivar (Lemos *et al.*, 2010).

There is great genetic variability within *A. squamosa* species in Mexico (Romero-Soler and Cuetzal-Ix, 2015). 'Seedless Cuban' was developed in Cuba and introduced to Florida in 1995 (Nakasone and Paull, 1998). Thailand and Taiwan have produced a few named cultivars of *A. squamosa* that are propagated vegetatively. Three old cultivars of Thailand are 'Fai' 'Nang' and 'Krang'. The main Taiwan cultivars are 'Ruan-Zhi', Da-mu', 'Xi-Lin', 'Tainung No. 1', 'Tainung No. 2' and 'Tainung No. 3' (Lee *et al.*, 2017; Thanarut and Ko, 2017).

Most *A. squamosa* trees in Florida are grown from seed and within a particular selection (e.g. 'Thai Lessard', 'Kampong Mauve'). The seedless types 'Cuban Seedless' and 'Brazilian Seedless' have been introduced. However, the seedless fruit split when nearing maturity and the fruit quality and yield were reported to be inferior to the seedy types (Crane *et al.*, 2016).

The Central Horticulture Experiment Station, ICAR, Godhra, Gujarat, India, is actively involved in collection and conservation of elite genotypes of *A. squamosa* from diversity-rich pockets of semi-arid areas of western India. The station maintains more than 50 genotypes in the field gene bank under hot semi-arid conditions (Yadav, 2018). A number of selections including 'Balanagar', 'Mammoth', 'Red Sitaphal', 'NMK-1', 'NMK-3', 'Annona-2', 'Phule Purandhar', 'Sindhan' and 'Raigarh' and introduced selections such as 'Washington PI 98797', 'Washington PI 107005', 'British Guiana' and 'Barbados' seedlings are being grown in different agroclimatic regions of India (Yadav, 2018). The Central Horticultural Experimental Station (ICAR-IIHR), Bhubaneswar, India, made a clonal selection named 'Arka Neelanchal Vikram', which bears good-quality fruit with high yield (15 kg per tree year$^{-1}$) and good shelf life (5–6 days). Another new selection named 'Thar Amrit' was selected from the diversity-rich population of *A. squamosa* from Panchmahal district, Gujarat, India (Yadav *et al.*, 2023). This cultivar is precocious, a regular bearer and a high yielder (24.8 kg per tree in year 11) under the rainfed semi-arid conditions of western Gujarat. A wide range of variability in physico-chemical attributes of *A. squamosa* genotypes was observed in Gujarat, Chhattisgarh, Maharashtra and Andhra Pradesh. Yadav *et al.* (2017) collected and evaluated 25 genotypes from different areas in Gujarat. They observed wide variability with respect to fruit weight (110.45–335.37 g), pulp weight (50.75–205.87 g), number of seeds per fruit (18–63), TSS (26.61–32.63°Brix), acidity (0.20–0.30%) and total sugar content (12.48–17.55%). The mineral content of the fruit also showed wide variations among the genotypes: 21.24–38.65 mg of magnesium, 257.26–295.74 mg of potassium, 4.26–15.27 mg of sodium and 15.35–21.43 mg of calcium (all 100 $g^{-1}$ of fruit pulp). The genotypes $GC_{12}$, $CG_{13}$ and $GC_{20}$ were found to be superior in terms of quality of fruit compared with other genotypes and could be used for further improvement.

The Bastar district of Chhattisgarh, India, is considered to have a large population of *A. squamosa* with wide variability. Jain *et al.* (2024) evaluated 16 genotypes and selected the superior genotypes. The variation in tree height (3.40–7.20 m), yield (8.56–24.37 kg per tree), fruit length (4.30–8.90 cm) and fruit breadth (4.67–4.46 cm) was recorded. They suggested that genotypes BCA-8, BCA-5 and BCA-7 could be exploited for future breeding programmes. Dikshit *et al.* (2008) collected 21 accessions of *A. squamosa* germplasm during an exploration programme from six districts of Maharashtra. Physico-chemical characterization of the fruit of different germplasms revealed variations in average fruit weight (90.8–375 g), total soluble sugar (19–26%), number of seeds per fruit (21.2–73), weight of pulp per fruit (44.4–188 g) and fruit:pulp ratio (37–54.2). The accessions IC415724, IC415725 and IC415732 were found to be superior to the other collections and could be used for future breeding programmes.

Soursop (*A. muricata*) germplasm diversity is found in the north of South America and is distributed in different tropical regions of the world (Love and Paull, 2011). The fruit is highly valued for its excellent flavour and nutraceutical and therapeutic properties (Ojezele *et al.*, 2016; Moreira *et al.*, 2020). Morphological characterization of soursop germplasm in Sri Lanka was carried out by Padmini *et al.* (2013). Forty-five morphological characteristics were recorded from 448 accessions collected from seven districts of three climatic zones and three national research centres, with the results revealing that five principal components affecting variation. The first principal component explained 26% of the variation and was associated with fruit characteristics such as length, diameter and weight of ripe fruit, and weight and number of seeds per fruit. The second explained 19% of the variation and was associated with trunk colour, leaf blade shape, average leaf width and peduncle length. The third explained 10% of the variation and was associated with trunk ramification, suckering tendency and the colour of young branches. The fourth and fifth principal components explained 7% variation each. The petal outer colour and location of ramification were associated with the fourth and pulp taste was associated with the fifth. Cluster analysis identified nine clusters with unique characteristics, the most prominent being the fruit weight:seed ratio, which is considered important in utilization of the fruit. It was concluded that clusters that have higher and lower fruit weight:seed ratio are important to promote for utilization in breeding programmes and conservation of germplasm (Padmini *et al.*, 2013).

In India, soursop is an underutilized fruit, mostly grown in home gardens in northern, southern and western India. However, few studies have surveyed and characterized the germplasm of soursop. Working with 100 germplasms in the Konkan region of Maharashtra, India, Ghavale *et al.* (2016) observed five principal components from 11 morphological qualitative, 36 morphological quantitative and 18 biochemical characteristics explained 150.85%, 75.17% and 97.98% of total variability, respectively in the germplasm. Cluster analysis identified 11 clusters of which the most prominent characteristic was pulp weight contributing the most towards variation, and this confirmed the pattern of clustering in the morphological quantitative dendogram. Other characteristics obtained in principle component analysis that determined clustering in the dendograms were uniformity of fruit size, fruit symmetry, total sugar, °Brix:total acidity ratio, ascorbic acid levels and shelf life. Variation among genotypes in fruit shape contributed a significant positive correlation with fruit symmetry, fruit symmetry with uniformity of fruit size and fruit colour, and uniformity of fruit size with fruit exocarp type and exocarp colour, while yield had a significant positive correlation with fruit length, diameter, girth, weight of fruit, pulp weight, pulp percentage and number of seeds. The authors concluded that clusters that have higher and lower pulp weight:seed ratio are important for utilization in breeding programmes.

Evaluating 16 years of fruit yield data in 71 soursop genotypes in Colombia to study genetic diversity among genotypes, Sanchez *et al.* (2018) used Tocher's optimization and the unweighted pair group method with arithmetic mean (UPGMA) methods. The Scott–Knott clustering test grouped the 71 soursop genotypes in five homogenous groups based on fruit yield. Group I was formed by the genotypes 59, 124 and 45, which represent the highest average during the measurement. Group II was formed by genotypes 1, 10, 15, 24, 47, 79, 81 and 170, which also presented a high average for fruit yield during the measurements. The other groups (III–V) were formed by genotypes with low average fruit yield. The authors suggested that crossing of genotypes 124 and 145 with genotypes 59 and 170 would have the potential to generate populations with large genetic variability and high average fruit yield.

The wild soursop population in Manabí, Ecuador, has high variability. Morphological variability of 60 accessions revealed 74% of variability in terms of morphological traits. Conglomerates of similarity were formed, in which the tree height, diameter of canopy and number of fruit per tree were found as the characteristics of greatest contribution for its conformation (Moreira *et al.*, 2020).

In Puerto Rico, the wide range of forms and types of seeding soursops were roughly divided into three general classifications, sweet, subacid and acid, then subdivided into round, heart-shaped, oblong or angular, and finally classified according to flesh consistency, which varies from soft and juicy to firm and comparatively dry. In El Salvador, two types of soursop are distinguished: guanabana azucaran (sweet), which is eaten raw and used for drinks; and guanaba acida (very sour), which is used only for drinks (Morton, 1987).

In China, 25 genera, 120 species and six varieties of Annonaceae are growing in Zhejiang, Jiangxi, Fujian, Hunan, Guangdong, Guangxi, Yunnan, Guizhou and Tibet (Lei *et al.*, 2022). The cherimoya cultivar 'African Pride' was introduced to China in 1981. This cultivar was found to be precocious and productive in the Guangzhou area of Guangdong; yields from 3–5-year-old trees produced 338% higher yields compared with other cultivars. The trees were winter hardy, withstanding a temperature of 2.7°C without injury. The average fruit weight was 250 g with a high soluble solids content (Zhushen *et al.*, 1999).

## 6.4 Hybridization

The long reproductive cycles, higher levels of heterozygosity and the costs associated with crosses limits annona breeding programmes. Existing commercial cultivars show considerable variations in growth, fruit set, fruit size and quality. No single cultivar has all the desirable characteristics (Paull and Durate, 2011).

Genetic improvement in *Annona* spp. has been very limited. Genetic improvement of annonas was first initiated in Florida and the Philippines (Wester, 1913) but did not result in any hybrids.

Current production of annonas in Australia is based mainly on atemoya (*A. squamosa* × *A. cherimola*) hybrids. A major breeding programme funded by the Australian Custard Apple Growers Association was initiated in 1998. The objective of this breeding programme was to select new cultivars of atemoya with smooth skin, symmetrical shape, low seed number (fewer than ten seeds 100 $g^{-1}$ of pulp), low susceptibility to blemishes and high levels of self-pollination. The breeding programme was initiated using a multi-pronged approach to achieve the breeding objectives. A conventional breeding programme using polycrosses of 12 of the best, advanced selections/cultivars ('Hillary, 'White', 'Martin' 'Paxton', 'MHRS Gold', 'MHRS Red', 'Palethrope', 'R11-T3', 'R11-T4', R11-T16' 'Ruby Queen', 'Bullock's Heart') was initiated (George *et al.*, 1999). Interspecific crosses have also been made between five different species: *A. cherimola*, *A. squamosa*, *A. reticulata*, *A. diversifolia* and atemoya. Most atemoya cultivars appear to be hybrids between *A. squamosa* and *A. cherimola* (George *et al.*, 2008). Four new *Annona* spp. hybrids with commercial potential have been selected: 'Maroochy Gold', 'Maroochy Star', 'Maroochy Tang' and 'KJ Pinks' as well as a bud sport of 'Pink's Mammoth' (George *et al.*, 2008). The progeny of *A. diversifolia* crosses inherited the strong apical dominance characteristic of *A. diversifolia* and its poor fruit-setting ability. The fruit quality was very good with the fruit being highly symmetrical and with excellent flavour, reminiscent of strawberry sherbet. Evaluation of the $F_2$ progeny, particularly with purple-skinned sugar apple, indicated that red skin colour may be carried by either a single- or double-recessive gene. Progeny of an atemoya × *A. reticulata* and *A. cherimola* × *A. reticulata* cross produced fruit that were late maturing (spring in Queensland), appearing to have inherited the flowering and fruiting characteristics of *A. reticulata*, which flowers in autumn and matures in late spring in the subtropical conditions of Australia. Various selections were made in both Australia and Florida but were discarded due to unacceptable fruit quality, as the progeny inherited the high level of internal grittiness from *A. reticulata* (George *et al.*, 1999).

Hort Innovation (Australia)-funded projects CU11000, carried out between 2012 and 2014 (Bignell and Brunn, 2014) and CU16002 carried out between 2018 and 2021 (Bignell, 2022) developed a number of hybrid selections. Some of the selections are as follows:

**Fig. 6.1.** Fruit of selection 453-1. Photograph courtesy of Dr Grant Bignell.

- 453-1 (Cross 22 (A) unknown × seedless sugar apple (S)): fruit size 500–800 g, green skin with red flecking in the interstices, very symmetrical fruit, good texture and only 4.5 seeds 100 $g^{-1}$ flesh (Fig. 6.1).
- 450-1 (Cross 22 (A) (unknown) × Maroochy Tank (Hillary White (A) × Fino de Jete (C)): medium to large fruit (600–900 g), spheroid shape, good flavour and taste; the fruit skin is lime green with yellow interstices and the fruit has 4.47 seeds 100 $g^{-1}$ of flesh.
- 464-2 (Cross 40 (A) (Martin × Maroochy Tank (Hillary White (A) × Fino de Jete (C)) × unknown): medium to large fruit (650–950 g), skin colour green with large lime green interstices, oval in shape, mildly tuberculate, good texture and flavour, and only 2.2 seeds 100 $g^{-1}$ of flesh.
- 464-1 (Cross 40 (A) (Martin × Maroochy Tank) (Hillary White (A) × Fino de Jete (C) × unknown): medium fruit size (500–600 g), green with large lime green interstices, white skin colour, middle tuberculate on the skin, oval in shape, good flavour and texture, flesh white with an average of 3.6 seeds 100 $g^{-1}$ of flesh.
- 430-1 (Cross 76 (A) (Maroochy Gold (A) × Maroochy Tank (A)) × seedless sugar apple (S): medium fruit (550–750 g), spheroid in shape, symmetrical, skin mostly tuberculate and good texture.
- 430-2 (Cross 76 (A) (Maroochy Gold (A) × Maroochy Tank (A)) × seedless sugar apple (S): moderately vigorous tree, medium-sized fruit (500–750 g), skin colour green with cream interstices, spheroid shape, symmetrical, white flesh, good texture and 3.6 seeds 100 $g^{-1}$ of flesh.
- 470-2 (KJ Pinks (A) × seedless sugar apple (S)): vigorous tree; large fruit (750–100 g), flesh colour white, skin strongly tuberculate, good texture, excellent flavour and 4.5 seeds 100 $g^{-1}$ of flesh (Fig. 6.2).
- 363 (Maroochy Tank (A) × seedless sugar apple (S)): vigorous tree, medium to large fruit (600–800 g), skin colour green with cream interstices, flesh white, oval fruit

**Fig. 6.2.** Fruit of selection 470-2. Photograph courtesy of Dr Grant Bignell.

shape, good texture and flavour, and has 4.4 seeds 100 $g^{-1}$ of flesh.

- Green Skin Selection 470-3: large (530–660 g) pale green fruit with pale yellow interstices (Fig. 6.3), moderately symmetrical, tuberculate, white flesh, a 'Pink's Mammoth' flavour and sweet (16.4–19.4°Brix); shelf life is approximately 6 days at ambient temperature.
- 649-1 ('KJ Pinks' × sugar apple 35-1): medium to large (378–626 g) bright red fruit (Fig. 6.4) with pale interstices, very symmetrical, moderately tuberculate, firm white flesh, flavour similar to 'Pink's Mammoth' and sweet (16.4–19.4°Brix); shelf life is approximately 5 days at ambient temperature.

Under the Annona Improvement Programme at Sangareddy, India, six hybrids were developed: Hybrid-1 (17/4 atemoya × 'Balanagar' (*A. squamosa*)), Hybrid-2 (15/2 'Red Sitaphal' × pond apple (*A. glabra*)), Hybrid-3 (15/3 'Red Sitaphal' × atemoya), Hybrid-4 (1/6 'British Guiana' × atemoya), Hybrid-5 and Hybrid-6 (15/3 'Red Sitaphal' × atemoya) (Girwani *et al.*, 2011). Evaluation of the hybrids 10 years after planting showed variability in fruit shape, fruit skin colour, pulp colour, areole shape, quality (TSS content of pulp) and shelf life (Table 6.2).

**Fig. 6.3.** Fruit of selection 470-3. Photograph courtesy of Dr Grant Bignell.

At the Indian Institute of Horticultural Research, Bangalore, a hybrid between atemoya 'Island Gem' × *A. squamosa* 'Mammoth' was selected from more than 800 highly variable inter- and intraspecific hybrid populations derived from 13 cross combinations. The hybrid showed consistently better performance in replicated field trials and was released as 'Arka Sahan'. This hybrid is characterized by small seeds, large flakes, a firm rind, a higher edible portion and high TSS (31°Brix), which is 41% more than 'Mammoth'. Six-year-old plants of 'Arka Sahan' yielded 17 kg of fruit per plant (Jalikop and Kumar, 1996). *A. glabra*, as a tetraploid, fails to produce viable progeny with diploids. However, atemoya has been found to cross freely with *A. reticulata* and *A. cherimola* but not with *A. glabra* (Jalikop and Kumar, 1995). An interspecific annona hybrid (21/15) was found to bear fruit with an excellent shelf life (6–7 days), fewer seeds (11.0 100 $g^{-1}$ fruit weight) and high TSS (>30°Brix) (Jalikop and Kumar, 1995). Interspecific hybrids involving *A. squamosa* and *A. reticulata* (as the female) and *A. cherimola* (as the pollen parent) can easily be produced as the fruit set in these crosses is fairly high, ranging from 37.5% to 52.5% (Thakur and Singh, 1964).

**Fig. 6.4.** Fruit of selection 649-1. Photograph courtesy of Dr Grant Bignell.

Jalikop (2010) stated that to incorporate desirable (novel) traits into a cultivar, *A. reticulata* can be used as a valuable source. However, the $F_1$ trihybrids showed reduced fertility with pronounced undesirable traits of *A. reticulata*. The survivability of the $F_2$ progeny was very poor. Jalikop (2010) suggested backcrossing instead of advancing trihybrid generation for transferring desirable genes from *A. reticulata*. Backcrossing the $F_1$ progeny of (*A. cherimola* × *A. squamosa*) × *A. reticulata* with *A. squamosa* showed productive introgressants in which the contribution of the *A. squamosa* genome was extended from 25% to 62.5% in the $F_1$ first backcross generation. 'Cutemoya', a hybrid of atemoya × *A. reticulata*, was selected and shows many desirable characteristics when compared with other *Annona* spp. cultivars and hybrids (Kumar and Jalikop, 2000).

The result of the studies conducted in Serdang, Selangor, Malaysia, by Mohd Khahid (2002) with six *Annona* spp. found that only

**Table 6.2.** Morphological and physico-chemical characteristics of six hybrids developed under the Annona Improvement Programme at Sangareddy, India. Data from Girwani *et al.* (2011).

| Hybrid | Fruit shape | Fruit colour | Areole shape | Pulp colour | Average fruit weight (g) | Total soluble solids (°Brix) | Shelf life (days) |
|---|---|---|---|---|---|---|---|
| Hybrid-1 | Conical | High green | Pointed | Pale white | 250 | 21.00 | 4 |
| Hybrid-2 | Round | Pale green | Rounded fused | Pale white | 136 | 28.25 | 2 |
| Hybrid-3 | Rounded | Greyish-green | Hexagonal | Creamy-white | 184 | 23.00 | 3 |
| Hybrid-4 | Cordate | Light green | Mammiform | White | 220 | 25.00 | 4 |
| Hybrid-5 | Rounded | Yellowish-green | Rounded | Creamy-white | 219 | 20.00 | 3 |
| Hybrid-6 | Rounded | Red | Hexagonal reticulated | White | 225 | 22.00 | 3 |

*A. muricata* crosses with *A. rollinia*, *A. glabra* and *A. montana* set fruit. The highest number of fruit formed was from the cross between *A. glabra* and *A. muricata*. Pollen from *A. muricata* set fruit for the pond apple and *Rollinia* spp. crosses, while pollen from *A. montana* produced fruit with *A. muricata*.

$F_1$ hybrids from an *A. muricata* × *A. montana* cross were intermediate between the two parents in height and plant growth habit, and flowered after about 2 years in the field. The flowers were yellow and very similar to soursop flowers in morphology and structure, but the hybrids failed to set fruit (Khalid, 2002).

The 'Brazilian Seedless' cultivar, originating from a natural change, is the only seedless cultivar of *A. squamosa* developed in Brazil, but has an irregular shape and low yield (Pereira *et al.*, 2019). The stenospermia in this mutant is due to the suppression of the *INO* genes, leading to a loss of the outer integument of the egg, which affects development of the seed (Lora *et al.*, 2011). The fruit of all individuals of $M_1$ and $M_2$ accessions and of the $F_1$ offspring ($P_1$ × M1 and $P_1$ × $M_2$) had seeds. The seedless characteristics was observed only in the fruit of the male parent $P_1$. These preliminary data suggest a monogenic inheritance with allelic interaction of the complete dominant type.

In Egypt, a hybrid between *A. squamosa* × *A. cherimola* is grown commercially (Ezzat *et al.*, 1974). 'Red Israel', a natural hybrid from *A. cherimota* × *A. reticulata*, has also been grown commercially (Leal and Paull, 2023).

## 6.5 Polyploidy

Genome duplication, or polyploidy, is one of the main specification factors in plants. It is especially frequent in hybrids and very valuable in many crops (Martin *et al.*, 2019). A limited number of species in the family Annonaceae belonging to two genera in the tribe Annoneae of the subfamily Annonoidea (*Annona* Lo and *Asimina* Adans, as *Rollinia* A. St-Hil. has been included in the genus *Annona* (Rainer, 2007)) produce edible fruits (Martin *et al.*, 2019). The *Annona* spp. that produce edible fruits include cherimoya (*A. cherimola* Mill.), sugar apple (*A. squamosa*), soursop (*A. muricata* L.) atemoya (*A. cherimola* × *A. squamosa*), custard apple (*A. reticulata*), ilama (*A. macroprophyllata* Donn. Sm.) and pond apple (*A. glabra*), as well as pawpaw (*Asimina triloba*) (Larranaga *et al.*, 2019). Most of the *Annona* spp. are diploid with the exception of *A. glabra* and *A. lutescens* (Morawetz, 1986), which have been reported as tetraploid as well as *Rollinia* spp. The hexaploid species include *Annona mucosa* and *A. pulchrinervis* (Walker, 1972; Morawetz, 1986).

Martin *et al.* (2019) studied ploidy in several. *Annona* spp.. They observed a ploidy level of 2× in *A. cherimola*, *A. squamosa*, *A. senegalensis*, *A. muricata*, *A. emarginata* and *Asimina triloba*. *Annona neosalicifolia*, which had a higher DNA content (4.82 pg), was found to be hexaploid (6×). The progeny from interspecific crosses of *A. cherimola* showed polyploidy genotypes that ranged from 2.5% to 33%, while hybrid atemoyas from the interspecific crosses had 35% triploids from a total of 186 genotypes analysed. With the aim of understanding the possible implications of the production of non-reduced gametes, pollen performance, pollen size and frequency distribution of pollen grains were quantified in the progeny of these crosses and the parents by Martin *et al.* (2019). They observed large polymorphisms in pollen grain size within the interspecific progeny, with higher production of unreduced pollen in triploids (38%) than in diploids (29%). Using polymerase chain reaction amplification of selected microsatellite loci, they noted two alleles in 13.7% of the pollen grains from the diploids, while 41.28% of the grains from the triploids amplified two alleles and 5.63% showed up to three alleles. This suggested that the larger pollen gains could correspond to diploid pollen and at a lower frequency to triploid pollen. Pollen performance was also affected, with lower pollen germination in the hybrid triploids than in both diploid parents. These results confirmed a higher percentage of polyploids in the interspecific cross, affecting pollen grain size and pollen performance. The occurrence of unreduced gametes in *A. cherimola*, *A. squamosa* and their interspecific progeny may cause the abnormalities of ploidy such as the triploids and tetraploids observed in this study (Martin *et al.*, 2019).

## 6.6 Molecular Characterization

The use of phenotypic characteristics has been the basis for the identification and characterization of species and cultivars. The combination of morpho-agronomic characteristics and DNA molecular markers constitutes a novel tool of great utility to characterize germplasm, estimate diversity levels and determine parental relationships among accessions and also to recommend genotypes with conservation and breeding potential (Nimisha *et al.*, 2013). Molecular markers can be used to discriminate between species and cultivars of annonas (Bharad *et al.*, 2009; Collevatti *et al.*, 2014; Larranga *et al.*, 2017).

Different molecular markers such as randomized amplified polymorphic DNA (RAPD) (Brown *et al.*, 2003; Ahmad *et al.*, 2010; Anuragi *et al.*, 2016; Abdelkawy and El-Nawman, 2017; Hasan *et al.*, 2017) amplified fragment length polymorphisms (AFLP) (Rahman *et al.*, 1998; Bharad *et al.*, 2009) and inter-simple sequence repeats (ISSR) (Ahmad *et al.*, 2010; Cota *et al.*, 2011; Thuy *et al.*, 2022) have been used to discriminate species and cultivars.

Analysis of isozyme patterns indicated that Spanish cherimoya cultivars were distinctly different from cultivars in California (Pascual *et al.*, 1993) and from atemoya (Ellstrand and Lee, 1987). The popularly grown Spanish cherimoya cultivars 'Fino de Jete' and 'Campa' exhibited identical banding patterns for 15 enzymes, indicating that they might be the same cultivar (Pascual *et al.*, 1993). In the South Coast Research and Extension Centre, California, approximately 30 cherimoya cultivars have been collected and verified by isozyme analysis. Twenty-five different isozyme loci have been identified, which are useful for determining the identity of a cultivar (Grossberger, 1999). Escribano *et al.* (2007) used 16 simple sequence repeat (SSR) loci to find molecular polymorphisms among 279 cherimoya accessions collected from Peru, Ecuador, Bolivia, Chile, Colombia, Costa Rice, Mexico, Spain, Portugal, Italy, the USA and Australia. A total of 79 fragments were amplified with 16 pairs of SSR primers, with an average of 4.9 bands per SSR. UPGMA cluster analysis indicated the relationships among the analysed assessions, showing some specific groups related to their geographical origins. Analysis of molecular variance (AMOVA) revealed the distribution of genetic variation of 148 accessions collected from putative cherimoya origin areas in Ecuador and Peru, showing that the major variations occurred within the valleys in each country. The results from this study indicated that SSRs are excellent codominant markers to improve cherimoya germplasm management, allowing researchers to distinguish synonymies and homonymics and to study cherimoya conserved diversity.

The genetic diversity of *Annona* spp. as determined by molecular markers shows variable levels of polymorphism. Ahmad *et al.* (2010) found that RAPD and ISSR primers detected 52 and 58% polymorphism among *Annona* spp. (*A. squamosa, A. muricata, A. reticulata and A. glabra*) grown in India, while Anuragi *et al.* (2016) reported polymorphism values of 83% and 93% for the corresponding methods. Guimaraes *et al.* (2013) found only 28.7% polymorphism among 64 accessions of sugar apple from south-eastern Brazil using RAPD markers.

A considerable degree of genetic diversity was found in soursop (*A. muricata*) germplasm accessions in Sri Lanka. In a study of 50 soursop accessions using 13 ISSR markers, Samaradiwakara *et al.* (2020) observed 139 amplified bands of which 118 (85%) were polymorphic. Cota *et al.* (2011) used 15 selected ISSR primers to evaluate the genetic diversity of *A. crassiflora*, and these primers generated 140 bands of which 123 (87.8%) were polymorphic. *Annona deceptrix* is an endemic tree of the province of Manabí, Ecuador, and is commonly known as cherimoya de monte o anonilla. This species produces conical, elongated fruit. However, natural populations have a high degree of vulnerability caused by human activities and disturbance. Pico-Mendoza *et al.* (2024) developed polymorphic microsatellite markers to evaluate its genetic diversity and population structure. Twenty-two polymorphic SSR loci were isolated using new-generation Nextera kit (Illumina MiSeq). The mean number of alleles per locus ranged from two to eight. All SSR markers developed were found to be promising candidates for analysing genetic variation within or between natural populations and could be transferable to other species of the genus *Annona*.

In Egypt, Abdelkawy and El-Nawman (2017) identified and distinguished the three annona cultivars, 'Abdelrazik' (*A. squamosa*), 'Senigalensis' (*A. senigalensis*) and 'Baladi' (*A. squamosa*) using RAPD fingerprinting. Almost identical fingerprinting for 'Abdelrazik' and 'Baladi' indicated high genetic similarity, while 'Senigalensis' was separated in an independent cluster.

DNA barcoding represents a significant advancement in the field of DNA-based species identification (Hebert *et al.*, 2003). This method used worldwide community efforts to construct large-scale public reference libraries, allowing reliable species identification (Coissac *et al.*, 2016). There are several cultivars and wild-type variants of annonaceous plants in Vietnam (Thang *et al.*, 2013). Thuy *et al.* (2022) used two DNA barcodes (*rbcL* and *rpoC1*) and two ISSR markers to assess the interspecies of 11 annona samples collected from different provinces in the Mekong Delta. The *rbcL* marker successfully amplified and sequenced *Annona* spp. for identification, while the *rpoC1* sequence required extensive editing and could not distinguish between *A. muricata*, *A. glabra* and *A. squamosa* samples in a BLAST analysis. The *rbcL* and *rpoC1* sequence properties of *A. squamosa* collected from Nag Bay, Hau Giang (NDL-NB) were analysed. Phylogenetic analysis revealed that sample NDL-NB, previously recognized as *A. squamosa* morphologically, grouped in the cluster of *A. muricata* based on *rpoC1* sequences. Among 33 bands generated from ISSR primers, 20 bands were polymorphic, accounting for 60.60%. Dendrogram analysis of ISSR data could classify seven DNA samples into three genotypes that were not geographically dependent. These results suggest that the performance of approved DNA barcodes and ISSR analysis appears to be species specific or genus specific.

## 6.7 Rootstock Breeding

Annonas are perennial fruit crops and so are subjected to biotic stresses such as pests and diseases and abiotic stresses such as soil and environmental factors. The focus on rootstock in annonas includes development of genotypes that can confer resistant to adverse soil conditions and affect the vigour of the scion. Other aspects include dwarfing characteristics for establishing high-density orchards, tolerance to calcareous soils, and tolerance to soil-borne problems like *Phytophthora* root rot, *Armillaria* rot, dieback and nematodes. However, very few research results are available on suitable rootstocks for use in different situations.

Different *Annona* spp. are used as rootstocks, but some species are not compatible; examples are *A. squamosa* with *A. cherimola*, and atemoya with *A. muricata* and *A. reticulata* (as discussed in Chapter 5, this volume). Seedlings of some *Annona* spp. such as *A. montana*, *A glabra* and *A. muricata* are tolerant to continuous flooding and periodic water logging. Commercial *Annona* spp. and cultivars grafted on to *A. glabra* rootstock have been shown to thrive in periodically flooded wetlands (Nunez-Elisea *et al.*, 1998).

In Australia, rootstock breeding work was initiated in the late 1990s. Rootstocks with potential dwarfing characteristics such as shortened internodes and a short, bushy stature have also been selected from seedling progeny (George *et al.*, 2005). The Hort Innovation-funded project CU16002 further evaluated seedling rootstocks of atemoya, sugar apple and other cherimoya selections to determine their suitability as an alternative to the industry standard cherimoya rootstock. In Mareeba, different rootstocks were tested: 550 (dwarf cherimoya selection), 404 (low/moderate vigour Taiwanese squamosa), APV2 (moderate vigour 'African Pride' seedling selections) and cherimoya industry standard rootstock. Dwarfing selections 404 and 550 reduced average tree vigour and shoot elongation of atemoya 'KJ Pinks' when compared with trees grafted on cherimoya and 'African Pride' (APV2) (Bignell, 2022).

## 6.8 Cultivars

### 6.8.1 Australia

The major atemoya cultivars grown in Australia are described below.

#### *6.8.1.1 'Pink's Mammoth'*

'Pink's Mammoth' was introduced from British Guiana in the early 1870s. It was named after Mr Pink, the manager of the Queensland Acclimatization Society, who introduced it. The fruit is large and irregularly shaped, with a knobby skin, which encourages mealybugs and makes postharvest cleaning difficult. It is considered to have better flavour and fewer seeds than other cultivars but it does have an unpleasant gritty texture near the skin. The average fruit weight is 700–900 g. The fruit has good flavour with a TSS content of 22°Brix (George *et al.*, 2005).

#### 6.8.1.2 'Hillary White'

This is a strain of 'Pink's Mammoth' selected from a bud sport by Mr Alan George of the Queensland Department of Primary Industries. The strain's cropping performance is between the standard 'Pink's Mammoth' and 'African Pride'. The fruit has a smoother skin than the standard 'Pink's Mammoth' but poor fruit shape. The average fruit weight is 500–700 g with five seeds 100 $g^{-1}$ of fruit. It has a medium texture (rating 6) and a TSS content of 22°Brix (George *et al.*, 2005).

#### 6.8.1.3 'African Pride'

The fruit is heart-shaped or conical, small to moderate in size (500–700 g) and usually contains more seeds (ten seeds 100 $g^{-1}$ of fruit) in relation to the flesh than 'Pink's Mammoth' (Sanewski, 1988; George *et al.*, 2005). The medium-thick skin is smoother than 'Pink's Mammoth'. It has a medium texture (rating 6), good flavour and is sweet (TSS 24°Brix).

#### 6.8.1.4 'Maroochy Gold' ('Hillary White' × red sugar apple (A. squamosa))

The average fruit weight is 600–800 g with five seeds 100 $g^{-1}$ of fruit. It has a medium texture (rating 7), good flavour (rating 8) and is medium sweet (TSS 22°Brix) (George *et al.*, 2005).

#### 6.8.1.5 'Maroochy Yellow'

The average fruit weight is 400–600 g with five seeds 100 g $fruit^{-1}$. It has a medium texture (rating 7) and is sweet (TSS 25°Brix) (George *et al.*, 2005).

#### 6.8.1.6 'Maroochy Smoothie'

The average fruit weight is 400–600 g with five seeds 100 $g^{-1}$ of fruit. It has a very good texture (rating 8) and is very sweet (TSS 28°Brix). It has a very good shelf life of 10 days (George *et al.*, 2005).

#### 6.8.1.7 'KJ Pinks'

This is a bud sport of 'Pink's Mammoth'. It has a natural fruit set of about 41%. It is highly productive (rating 10), with average fruit weight of 500–800 g with five seeds 100 $g^{-1}$ of fruit. It has a medium texture (rating 6) and good flavour (rating 7) but a poor shelf life (5 days) (George *et al.*, 2005).

#### 6.8.1.8 'Island Gem'

This produces small to medium-sized fruit of good flavour but with many seeds. This cultivar is susceptible to fruit splitting due to its thin skin.

### 6.8.2 Thailand

The main sugar apple (*Annona squamosa*) and atemoya cultivars grown in Thailand are described below. 'Pink's Mammoth' was introduced from Australia but is too low yielding and does not performing well in Thailand. Other available cultivars of *A. squamosa* include 'Thai Lessard', 'Kaset Green', 'NaDai' and 'Green Thai'.

#### 6.8.2.1 'Green Fai' and 'Purple Fai' (A. squamosa)

Fai is a local Thai type containing many seeds. The average fruit weight is 180–250 g. It has a green rind, is not easy to peel, and has white flesh and a sweet flavour. Two cultivars of the Fai type are available, 'Green Fai' and 'Purple Fai'.

#### 6.8.2.2 'Nang' (A. squamosa)

This was introduced from Vietnam. The average fruit weight is 180–300 g. It has a green rind, is easy to peel, and has white flesh and a sweet flavour. There are three types of 'Nang': 'Green Nang' 'Purple Nang' and 'Golden Nang'.

#### 6.8.2.3 'Krang' (A. squamosa)

The average fruit weight is 180–250 g. It has a purple rind, pink flesh and a sweet flavour.

#### 6.8.2.4 'Phet Pak Chong' (A. squamosa)

The fruit weight varies from 300 to 1000 g, and it has a green rind, white flesh and is easy to peel. The fruit is sweet (21°Brix) and has a good flavour (Komkuntod and Wanichkhun, 2014). This cultivar is drought tolerant, and it is easy to force flowering and off-season production.

#### 6.8.2.5 'Phetpackchong' (atemoya)

The fruit has a yellowish skin and the weight varies from 400 to 800 g. It is sweet (22°Brix), and has a firm texture and a more than acceptable flavour. When ready to eat, the fruit is firm enough to be sliced without collapsing, and the yield is as high as 50 kg per tree.

#### 6.5.2.6 'Mai Lee' (atemoya)

The fruit can be broken in half by hand when mature, and it is yellow skinned and not sticky. The yield is around 30 kg per tree.

### 6.8.3 Taiwan

Sugar apple is mostly grown in Taiwan. In the 1980s, local cultivars with fruit weights of 150–350 g were mainly grown. In the 1990s, the cultivar 'Soft Sticks', which varies from the local cultivars and has a larger fruit size (250–500 g) has become the main type grown. Taitung District Agricultural and Research Extension Station developed new cultivars such as 'Taitung No. 1' in 2001 with a fruit weight of 380–550 g, 'Taitung No. 2' in 2009 with a fruit weight of 500–1000 g and 'Taiung No. 3' in 2013, which produces fruit weighing 400–850 g. However, 'Taitung No. 2' accounts for 90% of production (Lee *et al.*, 2017).

### 6.8.4 Israel

'Kaspi' (or 'Caspi'), 'Geffner' and 'Red Annona' atemoya cultivars are available in Israel (Dr Uri Lavi, personal communication). However, the main cultivar is 'Geffner'.

#### 6.8.4.1 'Geffner'

This cultivar was selected by Mr Geffner and produces a good yield in Israel. The fruit is medium-sized (300–400 g), conical in shape and has small, pointed areoles (Fig. 6.5). It has excellent quality and flavour. The fertility is quite good even without artificial pollination (Dr Raffi Stern, personal communication).

**Fig. 6.5.** Fruit of the cultivar 'Gefner'. Photograph courtesy of Dr Raffi Stern.

### 6.8.5 Spain

Since the early days, the cherimoya growers of Almuñécar recognized what perhaps should be called races (known as 'castas' in Spain). The oldest recognized casta is 'Pinchudo', which is characterized by well-defined protuberances on the surface of the fruit. One strain or cultivar of this group, which is somewhat like pineapple in shape, is of unusually fine quality. It is aromatic, with a perfect combination of acids and sugars. Unfortunately, cherimoyas of the 'Pinchudo' group are thin-skinned and are subject to attack by Mediterranean fruit fly. Another race or casta that has been cultivated for a long time is the 'Negrito' or 'Negro'. Many people consider this to be the best flavoured of all cherimoyas. The flesh has the consistency of ice cream, and is sometimes rather gelatinous in character. There are only two commercial races of castas, 'Jete' or 'Fino de Jete' and 'Campas'. 'Campas' produces large fruit of uniformly good shape, but 'Fino de Jete' is the most important commercial cherimoya grown in Spain (Sarasola, 1960).

#### 6.8.5.1 'Fino de Jete'

This is a highly productive cultivar. The fruit can be round, oval, heart-shaped or kidney-shaped, with a symmetrical tendency concerning the peduncle axis. The skin consists of smooth, non-bulbous scales (impressa-type skin) (Fig. 6.6) and is relatively thick, which enhances its suitability for transport and handling. The fruit's skin colour is light green, while the pulp is ivory

**Fig. 6.6.** Fruit of the cultivar 'Fino de Jete'. Photograph courtesy of Dr Juan Jose.

white at the optimal harvesting stage. The seeds are enclosed in the carpels, meaning that detachment is not direct. The fruit quality is excellent, with a high sweet flavour blanched by a subacid taste, reaching a soluble sugar content of over 17°Brix.

### 6.8.6 New Zealand

In New Zealand, the cherimoya industry has developed from a few plantings in the mid-1980s to estimated plantings of 15,000 trees in Northland. Several cultivars were introduced and evaluated, and described in Table 6.3.

### 6.8.7 USA

A number of cultivars of cherimoya (*A. cherimola*) and a sugar apple cultivar (*A. squamosa*) are grown in the USA.

#### *6.8.7.1 'Bays' (A. cherimola)*

This was introduced in 1920 by William Weber in Ventura, California. The tree is broad and about 6.0 m tall. The fruit is round (Fig. 6.7), medium-sized, light green in colour and the skin shows fingerprint-like marks (impressa type). The flavour is good, almost lemony.

#### *6.8.7.2 'Big Sister' (A. cherimola)*

This was selected by James Neitzel in San Diego, California, in 1979. It is a sibling of 'Sabor'. The fruit is large, very smooth and impressa type (Fig. 6.8). It is often self-fertile.

#### *6.8.7.3 'Booth' (A. cherimola)*

This was selected by A.F. Booth in California in 1921. It is one of the hardiest cherimoya cultivars. The tree is 6.0–8.0 m in height, and the fruit is conical (Fig. 6.9), impressa type, medium-sized and seedy, with a papaya-like flavour.

#### *6.8.7.4 'Chaffey' (A. cherimola)*

This cultivar was selected by A.M. Chaffey in West Los Angeles, California in 1945. The tree has an open canopy and is fast growing and suitable for coastal areas. The fruit is small to medium-sized, round (Fig. 6.10) and impressa type, with a strong lemony flavour.

**Table 6.3.** Characteristics of cherimoya evaluated at Kerikeri Horticulture Research Station, New Zealand. Data from Anderson and Richardson (1990).

| Cultivar | Origin | Tree age (years) | Season | Skin type[a] | Fruit weight (g) | Seed number 100 $g^{-1}$ flesh | Shelf life (days) |
|---|---|---|---|---|---|---|---|
| 'Smoothey' | Tarranga, NZ | 5 | May–July | Intermediate[b] | 430 | 18 | 5–7 |
| 'Bays' | California, USA | 6 | June–Aug | Impressa | 450 | 20 | 5–7 |
| 'Spain' | California, USA | 6 | July–Aug | Intermediate[b] | 550 | 12 | 5–7 |
| 'Jete' | Canary Islands | 5 | July–Aug | Impressa | 550 | 15 | 5–7 |
| 'White' | California, USA | 6 | July–Sept | Intermediate[b] | 600 | 9 | 5–7 |
| 'McPherson' | Hawaii, USA | 5 | July–Sept | Impressa | 370 | 17 | 5–7 |
| 'PK2' | Tarranga, NZ | 5 | July–Sept | Impressa | 550 | 13 | 5–7 |
| 'Deliciosa' | California, USA | 5 | July–Oct | Tuberculasas | 630 | 21 | 3–5 |
| 'Burton's' | Kaitria, NZ | 5 | Aug–Sept | Impressa | 450 | 15 | 5–7 |
| 'Bronceada' | Chile | 5 | Aug–Oct | Intermediate[b] | 650 | 9 | 12–15 |
| 'Burton's Favourite' | Kaitria, NZ | 5 | Aug–Oct | Impressa | 470 | 19 | 5–7 |
| 'P52' | Peru | 5 | Aug–Oct | Impressa | 550 | 16 | 5–7 |
| 'Variety Q' | Tarranga, NZ | 5 | Aug–Oct | Impressa | 450 | 16 | 5–7 |
| 'Plna' | Chile | 5 | Aug–Oct | Deeply impressa | 540 | 21 | 5–7 |
| 'Chaffey' | California, USA | 5 | Aug–Oct | Impressa | 650 | 10 | 5–7 |
| 'Reretai' | Kaitria, NZ | 5 | Aug–Nov | Impressa | 400 | 21 | 5–7 |

[a]Skin type is described as either impressa (smooth), tuberculose (raised protuberances) or intermediate (a combination of impressa and tuberculose).
[b]Some cultivars produce different skin types on the same tree and/or some fruit.

#### 6.8.7.5 'Ecuador' (A. cherimola)

The tree of this cultivar has broad-limb, spreading branches and is very hardy. The fruit is medium-sized, dark green in colour (Fig. 6.11), mamillated and have a good flavour.

#### 6.8.7.6 'El Bumpo' (A. cherimola)

This was selected by Rudy Haluza in Villa Park, California, in 1986. The fruit is conical in shape (Fig. 6.12), medium-sized and mamillated, with very soft skin. It has an excellent flavour.

#### 6.8.7.7 'Pierce' (syn. 'Knight', 'Escondido White', 'Ryerson', 'Thomson-Spain' and 'Bayott') (A. cherimola)

This cultivar is believed to be from a group of scions imported from Mexico in the 1930s by Mr Knight of Orange Country. Dr H.F. Pierce planted a grove in Goleta made up largely of trees produced by Knight. This cultivar was

**Fig. 6.7.** Fruit of the cultivar 'Bays'. Photograph courtesy of Dr Gary S. Bender.

**Fig. 6.8.** Fruit of the cultivar 'Big Sister'. Photograph courtesy of Dr Gary S. Bender.

**Fig. 6.9.** Fruit of the cultivar 'Booth'. Photograph courtesy of Dr Gary S. Bender.

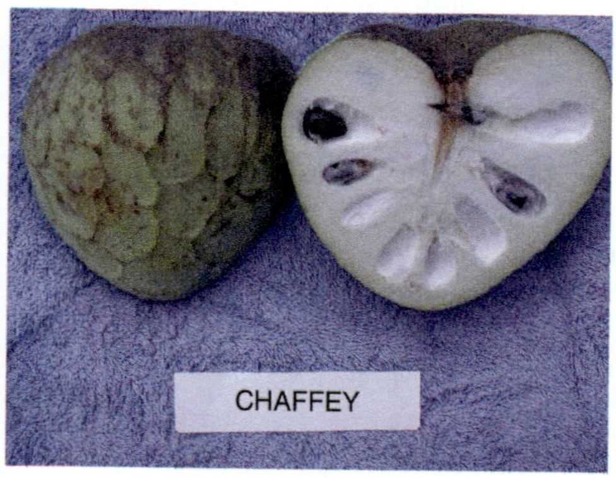


**Fig. 6.10.** Fruit of the cultivar 'Chaffey'. Photograph courtesy of Dr Gary S. Bender.

**Fig. 6.11.** Fruit of the cultivar 'Ecuador'. Photograph courtesy of Dr Gary S. Bender.

**Fig. 6.12.** Fruit of the cultivar 'El Bumpo'. Photograph courtesy of Dr Gary S. Bender.

Dr Pierce's favourite and was named 'Pierce' by him. The tree is vigorous with large, dark green leaves. The fruit is medium-sized, elongated and conically shaped with very smooth skin and a high sugar content.

#### *6.8.7.8 'White' (A. cherimola)*

This was selected by J.H. MacPherson in Lemon Grove, California, in 1928. The tree is tall, growing up to 10 m, and needs training. It grows and bears well near the coast. The fruit is large and conical, with superficial small lumps (umbonate). The flesh is juicy but has a weak flavour.

#### *6.8.7.9 'Whaley' (A. cherimola)*

This was selected in Hollywood, California, in 1924. The tree is moderately vigorous, and the fruit is medium to large in size, elongated, conical, tuberculate and light green in colour, with a good flavour. The seeds are enclosed in an obtrusive sac of flesh.

#### *6.8.7.10 'Leah Reese' (A. squamosa)*

This sugar apple cultivar was selected from a seedling population in Florida. The tree is medium in height. Hand-pollinated fruit are symmetric and broadly cordate, and the mean fruit weight is 638.73 g with an average of 90 seeds. The pulp is creamy-white with almost no stone cells, and has an excellent sweet, subacidic flavour (Crane *et al.*, 2021).

### 6.8.8 Ecuador

The Fruit Growing Programme of the Faculty of Agricultural Sciences, Central University of Ecuador, carried out an evaluation study of five preselected cherimoya genotypes in the province of Pichincha. The important fruit characteristics of these five genotypes are as follows (Herrera, 2006).

#### *6.8.8.1 'San Jose de Minas'*

The average fruit weight is 560.45 g and it has a yield of 42.0 t $ha^{-1}$ (500 plants $ha^{-1}$). The fruit has 60.1% pulp, 10.7% seeds and is sweet (24.1° Brix).

#### *6.8.8.2 'MAG – Tumbaco'*

The average fruit weight is 490.40 g, with a yield of 390.0 t $ha^{-1}$. The fruit contains 61.5% pulp and 7.2% seeds and has a sugar content of 21.6°Brix.

#### *6.8.8.3 'Loja'*

This has an average fruit weight of 337.0 g and a yield of 16.85 t $ha^{-1}$. The fruit contains 54.7% pulp and 5.8% seeds, and is sweet (26.6°Brix).

#### *6.8.8.4 'Fabulosa'*

The average fruit weight of this cultivar is 469.0 g and it has a yield of 23.5 t $ha^{-1}$. The fruit contains 65.5% pulp and 7.1% seeds, with a sugar content of 21.8°Brix.

#### *6.8.8.5 'Guideline'*

This has an average fruit weight of 407.27 and a yield of 20.36 t $ha^{-1}$. The fruit comprise 64.5% pulp and 7.9% seeds, with a sugar content of 24.0°Brix.

#### *6.8.8.6 'INIAP Fabulosa 2015'*

INIAP Fabulosa 2015 is another clone that has been selected for its high yield and good fruit quality. It has an average fruit weight of 491.55 ± 201.10 g. The fruit contains 55.87 ± 5.57% pulp, 9.45 ± 3.22% seeds, and has the following characteristics: 24.72 ± 5.19°Brix, 0.49 ± 0.15%

titratable acidity, 0.94 ± 0.01 g protein 100 $g^{-1}$, 1.82 ± 0.02 g fibre 100 $g^{-1}$ and 63.06 ± 13.63 mg vitamin C 100 $g^{-1}$.

### 6.8.9 Chile

In Chile, the main cultivars of cherimoya grown are 'Loma', 'Concha Lisa', 'Bronceada' and to a lesser extent 'Local Serena', 'Pina' and 'Santa Julia'. Gardiazabal and Cano (1999) evaluated ten cherimoya cultivars in Quillota, Chile. The important characteristics of the cultivars are as follows (Table 6.4).

#### 6.8.9.1 'Loma'

The fruit is round to round-conical in shape with a good taste and a high pulp content (78–83%). It contains 3.6–20.9 seeds 100 $g^{-1}$ of fruit. The TSS content of fruit varies between 15.1°Brix and 18.2°Brix.

#### 6.8.9.2 'Concha Lisa'

The fruit is light green in colour and medium to good in taste. The pulp content is 75–85% with 8.0–16.6 seeds 100 $g^{-1}$ of fruit. The fruit is medium sweet with a TSS of 14.1–16.5°Brix.

#### 6.8.9.3 'Bronceada'

The fruit is dark green in colour and round to round-conical in shape. The fruit contains 79–82% pulp and 7.3–16.8 seeds 100 $g^{-1}$ of fruit. It is medium sweet with a TSS of 13.8–15.5°Brix.

### 6.8.10 Egypt

The three main cultivars grown in Egypt are 'Balady', 'Abd El-Razik' and 'Finny'.

#### 6.8.10.1 'Balady'

The fruit set in this cultivar varies between 30.9% and 77.9% in different months (June, July and August) and according to the pollinator used (Nahla *et al.*, 2009), while fruit retention varies between 28.2% and 37.5%. The average fruit weight is 113.7–153.4 g (Fig. 6.13). The average number of seeds varies between 19.2 and 28.4 per fruit. The fruit is medium sweet (TSS 15.1–16.1°Brix) and the acidity content varies between 0.22% and 0.30%.

#### 6.8.10.2 'Abd El-Razik'

The fruit set and fruit retention vary from 26.7% to 78.1% and from 22.4% to 38.2%, respectively, in different months and according to the pollinator used (Nahla *et al.*, 2009). The average fruit weight is 125.6–225.79 g with 11.7–19.0 seeds per fruit. The TSS content of fruit varies between 14.1°Brix and 15.2°Brix and the fruit acidity is 0.21–0.30%.

**Table 6.4.** Physico-chemical characteristics of ten cherimoya cultivars grown in Chile. Data from Gardiazabal and Cano (1999).

| Cultivar | Fruit set (%) (self-pollination) | Fruit weight (g) (September harvest) | No. of seeds 100 $g^{-1}$ fruit | °Brix (September harvest) | Ascorbic acid in 100 ml juice (mg) |
|---|---|---|---|---|---|
| 'Cumbe' | 73.0 | 565.0 | 7.01 | 14.8 | 1.7 |
| 'Chiuna 2' | 85.0 | 497.0 | 7.65 | 15.4 | 2.0 |
| 'Fino de Jete' | 84.0 | 680.0 | 9.73 | 16.1 | 1.9 |
| 'Peru 66' | 90.0 | 696.0 | 7.06 | 14.5 | 1.6 |
| 'Peru 45' | 69.0 | 569.0 | 18.15 | 13.6 | 1.4 |
| 'Cholan' | 81.0 | 630.0 | 6.32 | 13.5 | 1.3 |
| 'Peru 41' | 76.0 | 629.0 | 10.51 | 13.2 | 1.4 |
| 'Peru 11' | 52.0 | 440.0 | 12.26 | 13.2 | 0.9 |
| 'Chiuna 3' | 81.0 | 199.0 | 18.95 | 20.0 | 3.4 |
| 'Loma' | 80.0 | 838.0 | 8.04 | 15.1 | 1.9 |

#### 6.8.10.3 'Finny'

This cultivar shows fruit set and fruit retention of 18.9–75.8% and 25.1–38.2%, respectively (Nahla *et al.*, 2009). The average fruit weight is 161.3–190.2 g and it has 16.1–23.4 seeds per fruit. The TSS:acidity ratio of the fruit varies between 52.0 and 64.0.

### 6.8.11 India

Sugar apple (*A. squamosa*, mostly known as custard apple in India) is grown in different agro-climatic regions of India. A number of selections and hybrids developed in India along with a few introduced cultivars are being evaluated.

#### 6.8.11.1 'Balanagar'

This cultivar is grown in more than 60% of areas under sugar apple in semi-arid regions of western India due to its wide adaptability, drought tolerance and suitability for difficult types of soil and high yielding characteristics under semi-arid conditions. The fruit has a high TSS (28–30°Brix) with a granular pulp texture and compact flakes (Yadav, 2018). The pulp is white (Fig. 6.14A), buttery and sweet, with moderate to plenty of juice and excellent flavour. The average fruit weight is about 200 g (Fig. 6.14B) but fruit has been recorded up to 620 g. The seeds are brown (Fig. 6.14C). This cultivar has a yield of about 6 t ha$^{-1}$ under the rainfed vertisol condition of Tamil Nadu (Rajadurai *et al.*, 2022).

**Fig. 6.13.** Fruit of the cultivar 'Balady'. Photograph courtesy of Dr Mona Arfa.

#### 6.8.11.2 'Mammoth'

The fruit is irregular to spherical in shape and weighs 170–250 g. The areoles are medium to large, very smooth and round, and the rind is lightly russet. The mesocarp is white with small to medium granules, and the pulp is white and translucent, juicy and sweet. The tree has a tendency to branch profusely and is prolific bearer. The rind becomes wrinkled on ripening, and the fruit has low acidity and good keeping quality (Parthasarathy *et al.*, 2021).

#### 6.8.11.3 'Pink's Mammoth'

The tree is about 3.5 m in height and the average fruit weight is 208.6 g. The fruit surface is mammillated, and the average number of seeds is 31.2 per fruit. The fruit has a strong aroma and is sweet (22.12°Brix), with 0.26% acidity and a short shelf life (5.40 days). The tree produces about 12.60 kg year$^{-1}$ (Kahar *et al.*, 2023).

**Fig. 6.14.** (A) Pulp, (B) fruit and (C) seeds of the cultivar 'Balanagar'. Photograph courtesy of Dr Vikas Yadav.

#### *6.8.11.4 'Washington 98797'*

The tree grows up to 3.0 m in height. The fruit is spheroid to cordate in shape, medium to big size. Average weight is 690 g, areoles medium, truncate, tuberculate, ridged, rind light russet-green; mesocarp with medium to big granules; pulp creamy-white, soft, moderately juicy, pleasant, moderately sweet. Seeds 50–60 fruit$^{-1}$. High yielder (Parthasarathy *et al.*, 2021).

#### *6.8.11.5 'Washington 107005'*

The tree grows up to 2.4 m. The fruit is spherical to short conical in shape, with an average size of 170–335 g, although up to 560 g has been recorded. The areoles are tuberculate, with small pits. The rind is greyish-green, and the mesocarp is white with small to medium granules. The pulp is soft, bright white and buttery, and is moderately to very juicy and sweet, with a delightful flavour and has 30–50 seeds per fruit (Parthasarathy *et al.*, 2021).

#### *6.8.11.6 'APK-1'*

The tree grows up to 4.0 m. The average fruit weight is 188.6 g with 34.0 seeds per fruit. The fruit is mamillated with a light green exocarp colour. This cultivar is a high yielder (a 7-year-old tree produces about 14.20 kg) and the fruit is medium sweet (21.0°Brix), with an acidity of 0.23% and a mild flavour (Kahar *et al.*, 2023).

#### *6.8.11.7 'Chittorgarh Local'*

The tree is tall, growing up to 5.60 m. The average fruit weight is 240.1 g with 41.2 seeds per fruit. The fruit surface is impressa type with a light green exocarp. The pulp is sweet (24.0°Brix), with 0.32% acidity and a mild flavour; it has a good shelf life of 9.20 days. The cultivar is a very high yielder, producing 15.2 kg per tree year$^{-1}$ (Kahar *et al.*, 2023).

#### *6.8.11.8 'Raydurg'*

This has a medium-sized tree (3.6 m). The average fruit weight is 149.55 g, with 37.85 seeds per fruit. It is a very high yielder producing an average of 16.87 kg per tree and 4.55 t ha$^{-1}$. The fruit is medium sweet with a TSS of 21.97°Brix (Rajadurai *et al.*, 2022).

#### *6.8.11.9 'T.P.-7'*

The tree is medium-sized (3.6 m). The fruit is small (178.0 g), with 31.6 seeds per fruit, an impressa fruit surface, and a light green exocarp. It has a mild flavour and is very sweet (25.60°Brix), with total sugar content of 24.84% and acidity of 0.124%. The average yield is 13.20 kg per tree year$^{-1}$ (Kahar *et al.*, 2023).

#### *6.8.11.10 'Barbados'*

The tree is dwarf (2.4 m). The fruit is spherical to cordate in shape and medium in size (230 g). The areoles are tuberculate and very rough. The rind is fine and the mesocarp has medium granules. The pulp is creamy-white and moderately juicy, with an agreeable flavour. This cultivar is a sweet, prolific bearer (Parthasarathy *et al.*, 2021).

#### *6.8.11.11 'Arka Sahan'*

The cultivar is a progeny of 'Island Gem' (atemoya) × 'Mammoth' (*A. squamosa*). The average fruit weight is about 210 g, and the fruit skin has a waxy bloom and is light green in colour with large flat eyes (Fig. 6.15). The creamy-white flesh is juicy with a mild and pleasant aroma, and is tender with few seeds (~9 seeds100 g$^{-1}$) and large segments. The edible pulp is very sweet (30.0°Brix). The fruit pulp (100 g) contains 2.49 g of crude protein, 42.29 mg of phosphorus and 225 mg of calcium. The yield is about 12.0 t ha$^{-1}$ and the tree is drought hardy (Parthasarathy *et al.*, 2021).

#### *6.8.11.12 'Arka Neelanchal Vikram'*

This high-yielding cultivar was developed through clonal selection. The average fruit weight is about 211 g, and it is sweet (TSS 23.5°Brix) with a sugar:acid ratio of 53.8 and a long shelf life. It has a relatively low incidence of mealybug. It was released by the Government of Odisha in 2016.

#### *6.8.11.13 'Thar Amrit'*

This was developed at the Central Horticultural Experiment Station, Godhra, Gujarat, and released in 2022. It is a high yielder (24.8 kg per tree in year 11), matures late in the season and is regular bearer (Fig. 6.16). It has an average fruit weight of 320.12 g, with a high pulp content (63.58%) and high TSS (29.12°Brix). The fruit is rich in ascorbic acid (45–50 mg 100 $g^{-1}$), iron (1.70 mg 100 $g^{-1}$), phosphorus (26.0 mg 100 $g^{-1}$), potassium (285.55 mg 100 $g^{-1}$) and calcium 918.21 mg 100 $g^{-1}$). It is drought hardy and able to provide a stable yield during aberrant agroclimatic conditions (Yadav *et al.*, 2023).

**Fig. 6.15.** Fruit of the cultivar 'Arka Shan'. Photograph courtesy of Dr Vikas Yadav.

#### 6.8.11.14 'Sindhan'

The tree is 4.0–4.5 m in height. The average fruit weight is 183.6 g, with 31.6 seeds per fruit. The fruit surface is impressa type and green in colour, and the fruit is sweet (22.0°Brix) with an acidity of 0.23%. It produces about 10.20 t $ha^{-1}$ (Fig. 6.17).

#### 6.8.11.15 'VNR Madhur'

This cultivar was developed by VNR Seeds, Raipur, Chattisgarh. It is high yielding and the fruit is medium-sized to very large (300–1000 g) (Fig. 6.18), with few seeds (4–5%). The pulp is creamy and the pulp content is high (>70%); it is very sweet (25–30°Brix).

## 6.9 Conclusion

Annonas are considered a commercial crop in about a dozen countries in the world. Each of these countries has developed/identified its own cultivars for commercial cultivation (Table 6.5).

**Fig. 6.16.** Fruit of the cultivar 'Thar Amrit'. Photograph courtesy of Dr Vikas Yadav.

Development of germplasm collections and *ex situ* evaluation of numerous accessions of cherimoya have been undertaken in Spain, Mexico, Ecuador, Peru, Chile, California and Portugal, of soursop in Ecuador, Colombia, Nigeria, Puerto Rico and Brazil, and of sugar apple in India, Taiwan, Malaysia, Brazil, Colombia, Cuba and Sri Lanka. Most of the atemoya selections were made in Australia. Researchers were able to identify promising types based on morpho-agronomic characterization and proposed that this as the basis for developing germplasm for commercial plantation.

The long reproductive cycles, high levels of heterozygosity and costs associated with crosses limits annona hybridization programmes. Major hybridization work on atemoya was carried out or is ongoing on in Australia and on sugar apple in India and Taiwan, which has resulted a few good hybrids with great market potential. Hybridization using wild *Annona* spp. shows promise for both biotic and abiotic stress tolerance, but very little research has been conducted to date.

**Fig. 6.17.** Fruit of the cultivar 'Sindhan'. Photograph courtesy of Dr Vikas Yadav.

**Fig. 6.18.** Fruit of the cultivar 'VNR Madhur'. Photograph courtesy of VNR Seeds, Chattisgarh.

**Table 6.5.** Annona cultivars available in different countries.

| Country | Crop | Cultivars |
|---|---|---|
| Australia | Atemoya | 'African Pride', 'Pink's Mammoth', 'KJ Pinks', 'Hilary White', 'Maroochy Gold' |
| Bolivia | Cherimoya | Local selections, 'Cumbe', 'Concha Lisa', 'Bronceada' |
| Brazil | Sugar apple | 'Brazilian Seedless' |
| | Cherimoya | 'Madeira', 'Crioula', 'Gigante das Alagoas' |
| | Atemoya | 'Gefner', 'Pink's Mammoth', 'African Pride' |
| Chile | | 'Bronceada', 'Concha Lisa', 'Cholan', 'Loma', 'Ju Niana', 'Local Serena', 'Calvo' |
| China | | 'Gefner', 'Damu', 'Luzuan', 'Soft Branch', 'Thin Scale', 'Purple', 'African 'Pride' |
| Colombia | Cherimoya | 'Concha Lisa', 'Amarilla', 'Blanca', 'Rugosa', 'Tocarema' |
| | Soursop | 'Elite' |
| Ecuador | Soursop | 'ICA', 'Giant' |
| | Cherimoya | 'Fino de Jete', 'Bays', 'White', 'Bronceada', 'Concha Lisa', 'Tumbaco', 'Loja', 'Fabulous' |
| Egypt | Sugar apple and | 'Balady', 'Abd El-Razik', 'Finny' |
| | Cherimoya | 'Benimazar', 'Cherimata' |
| India | Sugar apple | 'Balangar', 'Mammoth', 'Pink's Mammoth', 'Raydurg', 'Arka Sahan', 'Thar Amrit', 'VNR Madhur', 'Sindhan' |

*Continued*

**Table 6.5.** Continued.

| Country | Crop | Cultivars |
|---|---|---|
| Israel | Cherimoya | 'Gefner', 'Red Annona', 'Kapsi' ('Caspi') |
| Italy | Cherimoya | 'Campas', 'Chaffey', 'Fino de Jete', 'White', 'Daniela', 'Torre1', 'Torre2' |
| Mexico | Cherimoya | 'Bonita', 'Campas', 'White', 'Cortes-II-31' |
| | Soursop | 'Guanay-1', 'Guanay-2', 'Guanay-3', 'Acateno', 'Emilano Zapata' |
| New Zealand | Cherimoya | 'Asca', 'Ballinger', 'Canaria', 'Bays', 'McPherson', 'Smoothey', Burtons', 'Pina' |
| Peru | Cherimoya | 'Chiuna 1', 'Chiuna 2', 'Chiuna 3', 'Cumbre', 'Sinsemila', 'Asca', 'Chavez' |
| Philippines | Sugar apple | 'Aguinaldo', 'Lobo', 'Seedless' |
| Portugal | Cherimoya | 'Perry Vidal', 'Mateus I', 'Mateus III', 'Funchal' |
| Puerto Rico | Cherimoya | 'Bradley', 'Gefner', 'Priestly', 'Lisa', '47-18' |
| Spain | Cherimoya | 'Fino de Jete', 'Campas', 'Alboran', 'Jete' |
| Taiwan | Sugar apple | 'Da-Mo', 'Taitung 1', 'Taitung 2', 'Taitung 3' |
| Thailand | Sugar apple | 'Fai Keaw', 'Fai Krung', 'Nung Keaw', 'Nung Krung', 'Nun Thong', 'Petch Pakchong', 'Noi' |
| USA | Cherimoya | 'El Bumpo', 'Carmela', 'Bays', 'Big Sister', 'Booth', 'Chaffey', 'White', 'Whaley', 'Deliciosa', 'Lisa', 'Loma', 'Ryerson' |
| | Sugar apple | 'Red Sugar', 'Leah Reese' |
| | Custard apple | 'Camo Real', 'Fairchild Purpura', 'Dr Leon' |
| Venezuela | Soursop | 'Giant', 'Sugar-free soursop' |

## References


Abdelkawy, M.M.A. and El-Nawman S.M.A., E. (2017) Identification of three Egyptian *Annona* cultivars morphologically and biochemically using RAPD analysis. *Alexandria Journal of Agricultural Sciences* 62, 415–421. DOI: 10.21608/alexja.2017.67623

Agustín, J.A. (1999) Advances in research on genetic resources of cherimoya (*Annona cherimola* Mill.) in Michoacán State, Mexico. *Acta Horticulturae* 497, 189–200. DOI: 10.17660/ActaHortic.1999.497.10

Ahmad, I., Bhagat, T.V.R.S., Sharma, K.K., Simachalam, P. and Srivastava, R.C. (2010) ISSR and RAPD marker based DNA fingerprinting and diversity assessment of *Annona* spp. in South Andamans. *Indian Journal of Horticulture* 67, 147–151.

Anderson, P.A. and Richardson, A.C. (1990) Which cherimoya cultivar is best? *Orchardist of New Zealand* 63, 17–19.

Anderson, P.A. and Richardson, A.D. (1992) Cherimoya pruning essential for high quality fruit. *Orchardist of New Zealand* 65, 32–34.

Andrade, R. (2009) Morphological and molecular characterization of the custard apple (*Annona cherimola*) collection at the Tumbaco Experimental Farm INIAP – Ecuador. Biotechnology Engineering Thesis. ESPE, Universidad de las Fuerzas Armadas, Sangolquí, Ecuador.

Andrés-Agustín, J., González-Andrés, F., Nieto-Ángel, R. and Barrientos-Priego, A.F. (2006) Morphometry of the organs of cherimoya (*Annona cherimola* Mill.) and analysis of fruit parameters for the characterization of cultivars, and Mexican germplasm selections. *Scientia Horticulturae* 107, 337–346. DOI: 10.1016/j.scienta.2005.11.003

Anuragi, H., Dhaduk, H.L., Kumar, S., Dhruve, J.J., Parekh, M.J. and Sakure, A.A. (2016) Molecular diversity of *Annona* species and proximate fruit composition of selected genotypes. *3 Biotechnology* 6: 204. DOI: 10.1007/s13205-016-0520-9

Bharad, S.G., Kulwal, P.L. and Bagal, S.A. (2009) Genetic diversity study in *Annona squamosa* by morphological, biochemical and RAPD markers. *Acta Horticulturae* 839, 615–623. DOI: 10.17660/ActaHortic.2009.839.84

Bignell, G. (2022) *New Custard Apple Varieties and Enhanced Industry Productivity*. Final report. Hort Innovation, Sydney, Australia. Available at: https://era.dpi.qld.gov.au/id/eprint/9523/1/cu16002-final-report-complete%20%281%29.pdf (accessed 18 August 2025).

Bignell, G. and Bruun, D. (2014) *Accelerating the Development of the Australian Custard Apple Industry*. Final Report for HAL Project CU11000. Department of Agriculture, Fisheries and Forestry, Queensland. Australia. Available at: www.horticulture.com.au/globalassets/laserfiche/assets/project-reports/cu11000/cu11000-final-report-complete-2.pdf (accessed 28 August 2025).

Brown, J., Laurentín, H. and Dávila, M. (2003) Genetic relationships between nine *Annona muricata* L. accessions using RAPD markers. *Fruits* 58, 255–259. DOI: 10.1051/fruits:2003013

CHERLA (2008) *Inventory of Current* Ex situ *Germplasm Collections*. Deliverable 7, Project No. 015100, INCO Sixth Framework Programme of the European Commission, Brussels.

Coissac, E., Hollingsworth, P.M., Lavergne, S. and Taberlet, P. (2016) From barcodes to genomes: extending the concept of DNA barcoding. *Molecular Ecology* 25, 1423–1428. DOI: 10.1111/mec.13549

Collevatti, R.G., Telles, M.P.C., Lima, J.S., Gouveia, F.O. and Soares, T.N. (2014) Contrasting spatial genetic structure in *Annona crassiflora* populations from fragmented and pristine savannas. *Plant Systematics and Evolution* 300, 1719–1727. DOI: 10.1007/s00606-014-0997-9

Cota, L.G., Vieira, F.A., Melo Júnior, A.F., Brandão, M.M., Santana, K.N.O. *et al.* (2011) Genetic diversity of *Annona crassiflora* (Annonaceae) in northern Minas Gerais state. *Genetics and Molecular Research* 10, 2172–2180. DOI: 10.4238/vol10-3gmr1188

Crane, J.H., Balerdi, C.F. and Maguire, I. (2016) *Sugar Apple Growing in the Florida Home Landscape*. IFAS Extension HS38, University of Florida, Gainesville, Florida. Available at: https://edis.ifas.ufl.edu/publication/MG330 (accessed 18 August 2025).

Crane, J.H., Chagas, P.C. and Chagas, E.A. (2021) 'LeahReese': a new sugar apple (*Annona squamosa* L.) cultivar. *HortScience* 56, 1293–1294. DOI: 10.21273/HORTSCI16070-21

de Carvalho, P.S., Bezerra, J.E.F., Lederman, I.E., Alves, M.A. and Melo Neto, M.L. (2000) Avaliacao de genotipes de Pinheira (*Annona squamosa* L) no Vale do Rio Moxoto III – caracteristicas de crescimento Produca d 1993 a 1997. *Revista Brasileira de Fruticultura* 22, 27–30.

Dikshit, N., Bharad, S.G. and Badge, M.P. (2008) Diversity in custard apple germplasm collections from Maharashtra. *Indian Journal of Plant Genetic Resources* 21, 95–96.

Ellstrand, N.C. and Lee, J.M. (1987) Cultivar identification of cherimoya (*Annona cherimola* Mill.) using isozyme markers. *Scientia Horticulturae* 32, 25–31. DOI: 10.1016/0304-4238(87)90013-6

Escribano, P., Viruel, M.A. and Hormaza, J.I. (2007) Molecular analysis of genetic diversity and geographic origin within an *ex situ* germplasm collection of cherimoya by using SSRS. *Journal of the American Society for Horticultural Science* 132, 357–367. DOI: 10.21273/JASHS.132.3.357

Ezzat, A.H., Naguib, M. and Metwalli, S. (1974) Evaluation and determination of the maturity stage of the fruits of some *Annona* varieties. *Agriculture Research Review (Cairo)* 52, 7–17.

FAO (Food and Agriculture Organization of the United Nations) (2010) *The Second Report of the State of World's Plant Genetic Resources for Food and Agriculture*. FAO, Rome.

FAO (Food and Agriculture Organization of the United Nations) (2011) *Draft Updated Global Plan of Action for the Conservation and Utilization of Plant Genetic Resources for Food and Agriculture*. Fifth Session of the Intergovernmental Technical Working Group on Plant Genetic Resources for Food and Agriculture. FAO, Rome, Italy. Available at: https://openknowledge.fao.org/handle/20.500.14283/mb068e (accessed 18 August 2025).

Gardiazabal, F.I. and Cano, G.L. (1999) Characterization of 10 cherimoya (*Annona cherimola* Mill.) cultivars and their response to artificial pollination in Qillota, Chile. *Acta Horticulturae* 497, 225–254. DOI: 10.17660/ActaHortic.1999.497.12

George, A.P. and Nissen, R.J. (1987) The effects of root temperature on growth and dry matter production of *Annona* species. *Scientia Horticulturae* 31, 95–99. DOI: 10.1016/0304-4238(87)90111-7

George, A.P., Broadley, R.H., Nissen, R.J., Hamill, S.D. and Topp, B.L. (1999) Breeding strategies for atemoya and cherimoya. *Acta Horticulturae* 497, 255–268. DOI: 10.17660/ActaHortic.1999.497.13

George, A.P., Broadley, R., Nissen, R.J. and Hamill, S. (2005) Breeding and selecting new varieties and rootstocks of custard apple (*Annona* spp. hybrids) in subtropical Australia. *Acta Horticulturae* 694, 125–128. DOI: 10.17660/ActaHortic.2005.694.18

George, A.P., Broadley, R.H., Nissen, R.J., Bruun, D., Hormaza, J.I. *et al.* (2008) Pollination, breeding and selection of new varieties of custard apple (*Annona* spp. hybrids) in Australia. *Acta Horticulturae* 772, 215–218. DOI: 10.17660/ActaHortic.2008.772.31

Ghavale, S.L., Patil, R.S. and Dalvi, V.V. (2016) Morphological characterization of Bullock's heart (*Annonal reticulata* L.) germplasm in Konkan region of Maharashtra State (India). *International Journal of Tropical Agriculture* 34, 1013–1026.

Girwani, A., Madhavi, A., Surcsh Kumar, T. and Satyanarayan Reddy, G. (2011) Evaluation of custard apple hybrids for fruit yield and quality attributing characters. *Acta Horticulturae* 890, 251–254. DOI: 10.17660/ActaHortic.2011.890.35

Grossberger, D. (1999) The California cherimoya industry *Acta Horticulturae* 497, 119–142. DOI: 10.17660/ActaHortic.1999.497.6

Guimaraes, J.F.R., Nietsche, S., Costa, M.R., Moreira, C.B.R., Pereira, M.C.T. *et al.* (2013) Genetic diversity in sugar apple (*Annona squamosa*) by using RAPD markers. *Revista Ceres* 60, 428–431.

Hasan, A.E.Z., Bermawie, N., Julistiono, H., Riyanti, E.I., Hasim *et al.* (2017) Genetic diversity analysis of soursop (*Annona muricata* L.) in West Java region of Indonesia using RAPD markers. *Annual Research & Review in Biology* 14, 1–7. DOI: 10.9734/ARRB/2017/34354

Hebert, P.D.N., Cywinska, A., Ball, S.L. and deWaard, J.R. (2003) Biological identifications through DNA barcodes. *Proceedings of the Royal Society of London. B* 270, 313–321. DOI: 10.1098/rspb.2002.2218

Hermoso González, J.M., Pérez de Oteyza, M.A., Ruiz Nieto, A. and Farré Massip, J.M. (1999) The Spanish germplasm bank of cherimoya (*Annona cherimola* Mill.). *Acta Horticulturae* 497, 201–224. DOI: 10.17660/ActaHortic.1999.497.11

Herrera, G. (2006) Study of vegetative, floral and fruit development in five genotypes of cherimoya (*Annona cherimola*). Thesis, Central University of Ecuador, Tumbaco-Pichincha, Ecuador.

Higuchi, H., Utsunomiya, N. and Sakuratani, T. (1998) Effects of temperature on growth, dry matter production and $CO_2$ assimilation in cherimoya (*Annona cherimola* Mill.) and sugar apple (*Annona squamosa* L.) seedlings. *Scientia Horticulturae* 73, 89–97. DOI: 10.1016/S0304-4238(97)00142-8

Jain, D.K., Dewangan, R.K., Nag, J.L., Nag, G.P., Sing, D.P. *et al.* (2024) Biodiversity of custard apple genotypes in Bastar District of Chhattisgarh. *Pharma Innovation Journal* 13, 149–152.

Jalikop, S.H. (2010) Tree and fruit traits of progenies from the cross between (*A. cherimola* Mill. × *A. squamosa* L.) and *A. reticulata* L. and approaches for the introgression of valuable genes from *A. reticulate* L. *Euphytica* 173, 161–171.

Jalikop, S.H. and Kumar, P.S. (1995) Improvement of some semi-arid fruits through selection and hybridization. *Annual Report 1994–95*. Indian Institute of Horticultural Research, p. 5.

Jalikop, S.H. and Kumar, P.S. (1996) *Improvement of Semi-arid Fruits Through Selection and Hybridization*. Annual Report 1994–1995, Indian Institute of Horticultural Research, Bengaluru, India.

Ojezele, O.J., Ojezele, M.O. and Adeosun, A.M. (2016) Comparative phytochemistry and antioxidant activities of water and ethanol extract of *Annona muricata* linn leaf, seed and fruit. *Advances in Biological Research* 10, 230–235. DOI: 10.5829/idosi.abr.2016.10.4.10514

Kahar, C., Mahawer, L.N., Singh, V., Bairwa, H.L., Babu, R. *et al.* (2023) Varietal evaluation of custard apple to find suitable variety in Mewar Region of Rajasthan. *Frontiers in Crop Improvement* 11, 3195–3203.

Khalid, M.Z.M. (2002) Hybridization between selected *Annoaceae* species. *Acta Horticulturae* 575, 367–369.

Komkuntod, R. and Wanichkhun, K. (2014) Sugar apple "Phet Pak Chong" Variety. Chansatit Organic Institute, Kasetsart University, Thailand.

Kumar, P.S. and Jalikop, S.H. (2000) Cross-compatibility among *Annona* species. *Indian Journal of Horticulture* 57, 309–313.

Larranaga, N., Albertazzi, F.J., Fontecha, G., Palmieri, M., Rainer, H. *et al.* (2017) A Mesoamerican origin of cherimoya (*Annona cherimola* Mill.): implications for the conservation of plant genetic resources. *Molecular Ecology* 26, 4116–4130. DOI: 10.1111/mec.14157

Larranaga, N., Albertazzi, F.J. and Hormaza, J.I. (2019) Phylogenetics of *Annona cherimola* (Annonaceae) and some of its closest relatives. *Journal of Systematics and Evolution* 57, 211–221. DOI: 10.1111/jse.12473

Leal, F. and Paull, R.E. (2023) The genus *Annona*: botanical characteristics, horticultural requirements and uses. *Crop Science* 63, 1030–1049. DOI: 10.1002/csc2.20833

Lee, W.-L., Huang, C.C., Kuan, C.S., Jiang, S.W. and Hsieh, H.Y. (2017) Development of $GA_3$ tropical fruit varieties and cultivation techniques in Taiwan. *Acta Horticulturae* 1166, 47–54. DOI: 10.17660/ActaHortic.2017.1166.7

Lei, J.M., Liang, Z.R., Zhang, H., Li, Y.Q. and Xue, B. (2022) Advances in exploitation and utilization of wild fruit resources of Annonaceae in China. *Journal of Fruit Science* 39, 121–130.

Lemos, E.E.P., Barros, P.G., Campos, R.S., Salvador, T.L. and Santos, M.Q.C. (2010) Essential descriptors for sugar apple (*Annona squamosa* L.) cultivars. In: *28th International Horticultural Congress, Lisbon, Portugal*. International Society for Horticultural Science, Lisbon, p. 175 (abstract).

Lora, J., Hormaza, J.I., Herrero, M. and Gasser, C.S. (2011) Seedless fruits and the disruption of a conserved genetic pathway in angiosperm ovule development. *Proceedings of the National Academy of Sciences USA* 108, 5461–5465. DOI: 10.1073/pnas.1014514108

Love, K. and Paull, R.E. (2011) Fruit, nut and beverage crops series FM-22. In: *Soursop*. Honolulu, University, University of Hawaii, p. 6.
Mahdeem, H. (1990) Best of the Annnonas. *Tropical Fruit World* 1, 110–114.
Martin, C., Viruel, M.A., Lora, J. and Hormaza, J.L. (2019) Polyploidy in fruit tree crops of the genus Annona (Annonaceae). *Frontiers in Plant Science* 10: 99. DOI: 10.3389/fpls.2019.00099
Monastra, F. (1997) Current status of Annonaceae in Italy. *Mesfin Newsletter* 3, 11–12.
Morawetz, W. (1986) Remarks on Karyological differentiation patterns in tropical woody plants. *Plant Systematics and Evolution* 152(1–2), 49–100. DOI: 10.1007/BF00985351
Moreira, R., Morell, H.R., Ardisana, E.F.H. Feican, C.G., Mestanza, S. and Viera-Arroyo, W. (2020) *In situ* morphological characterization of soursop (*Annona muricata* L.) plants in Manabí, Ecuador. *Enfoque UTE* 11, 58–70. DOI: 10.29019/enfoque.v11n2.536
Morton, J.F. (1987) Soursop. *Fruits of Warm climates. Republished in 2013*. Echo Point Books and Media, p. 550.
Nahla, A.A., Farida, A.A. and Galila, G.S. (2009) Studies on the compatibility among some *Annona* cultivars as pollinators and their effects on fruit set and fruit quality. *Egyptian Journal of Horticulture* 36, 189–200.
Nakasone, H.Y. and Paull, R.E. (1998) Annonas. In: Nakasone, H.Y. and Paull, R.E. (eds) *Tropical Fruits*. CAB International, Wallingford, UK, pp. 45–75.
Nimisha, S., Kherwar, D., Ajay, K.M., Singh, B. and Usha, K. (2013) Molecular breeding to improve guava (*Psidium guajava* L.): current status and future prospective. *Scientia Horticulturae* 164, 578–588. DOI: 10.1016/j.scienta.2013.DOI: 10.017
Nunes, R.E. de. F. (1997) The actual status of cherimoya cultivation in Madeira Island. In: *Second MESFIN on Plant Genetic Resources, Madeira, Portugal*, pp. 133–151.
Nunez-Elisea, R., Schaffer, B., Crane, J.H. and Colls, A.M. (1998) Impacts of flooding on *Annona* species. *Proceedings of the Florida State Horticultural Society* 111, 317–319.
Padmini, S.M.P.C., Pushpakumar, D.K.N.G. and Samarasekera, R. (2013) Morphological characterization of soursop (*Annonas muricata* L.) germplasm in Sri Lanka. *Tropical Agricultural Research* 24, 362–374.
Parthasarathy, V.A., Sakthivel, T. and Rathore, D.S. (2021) Annonas. In: Parthasarathy, V.A., Base, T.K., Mitra, S.K., Ghosh, B., Chakorty, I. *et al* (eds) *Fruits: Tropical and Subtropical*, Vol. 1. Daya Publishing House, New Delhi, pp. 443–445.
Pascual, L., Perfectti, F., Gutierrez, M. and Vargas, A.M. (1993) Characterizing isozymes of Spanish cherimoya cultivars. *HortScience* 28, 845–847. DOI: 10.21273/HORTSCI.28.8.845
Paull, R.E. and Durate, O. (2011) Annonas: cherimoya, atemoya and sweetsop. In: Paull. R.E. and Duarte, O. (eds) *Tropical Fruits*, Vol. 1, 2nd edn. CAB International, UK, pp. 123–152.
Pereira, M.C.T., Crane, J.H., Nietsche, S., Amounts, W. and Santos, M.A. (2014) Growth regulators in effective fruiting and quality of fruits of atemoya 'Gefner'. *Brazilian Agriculture Research* 49, 281–289.
Pereira, M.C.T., Nietsche, S., São José, A.R., Lemos, E.E.P., Mizubutsi, G. *et al.* (2019) Anonáceas: Pinha (*Annona squamosa* L.), Atemóia (*Annona squamosa* x *Annona cherimola* Mill.) e graviola (*Annona muricata*). In: Paula Junior, T.J. and Venzon, M. (eds). *101 Culturas: Manual de Tecnologias Agrícolas*. Empresa de Pesquisa Agropecuária de Minas Gerais (EPAMIG), Belo Horizonte, Brazil, pp. 111–123.
Pico-Mendoza, J., Madrid, L., Pinoargote, M., Limongi, R., Peña, G. *et al.* (2024) Simple sequence repeat markers (SSR) in *Annona deceptrix* Westra H. Rainer, an endangered species of the Ecuadorian Coast. *Journal of Tropical Forest Science* 36, 132–138. DOI: 10.26525/jtfs2024.36.1.132
Pinto, A. and Silva, E.M. (1996) *Graviola para Exportacado Aspectos Tecnicos da Producao*. Embrapa/SPI, Brasilia.
Pinto, A.C. de Q., Ramos, V.H.V. and Rodriques, A.A. (2001) Formacao do pomar. In; Oliveria, G.P. (ed.) *Graviola Producado*. Embrapa Informacao Technolgica, Brasilia, pp. 22–25.
Pinto, A.C. de. Q., Corderio, M.C.R., Andrade, S.R.M. de., Ferreira, F.R., Filgueiras, H.A. de C. *et al.* (2005) Annona *Species*. International Centre for Underutilized Crops, University of Southampton, UK.
Rahman, S.M.M., Shimada, T., Yamamoto, T., Yonemoto, J. and Yoshida, M. (1998) Genetic diversity of cherimoya cultivars revealed by amplified fragment length polymorphism (AFLP) analysis. *Breeding Science* 48, 5–10.
Rainer, H. (2007) Monographic studies in the genus *Annona* L. (Annonaceae): inclusion of the genus Rollinia A. St.-Hil. *Annalen Des Naturhistorischen Muscums in Wien A* 108. pp. 191–205.
Rajadurai, K.R., Rajkumar, J. and Rajababu, C. (2022) Studies on the performance of custard apple (*Annona squamosa* L.) cultivars and varieties under rainfed vertisol condition of Tamil Nadu. *Journal of Scientific Research and Reports* 28, 7–13. DOI: 10.9734/jsrr/2022/v28i930543

Rasai, S., George, A.P. and Kantharajah, A.S. (1995) Tissue culture of *Annona* spp. (cherimoya, atemoya, sugar apple and soursop): a review. *Scientia Horticulturae* 62, 1–14. DOI: 10.1016/0304-4238(94)00758-8

Richardson, A. and Anderson, P. (1993) A detailed evaluation of cherimoya cultivars. *Orchardist of New Zealand* 66, 28–32.

Romero-Soler, K.J. and Cetzal-Ix, W. (2015) Cultivated species of the genus *Annona* (Annonaceae) from the Yucatan Peninsula. *Mexico CICY Herbarium* 7, 147–153.

Samaradiwakara, S.H.M.R.N.P., Samarasinghe, W.L.G., Shantha, P.G.S., Jayarathna, K.G.C.N., Dehigaspitiya, P. and Ubeysekera, N.M. (2020) Genetic diversity assessment of selected *Annona muricata* L. germplasm in Sri Lanka. *Asian Journal of Agricultural and Horticultural Research* 7, 1–13. DOI: 10.9734/ajahr/2020/v7i130084

Sanchez, C.F.B., Lopes, B.E., Teodoro, P.E., Garcia, A.D.P., Peixoto, L.D.A. *et al.* (2018) Genetic diversity among soursop genotypes based on fruit production. *Bioscience Journal* 34, 122–128.

Sanewski, G. (1988) *Growing Custard Apples*. Queensland Department of Primary Industries, Brisbane, Australia.

Santos, R.C.D., Ribeiro, L.M., Mercadante-Simões, M.O., Costa, M.R., Nietsche, S. *et al.* (2014) Stenospermy and seed development in the "Brazilian seedless" variety of sugar apple (*Annona squamosa*). *Anais da Academia Brasileira de Ciências* 86, 2101–2108. DOI: 10.1590/0001-3765201420140206

Sarasola, L. (1960) Cherimoya culture in Spain. In: *California Avocado Society Yearbook*, Vol. 44, pp. 47–53. Available at: www.avocadosource.com/CAS_Yearbooks/CAS_44_1960/CAS_1960_PG_47-53.pdf (accessed 18 August 2025).

Scheldeman, X. and van Damme, P. (1999) Promising cherimoya (*Annona cherimola* Mill.) accessions in Loja province, Southern Ecuador. *Acta Horticulturae* 497, 173–180. DOI: 10.17660/ActaHortic.1999.497.9

Senthilkumar, T., Prabhavathi, S.J., Alagesan, A., Punitha, A., Vidhya, D. *et al.* (2024) Evaluation, flowering behavior physico-chemical and keeping quality of custard apple: a review. *Journal of Experimental Agriculture International* 46, 96–104. DOI: 10.9734/jeai/2024/v46i32329

Thakur, D.R. and Singh, R.N. (1964) Studies on pollen morphology, pollination and fruit set in some *Annonas*. *Indian Journal of Horticulture* 22, 10–18.

Thanarut, C. and Ko, J.-C. (2017) Guava, wax apple, sugar apple and pineapple productions in Thailand. *Acta Horticulturae* 1166, 7–13. DOI: 10.17660/ActaHortic.2017.1166.2

Thang, T.D., Dai, D.N., Hoi, T.M. and Ogunwande, I.A. (2013) Study on the volatile oil content of *Annona glabra* L., *Annona squamosa* L. *Annona muricata* L. and *A. reticulata* L. from Vietnam. *Natural Product Research* 27, 1232–1236. DOI: 10.1080/14786419.2012.724413

Thuy, N.P., Trai, N.N., Tuu, N.T., Thu, N.K., Luan, N.T. *et al.* (2022) The analysis of DNA barcoding and the evolutionary divergence of *Annona* by ISSR markers. *Journal of Southwest Jiaotong University* 57, 41–57. DOI: 10.35741/issn.0258-2724.57.4.4

Viteri, P.F., León, J.F. and Ortega, C. (1999) Evaluation of shoot inducers in cherimoya (*Annona cherimola* Mill.). *Acta Horticulturae* 497, 347–354. DOI: 10.17660/ActaHortic.1999.497.19

Walker, J.W. (1972) Chromosome numbers, phylogeny, phytogeography of the Annonaceae and their bearing on the (original) basic chromosome number of angiosperms. *Taxon* 21, 57–65. DOI: 10.2307/1219222

Wester, P.J. (1913) Annonaceous possibilities for the plant breeder. *Philippine Agriculture Review* 6, 312–321.

Yadav, V. (2018) Cultivation of custard apple fetches more. *Indian Horticulture* 63, 42–46.

Yadav, V., Singh, A.K., Singh, S. and Rao, V.V.A. (2017) Variability in custard apple (*Annona squamosa*) genotypes for quality characters from Gujarat. *Indian Journal of Agricultural Sciences* 87, 1627–1632. DOI: 10.56093/ijas.v87i12.76493

Yadav, V., Singh, A.K., Hiwale, S.S., Singh, S. and Sharma, B.D. (2023) Thar amrit: a new high yielding custard apple variety for dryland. *Indian Horticulture* 68, 11–13.

Zhushen, X., XiDong, X. and JingYue, L. (1999) The performance of AP custard apple variety and its cultural techniques in Guangzhou area. *South China Fruits* 28, 24–26.

van Zonneveld, M., Scheldeman, X., Escribano, P., Viruel, M.A., van Damme, P. *et al.* (2012) Mapping genetic diversity of cherimoya (*Annona cherimola* Mill.): application of spatial analysis for conservation and use of plant genetic resources. *PLOS ONE* 7: e29845. DOI: 10.1371/journal.pone.0029845

# 7 Biotechnology

**Sisir Mitra***
*Former Professor, Faculty of Horticulture, Bidhan Chandra Krisi Viswavidyalaya (State Agricultural University), Mohanpur, West Bengal, India*

## 7.1 Introduction

The application of biotechnology to plant breeding includes the development of tools and knowledge to increase accuracy and efficiency for pre-breeding and selection for crop improvement and has multiple advantages over traditional technologies. Biotechnology is also used for characterization of an organism at the cellular or molecular level to accurately genotype for authentication and monitoring. This chapter discusses the various *in vitro* methodologies that have been investigated in cherimoya and other annonas used in breeding programmes, including (i) adventitious organogenesis and regeneration from cell culture; (ii) ploidy manipulation to obtain haploid, tetraploid and triploid plants (seedless); (iii) genetic transformation for introduction of genes to control postharvest processes and genes to provide resistance to pathogens and insects; and (iv) micropropagation and regeneration of wild *Annona* spp. or related Annonaceae (Encina *et al.*, 2014).

## 7.2 *In Vitro* Culture

Researchers have successfully micropropagated cherimoya (*Annona cherimola*), sugar apple (*Annona squamosa*), soursop (*Annona muricata*) and atemoya (*A. cherimola* × *A. squamosa*) (Pinto *et al.*, 2005).

Encina *et al.* (1994) developed a protocol for micropropagation from nodal explants of juvenile cherimoya cultivar 'Fino de Jete'. Murashige and Skoog (MS) medium supplemented with either 0.66 μM BA or 1.36 μM zeatin was used. After sprouting and multiplication, 95% rooting efficiency was achieved by pretreatment of *in vitro* shoos for 7 days on semi-solid MS medium containing activated charcoal (0.1% w/v). Upon transfer of the shoots into semi-solid MS medium supplemented with 500 μM indole-3-butyric acid (IBA), 58.4 mM sucrose and 200 mg citric acid $l^{-1}$ for 10 days (7 days in darkness, followed by 3 days of continuous light), rooting was induced. Elongation of the roots was carried out in a semi-solid half-strength MS medium. Following rooting, the acclimatization rate reached 100% (Encina *et al.*, 1994). Inoculation of micropropagated plantlets with arbuscular mycorrhizal fungi (*Glomus* spp.) improves vegetative growth and development of plantlets in the glasshouse and substantially reduces acclimatization time in juvenile and mature plants (Padilla and Encina, 2005).

Plantlet differentiation in soursop was described by Bejoy and Hariharan (1992). They found a mean of 4.8 shoots per hypocotyl

*Email: sisirm55@gmail.com


DOI: 10.1079/9781800627628.0007

explant growing in MS medium containing 6-benzylaminopurine (BAP) and 1-naphthaleneacetic acid (NAA). However, only relative success in rooting and 35% survival was achieved. Lemos and Blake (1996b) developed a micropropagation protocol for both juvenile and adult soursop. Axillary shoots on nodal explants were multiplied on woody plant medium (WPM) containing 0.5 μM NAA and 8.9 μM BAP. Shoots were preconditioned on WPM with 3% activated charcoal and rooted in liquid WPM containing 21.5 μM NAA and 1% galactose. Approximately 70% rooting occurred under these conditions, and all rooted plantlets were acclimatized. Adventitious shoot regeneration was reported from 67% of internodal explants of the mature phase on semi-solid Nitsch medium supplemented with 8.9 μM BAP, 2.7 μM NAA, 0.1% polyvinylpyrrolidone, 0.2% casein hydrolysate and 2% sorbitol (Lemos and Baker, 1998). Abubacker and Deepalakshmi (2017) developed a protocol for *in vitro* plantlet regeneration of *A. muricata* through direct shoot regeneration and rooting. Callus production was achieved in leaf base explants on MS medium supplemented with 1.5 g BAP $l^{-1}$ and 2.0 mg 2,4-dichlorophenoxyacetic acid (2,4-D) $l^{-1}$. Callus cell morphogenesis indicated the establishment of a pro-embryo, perfect dichotomy and shoot apices. Axillary shoot formation from nodal explants was achieved on MS medium with 1.0 mg BAP$l^{-1}$, 0.5 mg kinetin $l^{-1}$ and 1.5 mg IBA $l^{-1}$. The explants rooted in MS medium with 2.0 mg $l^{-1}$ IBA along with 0.5 mg NAA $l^{-1}$ and 1.0 mg BAP $l^{-1}$. The plants were acclimatized in a greenhouse and then under natural environmental conditions.

Ba *et al.* (2021) reported the best regeneration from cotyledonary nodes of juvenile soursop on MS medium supplemented with 2 mg BAP $l^{-1}$. Addition of 0.1 and 0.2 mg NAA $l^{-1}$ on MS medium caused the best elongation and average number of nodes for the new shoots (9.11 cm and 5.62 nodes, respectively). In the presence of 50 mg IBA $l^{-1}$, best rooting (83.33%) was obtained when incubated for 5 days.

Shoot proliferation of sugar apple was achieved with hypocotyls and nodal cuttings in WPM supplemented with BAP and silver thiosulphate to control leaf abscission (Lemos *et al.*, 1994; Lemos and Blake, 1996a,b). Rooting was obtained when shoots were preconditioned in medium with activated charcoal (10 g $l^{-1}$) and then treated with NAA (43 μM) or IBA (39 μM) (Lemos and Blake, 1996b). The rooting percentage improved if sucrose was replaced with 47% galactose. Following acclimatization, 80% of plants survived and were transferred to soil. Amin *et al.* (2002) successfully carried out micropropagation from field-grown mature sugar apple plants using shoot tips and nodes as explants. For axillary shoot proliferation, nodal explants showed the best results on modified MS medium containing 2 mg BAP $l^{-1}$ and 0.1 mg NAA $l^{-1}$. The highest percentage of rooting was obtained with 0.1 mg IBA $l^{-1}$. The regenerated shoots with well-developed roots were gradually acclimatized and successfully transferred to soil under natural conditions. Ba *et al.* (2021) obtained the best shoot formation from the juvenile plant by using terminal apices of sugar apple in modified MS medium supplemented with 2 mg BAP $l^{-1}$. However, the best rooting was obtained using cotyledonary nodes in a 3-day rhizogenic induction period in the dark with IBA at 50 mg $l^{-1}$ and activated charcoal at 2 g $l^{-1}$.

Multiple shoot formation was induced in different media from excised hypocotyls and nodal cuttings from 5-day-old trees of atemoya 'African Pride' (Rasai *et al.*, 1994). The highest number of shoots was obtained from the distal part (near the root) of the hypocotyl cultured at a photosynthetic photon flux (PPF) of 36 μmol $m^{-2}$ $s^{-1}$ on MS medium supplemented with 2% sucrose, 2 mg BAP $l^{-1}$, 0.5 mg 6-furfurylaminopurine (kinetin) $l^{-1}$, 0.1 mg biotin $l^{-1}$ and 0.1 mg calcium pantothenate $l^{-1}$ and solidified with 0.8% agar. The number of shoots of mature nodal cuttings cultured on modified MS medium, supplement with ammonium nitrate at 1 g $l^{-1}$ and the abovementioned growth regulators and addenda, was higher than on MS basal medium or WPM medium. The shoots were incubated in liquid MS medium containing 50 mg IBA $l^{-1}$ in the dark and after 3 days were transferred to MS basal medium in the light, supplemented with 0.25% activated charcoal and solidified with 8% agar. Overall, 40% of the shoots rooted. The rooted plantlets were successfully transferred to soil and planted in a glasshouse and 70% showed rapid growth (Rasai *et al.*, 1994).

Both *A. cherimoya* and *A. muricata* are phenol-reactive plants and are sensitive to wounding. Phenol metabolism is stimulated immediately after the cells of the isolated explant are separated from the mother plant. The phenolic compounds diffuse quickly in the tissue-culture medium and the concentrations produced are toxic for the explant (Debergh and Read, 1991). Addition of polyvinylpyrrolidone in the tissue-culture medium prevented blackening due to phenolic exudation in the media. Cutting explants from semi-woody branches of *A. cherimola* and *A. muricata* promoted direct bud sprouting in an initial optimal culture medium. Nitsch and Nitsch (1969) medium supplemented with 2.32 μM kinetin and 1.36 μM zeatin increased proliferation of shoots in both species, and these were elongated by adding 1.44 μM gibberellic acid ($GA_3$). Both species formed eight new shoots in 60 days of culture, with one subculture after 30 days (Bridg, 2000). Supplementation with 20.6 mM $NH_4^+$ and 39.4 mM $NO_3^-$ promoted good-quality green shoots available for rooting. For rooting, one-quarter strength Nitsch and Nitsch (1969) macro-salts medium supplemented with 1% sucrose and 3% Gelrite was found to improve the rooting. The rooted plants were successfully established in quartz-sand substrate in a greenhouse with 90% relative humidity (Figs 7.1 and 7.2) (Bridg, 2000).

**Fig. 7.1.** *In vitro* response of *Annona cherimola* during multiplication stages and morphogenic events related to root organogenesis. (A, B) Shoot proliferation on Nitsch and Nitsch (1969) medium with 2.32 μM potassium and 1.6 μM zinc. (B) *In vitro* rooting after 70 days on half NN-69 salts supplemented with 10 % sucrose. (C) *Ex vitro* root formation of induced by 1% IBA (Rhizopon) after 50 days. (D) *Ex vitro* adaptation. Photographs courtesy of Dr Hannia Bridg.

**Fig. 7.2.** *In vitro* response of *Annona muricata* during multiplication stages and morphogenic events related to root organogenesis. (A, B) Shoot elongation. (C) *In vitro* rooting after 70 days on half NN-69 salts supplemented with 10 % sucrose. (D) *Ex vitro* root formation of induced by 1% IBA (Rhizopon) after 50 days. (E) *Ex vitro* development. (F) Regeneration after 4 months. (G) Sequence of *ex vitro* growth in quartz-sand pots of *in vitro* regenerant. Photographs courtesy of Dr Hannia Bridg.

## 7.3 Ploidy Manipulation

### 7.3.1 Haploid recovery

Haploid induction from cultured anthers of *A. squamosa* was described by Nair *et al.* (1983). Dissected anthers from 1–1.5 cm flower buds, containing mid-uninucleate and uninucleate pollen grains, formed callus on Nitsch and Nitsch (1969) medium supplemented with 28.6 μM NAA. Callus formation from anthers required a preliminary incubation period of 7 days in darkness. Regeneration of single shoots from callus occurred on Nitsch medium supplemented with 8.9 μM BAP and 5.4 μM NAA. Multiple shoot development occurred on Nitsch medium supplemented with 8.9 μM BA and 0.6 μM indole-3-acetic acid (IAA). The haploid number of regenerated plantlets ($n = x = 7$) was verified.

### 7.3.2 Triploid recovery

Attempts to regenerate triploid ($3n = 3x = 21$) plants from endosperm tissue have had limited success. Shoots or roots but not complete plantlets have been regenerated from mature endosperm of *A. squamosa* (Nair *et al.*, 1983). Endosperm was removed from sterilized mature seeds and explanted 2–4 days after germination. Seeds were pre-treated for 1 h with 0.3 μM $GA_3$ and incubated on White's basal medium supplemented with 2.9 μM $GA_3$, 5.4 μM NAA, 0.9 μM BAP and 0.5 μM kinetin. Triploid shoots were induced from endosperm callus on Nitsch medium supplemented with 8.9 μM BAP and 2.7 μM NAA, and triploid root induction occurred on medium supplemented with 28.6 μM IAA.

### 7.3.3 Tetraploid recovery

Few tetraploid genotypes have been developed from cherimoya and atemoya (e.g. 'Fino de Jete' and '1-464'), using colchicine treatments to duplicate the chromosome number of diploid cherimoya (Encina *et al.*, 2014). This treatment recovered triploid and tetraploid plants of cherimoya and some tetraploid lines from atemoya after evaluation by flow cytometry of the ploidy level of the resultant shoot tissues after antimitotic shocks and treatments.

## 7.4 Flowering and Fruit Development

Flowering is a vital development process in the life cycle of plants. A coordinated transition from vegetative growth to reproductive development is required for successful reproduction in higher plants (Li *et al.*, 2011). As an intricate biological and morphological process, flower development is regulated by a large number of genes (Huang *et al.*, 2013). In some flowering plants, MADS-box genes work as organ identity genes during flower development, and loss-of-function mutations in these genes cause homeotic transformations of floral organs (Schwarz-Sommer *et al.*, 1990). Liu *et al.* (2017) undertook comprehensive characterization and expressing profiling analysis of seven sugar apple MADS-box genes using the method of rapid amplification of cDNA ends. Domain and phylogenetic analyses grouped these seven MADS-box genes into six different clades, which showed high similarity with orthologues in *Arabidopsis*. Expression patterns of these MADS-box genes were investigated during different flower development stages and in various reproductive organs, including petals, stamens, sepals and pistil. Most of the MADS-box genes were found to be least expressed in the sepal, and *AsAGL67* and *AsAGL80* expression was weak in all tissues. *AsSEPI* and *AsAGAMOUS* showed highest expression levels in the stamen and pistil, while *AsAGL12* showed stamen-specific expression. Dynamic expression patterns of MADS-box genes in different reproductive stages suggest involvement in flower development. A number of these MADS-box genes (*AsAGL12*, *AsAGMOUS*) were regulated by the phytohormones gibberellin and abscisic acid (ABA).

A full-length cDNA sequence of the homologous *AGAMOUS* (*AG*) gene was cloned from sugar apple (Liu *et al.*, 2016a). Sequence analysis showed that the *AsAG* gene contains a 669 bp open reading frame encoding 222 amino acids (GenBank accession no. KT159768). The predicted secondary structure showed that the AsAG protein is made up of 59.46% α-helices, 8.11 β-turns, 14.41% extended strands and 18.02% random coils. The deduced amino acid sequence contained a highly conserved MADS domain and a secondary conserved K domain. Subcellular localization showed that the AsAG protein was located in the nucleus. The quantitative real-time polymerase chain reaction (qRT-PCR) results suggested that the *AsAG* gene showed a high transcription level in flowers and could be detected during the entire period of flower development with a high transcription level in the stage of flower bud IV. The expression level in the stamen was higher than that in the pistil, sepal and petal. The expression level in a perfect flower was higher than in a deformed flower. After 2 and 4 h treatment with GA and ABA signal molecules in flower buds, the expression pattern of *AsAG* showed negative regulation by GA and positive regulation by ABA (Liu *et al.*, 2016a). The expression levels of genes involved in auxin transport or signalling were altered in buds during floral initiation in sugar apple. Several known ARFs (auxin response factors) and auxins/IAA were identified, suggesting, that the ARF–auxin/IAA regulatory pathway is vital for flower development in sugar apple (Liu *et al.*, 2016b). In sugar apple, expression of *IAA3* increased during floral transition and flower development, especially in the FL2 flowering stage (Liu *et al.*, 2016b).

In sugar apple, some cultivars have fewer seeds although some of the popular cultivars such as 'Sitaphal' have more seeds per fruit. Presequencing (Roche, 454GS FLX+) of the transcriptome from the early stages of fruit development (0, 4, 8 and 12 days after pollination) in two sugar apple cultivars, 'Sitaphal' and 'NMK-1', was studied by Gupta *et al.* (2015). Approximately 1.9 million high-quality clean reads were obtained in the cDNA library from the developing fruits of both cultivars, with average reads of about 568 bp. Quality reads were assembled *de novo* into 2074–11,004 contigs in the developing fruit samples at different stages of development. The contig sequence data of all four stages of each cultivar were combined into larger units resulting into 14,921 ('Sitaphal') and 14,178 ('NMK-1') unigenes, with a mean size of more than 1 kb. Assembled unigenes were functionally annotated by querying against the protein sequence of five public databases (NCBI non-redundant, *Prunus persica*, *Vitis vinifera*, *Fragaria vesca* and *Amborella trichopoda*), with an E-value cut-off of $10^{-5}$. A total of 4588 ('Sitaphal') and 2502 ('NMK-1') unigenes did not match any known protein in the NCBI non-redundant database. These sequences could be genes specific to *Annona* spp. or belong to untranslated regions. Several of the unigenes representing pathways related to primary and secondary metabolism and seed and fruit development were expressed at a higher level in 'Sitaphal', which is a densely seeded cultivar in comparison to the poorly seeded 'NMK-1'. A total of 2629 ('Sitaphal') and 3445 ('NMK-1') single-sequence repeat motifs were identified. These may be potential candidates for transcript-based microsatellite analysis in *A. squamosa* (Gupta *et al.*, 2015). Rodrigues *et al.* (2024) recently stated that inheritance of the presence or absence of seeds in *A. squamosa* is mediated by a singly fully recessive gene and is caused by a deletion of the *INNER NO OUTER* (*INO*) locus. Segregation analysis of $F_2$ and backcross descendants of 'Brazilian seedless' (Bs) to fertile wild-type varieties in this species with a multi-year generation time showed that seedlessness was controlled by a single locus. Comparison of whole-genome sequences of a wild-type plant and a third accession, 'Hawaiian Seedless', identified a 16 kb deletion including *INO* in this line. Thai Seedless and Bs lines were shown to have an identical deletion, indicating a common origin from a single deletion event. The observed deletion is the only significant difference between the wild-type and 'Hawaiian Seedless' line over 587 kb, probably encompassing more than 3.5 cM, showing that this deletion is the cause of the seedless trait.

## 7.5 Ripening Regulation

Expansions of plant cell-wall proteins disrupts hydrogen bonds within the cell-wall matrix, contributing to cell-wall degradation, and leading to a rapid decrease in fruit flesh firmness (Marowa *et al.*, 2016). The genes for expansin (*EXP*), aldehyde dehydrogenase (*ALDH*) and succinyl CoA ligase (*Succ-CoA*) are involved in important biochemical processes during fruit ripening. Berumen-Varela *et al.* (2020) evaluated the expression profile of genes related to fruit softening, the tricarboxylic acid cycle and acetaldehyde metabolism during ripening of soursop fruit stored at 28 ± 2°C and 15 ± 2°C. The fruit stored at 15 ± 2°C prolonged the postharvest shelf life up to 9 days and showed higher firmness at 3 days compared with those stored at 28 ± 2°C. Postharvest storage at 15 ± 2°C induced expression of the *EXP*, *Succ-CoA* and *ALDH* genes at 3 days, while an increase in expression of *EXP* and *Succ-CoA* was seen at 6 days in fruit stored at 28 ± 2°C. Based on the different gene expression patterns, Berumen-Varela *et al.* (2020) concluded that postharvest storage at 15 ± 2°C triggers expression of the *EXP*, *Succ-CoA* and *ALDH* genes at the early stages of soursop ripening.

The expression patterns of several cell-wall-related genes in cherimoya during postharvest ripening were analysed by qRT-PCR (Liu *et al.*, 2016c). These authors observed highly diversified expression patterns of one *AcPE*, three *ACXET* and three *ACEXP* genes during fruit softening. Postharvest treatment with citric acid and chitosan coating delayed the accumulation of *AcXEP1*, *AcEXP1* and *AcPE* by reducing their mRNA levels. Atemoya fruit has a short shelf life, and splitting of fruit occurs frequently during the postharvest period. Li *et al.* (2019) prepared various cDNA libraries independently for Illumina sequencing. A total of 17,860 genes, including 16,190 upregulated and 1670 downregulated genes, were identified as differentially expressed genes (DEGs). A number of DEGs have been assigned to most of the key components involved in different hormonal signalling pathways, indicating that the splitting process of atemoya fruit may be controlled by an intricate hormonal signalling network. The expression levels of most polygalacturonase, pectinesterase and cellulose-encoding genes increased in the split fruits compared with non-split fruits, which probably contributed to the reduction in skin elasticity. Identification of the DEGs related to phytohormones and cell walls from this study will facilitate our understanding of the regulatory mechanisms underlying the fruit splitting process of atemoya.

## 7.6 Postharvest Microbial Contamination

Annona fruits are highly prone to attack by several pathogens due to their high moisture content and this leads to massive economic loss. Metagenomic analysis of genomic DNA purified from *A. muricata* and *Malus domestica* showed that several bacterial species are associated with the fruit, and this leads to deterioration, rotting or discoloration, which may affect their preservation leading to postharvest fruit losses (Moore *et al.*, 2022). Clustering in the Greengenes database (http://greengenes.lbl.gov, accessed 1 September 2025) revealed 570 operational taxonomic units (OTUs), and α-diversity indices indicated a high diversity and abundance of the microbial community. Taxonomic analysis grouped the bacterial community into 24 phyla and 455 genera. The microbiome of the samples found was dominated by distinct populations of four phyla: Proteobacteria (54.81%), Bacteroidetes (18.59%), Actinobacteria (11.13%) and Firmicutes (7.29%). Functional genes were predicated for 16S rRNA gene sequences based on information from the Kyoto Encyclopedia of Genes and Genomes (KEGG), which indicated amino acid metabolism, carbohydrate metabolism, xenobiotic biodegradation, lipid metabolism and biosynthesis of other secondary metabolites as predominant metabolic categories. This study revealed that *A. muricata* fruit can host several human, plant and soil pathogens, and proper identification of bacterial organisms using metagenomics is a step towards proffering ways to prevent or control microorganisms on the fruit (Moore *et al.*, 2022).

## 7.7 Reference Genome

Annonaceae is an important family in the Magnoliidae clade, presenting morphological diversity with high economic value, and distributed widely in tropical and subtropical forests. Recently published results strongly disagree on the position of magnoliids, supporting either a sister relationship to eudicots and monocots (Chen *et al.*, 2019; Hu *et al.*, 2019; Rendón-Anaya *et al.*, 2019) or to eudicots alone (Chaw *et al.*, 2019).

Chromosome-level genome assembly of soursop was reported by Strijk *et al.* (2021), who assembled the genome using 444.32 Gb of DNA sequences (676× sequencing depth) from PacBio and Illumina short-reads, in combination with 10× Genomics and Bionano data (v.1). A total of 949 scaffolds were assembled with a final size of 656.77 Mb (v.1: scaffold N50 of 3.43 Mb) and then further improved to seven pseudo-chromosomes using Hi-C sequencing data (v.2: scaffold N50 of 93.2 Mb, total size in chromosomes: 639.6 Mb). They observed very low heterozygosity (0.06%), while repeat sequences accounted for 54.87% of the genome, and 23,375 protein-coding genes with an average of 4.79 exons per gene were annotated using *de novo*, RNA-sequencing and homology-based approaches. Reconstruction of the historical population size showed a slow continuous contraction, probably related to Cenozoic climate changes. Soursop was the first genome in the Annonaceae to be assembled, supporting further studies of floral evolution in magnoliids, providing an essential resource for delineating relationships of ancient angiosperm lineages (Strijk *et al.*, 2021).

Chromosome-level genome assembly of cherimoya 'Fino de Jete' was developed by Talavera *et al.* (2023). A combination of different sequencing technologies (Illumina and Pacific Biosciences) was implemented in order to assemble the reference genome. The reference genome resulted in an assembly of 1.13 Gb and N50 of 170.86 Mb, anchored into seven pseudo-molecules and with a completeness of 95.6%. A total of 41,413 protein-coding genes were identified, many of which were found to be related to secondary metabolism, defence mechanisms, the stress response and development.

A chromosome-level genome assembly for cherimoya 'Booth' was developed by Li *et al.* (2024). The genome assembly has a size of 794 Mb with an N50 of 97.59 Mb. The seven longest scaffolds account for 87.6% of the total genome length, which corresponds to the seven pseudochromosomes. A total of 45,272 protein-coding genes (≥30 amino acids) were predicted with 92.9% gene content completeness. Phylogenetic analysis suggests that eudicots and magnoliids are more closely related to each other than to monocots. Moreover, the order Magnoliales has been found to be more closely related to the order Laurales than to the order Piperales. Genome comparison revealed that cherimoya 'Booth' has 200 Mb repeats than the Spanish cultivar 'Fino de Jete', despite their highly similar (>99%) genome sequence identify and collinearity. The high heterozygosity rate observed in the 'Booth' (1.14%) and 'Fino de Jete' (1.05%) genomes suggest that a high-quality phased haploid genome will be needed in the future. These two cultivars diverged during the early Pleistocene (1.93 Mya), which suggests a different origin and domestication of the cherimoya.

The chloroplast genome of soursop has been sequenced, assembled and annotated (Niu *et al.*, 2020). It was found to be a double-stranded ring structure with a size of 196,038 bp consisting of four regions: a large single-copy region of 75,339 bp, a small single-copy region of 3105 bp and two inverted-repeat regions of 58,797 bp. The GC content of the whole chloroplast genome was 39.92%. A total of 111 protein-coding genes, one pseudogene, 38 tRNA genes and eight rRNA genes were annotated in the chloroplast genome, and the total number of genes was 158. DNA sequences of the chloroplast genomes of 19 species, which belong to three families in the order Magnoliales, were analysed and a phylogenetic tree was constructed. The results indicated that *A. muricata*, *A. cherimola*, *Uvaria macrophylla*, *Greenwayodendron suaveolens* and *Cheirodendron hainanense* have a close phylogenetic relationship (Niu *et al.*, 2020).

The mountain soursop (*Annona montana*) is found mainly in western South America and also grows in China and India. Tang *et al.* (2023) reported a high-quality genome of *A. montana* obtained using the Pacific Biosciences sequencing platform and High-resolution Chromosome Conformation Capture (Hi-C) technology. These authors observed a low level of heterozygosity in the *A. montana* genome corresponding to a genome size of 1.09 Gb. They assembled 974.35 Mb of the genome with a contig N50 value of 7.89 Mb. The completeness of the assembled genome was 90.80% based on analysis of Benchmarking Universal Single-Copy Orthologues (BUSCO), and 92.12 Gb of clean data was obtained using a sequencing library of genome-wide chromosome conformation capture (Hi-C). A total of 973.54 Mb (99.92%) of the genome was anchored, which was mapped to seven pseudochromosomes, the length of which ranged from 90.58 to 188.31 Mb. They recognized 32 genes involving starch degradation and 193 genes encoding ten key enzymes related to cell-wall degradation. These genes may be related to the softening of *A. montana* fruit.

## 7.8 Cryopreservation

Cherimoya pollen has a short viability and high sensitivity to desiccation and consequently conservation is problematic. Due to the protogynous dichogamy of cherimoya and the absence of proper pollinating vectors, hand pollination with fresh pollen is a common practice. Lora *et al.* (2006) studied the conservation of cherimoya pollen at −20°C, −80°C and −196°C for up to 3 months. *In vitro* germination of fresh pollen was 57.1% (4°C) and progressively reduced with conservation time at the three temperatures, reaching a minimum after 3 months of storage of 10.4%, 14.2% and 13.6% at −20°C, −80°C and −196°C, respectively. Differences in germination among temperatures were only significant during the first 2 weeks of storage. Field pollination with pollen stored for up to 3 months at the three temperatures showed no yield differences compared with pollination performed with fresh pollen. The results indicate that pollen collected and stored at subzero temperatures at the start of the cherimoya blooming season can be used throughout the blooming season, avoiding the need to collect fresh pollen daily. Chander *et al.* (2019) observed that pollen of *A. squamosa* 'Balanagar' (collected on the second day of anthesis) showed maximum germination (43.47%) in 5% sucrose and 100 p.p.m. boric acid, while that collected on the first day of anthesis did not show any germination. Pollen stored under varying temperatures lost viability within 1 month, while pollen stored at −196°C in liquid nitrogen retained germination for 2 months. Mountain soursop (*A. montana*) seeds, which is intermediate in nature, having 6–22% moisture content, show 100% seed viability, but this decreased to 80% after stepwise freezing. In stepwise freezing, seeds were immersed in cryotubes containing vitrification solution composed of 30% glycerol, 0.4 M sucrose, 15% ethylene glycol and 15% dimethyl sulfoxide per litre of MS solution and placed in an ice bath for 30 min. Finally, the solution was rapidly replaced with pre-cooled fresh vitrification solution and the tube was immediately transferred into liquid nitrogen for ultra-low-temperature storage (Zeng *et al.*, 2024).

## 7.9 Conclusions

Annonas show a high degree of heterozygosis, and to obtain homogeneous and productive orchards, it is necessary to avoid propagation by seed. *In vitro* tissue-culture methods can be applied to *Annona* spp. to overcome this problem.

The MADS-box gene family plays an important role in several plant processes, such as floral development, flowering time control and fruit ripening regulation. Forty-six MADS-box genes have been identified in the *A. montana* genome. The progress in developing genomic resources for *Annona* spp. lags behind what has been achieved for temperate fruit trees and annual crops (Li *et al.*, 2024). While the first sequenced genome in the genus *Annona*, *A. muricata*, was published in 2021 (Strijk *et al.*, 2021), only raw DNA reads are available in GenBank, and its genome assembly is still unavailable. Recently, the *A. cherimola* genome was analysed from the Spanish cultivar 'Fino de Jete' (Talavera *et al.*, 2023) and the American cultivar 'Booth' (Li *et al.*, 2024). The work of Li *et al.* (2024) with 'Booth' has revealed insights into the intraspecific divergence and the evolution of the Magnoliales. Both the *A. muricata* and *A. cherimola* genomes are crucial resources for boosting breeding and identifying new markers associated with traits of interest.

## References


Abubacker, N. and Deepalakshmi, T. (2017) *In vitro* direct regeneration of *Annona muricata* L. from nodal explant. *Biosciences Biotechnology Research Asia* 14, 123–128. DOI: 10.13005/bbra/2426

Amin, M.N., Nahar, K., Ahmed, F. and Ahmad, S. (2002) Micropropagation of *Annona squamosa* L. using explants (shoot tip and node) of field grown mature plants. *Pakistan Journal of Biological Sciences* 5, 394–397. DOI: 10.3923/pjbs.2002.394.397

Ba, O., Dieme, A., Ndoye, A.L. and Sy, M.O. (2021) *In vitro* clonal propagation from juvenile and different explant types of two edible Annonaceae species: *Annona muricata* L. and *Annona squamosa* L. *Advances in Bioscience and Biotechnology* 12, 458–480. DOI: 10.4236/abb.2021.1212029

Bejoy, M. and Hariharan, M. (1992) *In vitro* plantlet differentiation in *Annona muricata*. *Plant Cell Tissue and Organ Culture* 31, 245–247. DOI: 10.1007/BF00036231

Berumen-Varela, G., Martínez-González, E.M., Palomino-Hermosillo, Y.A., Jiménez-Zurita, J.O., Pena-Sandoval, G.R. *et al.* (2020) Expression profile of *EXP*, *Succ-CoA*, and *ALDH* genes in soursop (*Annona muricata* L.) fruits during ripening in response to refrigeration conditions. *Indian Journal of Biotechnology* 19, 254–262.

Bridg, H. (2000) Micropropagation and determination of the *in vitro* stability of *Annona cherimola* Mill. and *Annona muricata* L. PhD thesis, Humboldt University of Berlin, Germany.

Chander, S., Rajasekharan, P.E. and Kurian, M. (2019) Pollen storage studies in sugar apple (*Annona squamosa* L.) cv. Balanagar. *Israel Journal of Plant Sciences* 66, 196–202. DOI: 10.1163/22238980-20191080

Chaw, S.-M., Liu, Y.-C., Wu, Y.-W., Wang, H.-Y., Lin, C.-Y.I. *et al.* (2019) Stout camphor tree genome fills gaps in understanding of flowering plant genome evolution. *Nature Plants* 5, 63–73. DOI: 10.1038/s41477-018-0337-0

Chen, J., Hao, Z., Guang, X., Zhao, C., Wang, P. *et al.* (2019) *Liriodendron* genome sheds light on angiosperm phylogeny and species-pair differentiation. *Nature Plants* 5, 18–25. DOI: 10.1038/s41477-018-0323-6

Debergh, P.C. and Read, P.E. (1991) *Micropropagation*. In: Debergh, P.C. and Zimmerman, R.H. (eds) *Micropropagation: Technology and Application*. Kluwer Academic Publisher, Netherlands, pp. 1–13.

Encina, C., Barcelo-Munoz, A., Herrero-Castano, A. and Pliego-Alfaro, F. (1994) *In vitro* morphogenesis of juvenile *Annona cherimola* Mill. bud explants. *Journal of Horticultural Science* 69, 1053–1109. DOI: 10.1080/00221589.1994.11516544

Encina, C.L., Martin, E.C., Lopez, A.A. and Padilla, I.M.G. (2014) Biotechnology applied to *Annona* species: a review. *Revista Brasileira de Fruticultura* 36, 17–21. DOI: 10.1590/S0100-29452014000500002

Gupta, Y., Pathak, A.K., Singh, K., Mantri, S.S., Singh, S.P. *et al.* (2015) *De novo* assembly and characterization of transcriptomes of early-stage fruit from two genotypes of *Annona squamosa* L. with contrast in seed number. *BMC Genomics* 16: 86. DOI: 10.1186/s12864-015-1248-3

Hu, L., Xu, Z., Wang, M., Fan, R., Yuan, D. *et al.* (2019) The chromosome-scale reference genome of black pepper provides insight into piperine biosynthesis. *Nature Communications* 10: 4702. DOI: 10.1038/s41467-019-12607-6

Huang, Y.J., Liu, L.L., Huang, J.Q., Wang, Z.J., Chen, F.F. *et al.* (2013) Use of transcriptome sequencing to understand the pistillate flowering in hickory (*Carya cathayensis* Sarg.). *BMC Genomics* 14, 1471–2164. DOI: 10.1186/1471-2164-14-691

Lemos, E.E.P. and Baker, D.A. (1998) Shoot regeneration in response to carbon source on internodal explants of *Annona muricata* L. *Plant Growth Regulation* 25, 105–112. DOI: 10.1023/A:1006058301707

Lemos, E.E.P. and Blake, J. (1996a) Micropropagation of juvenile and adult *Annona squamosa*. *Plant Cell, Tissue and Organ Culture* 46, 77–79. DOI: 10.1007/BF00039698

Lemos, E.F.P. and Blake, J. (1996b) Micropropagation of juvenile and mature *Annona muricata* L. *Journal of Horticultural Science* 71, 395–403. DOI: 10.1080/14620316.1996.11515420

Lemos, E.E.P. and Blake, J. Lumsden, P.J., Nicholas, J.R. and Davies, W.J. (eds). (1994) *Leaf Abscission in Micropropagated Sugar Apple (Annona squamosa L.)*. In: Lumsden, P.J., Nicholas, J.R. and Davies, W.J. (eds) *Physiology, Growth and Development of Plants in Culture*. Springer, Dordrecht, Netherlands, pp. 227–223.

Li, C., Gu, M., Shi, N., Zhang, H., Yang, X. *et al.* (2011) Mobile FT mRNA contributes to the systemic florigen signalling in floral induction. *Scientific Reports* 1: 73. DOI: 10.1038/srep00073

Li, H., Li, W., Zhang, T., Zhong, J., Liu, J. *et al.* (2019) Comparative transcriptomic analysis of split and non-split atemoya (*Annona cherimola* Mill. × *Annona squamosa* L.) fruit to identify potential genes involved in the fruit splitting process. *Scientia Horticulturae* 248, 216–224. DOI: 10.1016/j.scienta.2019.01.017

Li, T., Zheng, J., Nousias, O., Yan, Y., Meinhardt, L.W. *et al.* (2024) The American cherimoya genome reveals insights into the intra-specific divergence, the evolution of *Magnoliales*, and a putative gene cluster for acetogenin biosynthesis. *Plants* 13: 636. DOI: 10.3390/Plants13050636

Liu, K., Li, H., Zhong, S., Yuan, C., Chen, Y. *et al.* (2016a) Cloning, subcellular localization and expression analysis of *AsAG* from sugar apple (*Annona squamosa* L.). *Scientia Agricultura Sinica* 49, 142–154.

Liu, K., Feng, S., Pan, Y., Zhong, J., Chen, Y. *et al.* (2016b) Transcriptome analysis and identification of genes associated with floral transition and flower development in sugar apple (*Annona squamosa* L.). *Frontiers in Plant Science* 7: 1695. DOI: 10.3389/fpls.2016.01695

Liu, K., Liu, J., Li, H., Yuan, C., Zhong, J. and Chen, Y. (2016c) Influence of postharvest citric acid and chitosan coating treatment on ripening attributes and expression of cell wall related genes in cherimoya (*Annona cherimola* Mill.) fruit. *Scientia Horticulturae* 198, 1–11. DOI: 10.1016/j.scienta.2015.11.008

Liu, K., Feng, S., Jiang, Y., Li, H., Huang, S. *et al.* (2017) Identification and expression analysis of seven MADS-box genes from *Annona squamosa*. *Biologia Plantarum* 61, 24–34. DOI: 10.1007/s10535-016-0688-1

Lora, J., Oteyza, M.A.P., Fuentetaja, P. and Hormaza, J.I. (2006) Low temperature storage and *in vitro* germination of cherimoya (*Annona cherimola* Mill.) pollen. *Scientia Horticulturae* 108, 91–94. DOI: 10.1016/j.scienta.2005.12.003

Marowa, P., Ding, A. and Kong, Y. (2016) Expansins: roles in plant growth and potential applications in crop improvement. *Plant Cell Reports* 35, 949–965. DOI: 10.1007/s00299-016-1948-4

Mathur, N., Ramawat, K.G. and Dilip, N. (1995) Rapid in vitro multiplication of jujube through mature stem explants. *Plant Cell, Tissue and Organ Culture* 43, 75–77. DOI: 10.1007/bf00042675

Moore, O., Ataga, A. and Ogbuji, N. (2022) Comparative bacterial metagenomics of soursop (*Annona muricata* L.) and apple (*Malus domestica* B.). *Plant Science Today* 9, 891–899. DOI: 10.14719/pst.1749

Nair, S., Gupta, P.K. and Mascarenhas, A.F. (1983) Haploid plants from *in vitro* anther culture of *Annona squamosa* L. *Plant Cell Reports* 2, 198–200. DOI: 10.1007/BF00270103

Nitsch, J.P. and Nitsch, C. (1969) Haploid plants from pollen grains. *Science* 163, 85–87. DOI: 10.1126/science.163.3862.85

Niu, Y.F., Li, K.X. and Liu, J. (2020) The complete chloroplast genome of *Annona muricata* L.: a tropical fruit with important medicinal properties. *Mitochondrial DNA B: Resources* 5, 3330–3332. DOI: 10.1080/23802359.2020.1820394

Padilla, I.M.G. and Encina, C.L. (2005) Changes in root morphology accompanying mycorrhizal alleviation of phosphorus deficiency in micropropagated *Annona cherimola* Mill. plants. *Scientia Horticulturae* 106, 360–369. DOI: 10.1016/j.scienta.2005.05.001

Pinto, A.C. de. Q., Corderio, M.C.R., de Andrade, S.R.M., Ferreira, F.R., Filgueiras, H.A. de C. *et al.* (2005) Annona *Species*. International Centre for Underutilized Crops, University of Southampton, UK.

Rasai, S., Kantharajah, A.S. and Dodd, W.A. (1994) The effect of growth-regulators, source of explants and irradiance on *in vitro* regeneration of atemoya. *Australian Journal of Botany* 42, 333–340. DOI: 10.1071/BT9940333

Rendón-Anaya, M., Ibarra-Laclette, E., Méndez-Bravo, A., Lan, T., Zheng, C. *et al.* (2019) The avocado genome informs deep angiosperm phylogeny, highlights introgressive hybridization, and reveals pathogen-influenced gene space adaptation. *Proceedings of the National Academy of Sciences* 116, 17081–17089. DOI: 10.1073/pnas.1822129116

Rodrigues, B.R.A., Gasser, C.S., Pimenta, S., Pereira, M.C.T. and Nietsche, S. (2024) Seedless fruit in *Annona squamosa* L. is monogenic and conferred by *INO* locus deletion in multiple accessions. *Plant Reproduction* 37, 71–84. DOI: 10.1007/s00497-023-00464-9

Schwarz-Sommer, Z., Huijser, P., Nacken, W., Saedler, H. and Sommer, H. (1990) Genetic control of flower development by homeotic genes in *Antirrhinum majus*. *Science* 250, 931–936. DOI: 10.1126/science.250.4983.931

Strijk, J.S., Hinsinger, D.D., Roeder, M.M., Chatrou, L.W., Couvreur, T.L.P. *et al.* (2021) Chromosome-level reference genome of the soursop (*Annona muricata*): A new resource for magnoliid research and tropical pomology. *Molecular Ecology Resources* 21, 1608–1619. DOI: 10.1111/1755-0998.13353

Talavera, A., Fernandez-Pozo, N., Matas, A.J., Hormaza, J.I. and Bombarely, A. (2023) Genomics in neglected and underutilized crops: a chromosome-scale genome sequence of cherimoya (*Annona cherimola*). *Plants, People, Planet* 5, 408–423. DOI: 10.1002/ppp3.10366

Tang, G., Chen, G., Ke, J., Wang, J., Zhang, D. *et al.* (2023) The *Annona montana* genome reveals the development and flower formation in mountain soursop fruit. *Ornamental Plant Research* 3: 14. DOI: 10.48130/OPR-2023-0014

Zeng, L., Sun, Z., Fu, L., Gu, Y., Li, R. *et al.* (2024) Cryopreservation of medicinal plant seeds: strategies for genetic diversity conservation and sustainability. *Plants* 13: 2577. DOI: 10.3390/plants13182577

# 8 Orchard Management

**Sisir Mitra***
*Former Professor, Faculty of Horticulture, Bidhan Chandra Krisi Viswavidyalaya (State Agricultural University), Mohanpur, West Bengal, India*

## 8.1 Introduction

The performance of an orchard depends on its management, which includes water and nutrient management, selection of the correct cultivars, planting technique, plant density, canopy management for flowering and fruiting, and improved light efficiency through pruning to optimize the size and shape of the trees. Under normal planting density spacing (5–10 × 7–12 m), the interspace may be utilized by growing short-duration intercrops to make the orchard productive in the early years (years 2–4) of planting. Annonas are grown mainly as a rainfed crop in most places. In such situations, mulching is an important and useful practice. Mulches are important in conserving soil moisture, maintaining root-zone temperature, controlling weeds and mitigating soil erosion (Mitra and Pathak, 2021).

## 8.2 Soil

All *Annona* spp. are capable of growing in a wide range of soil types, from sandy soil to clay loams. Nevertheless, they differ in their soil preferences. Higher yields occur on more well-drained sandy to sandy-loam soils, except for *Annona glabra*, which can succeed in shallow ponds. Drainage is essential to avoid root-rot diseases, hence the interest in *A. glabra* as a rootstock because of its tolerance of wet soils (Nakasone and Paull, 1998). In Spain, *Annona cherimola* (cherimoya) is cultivated in deep, alluvial soil with a high lime content or in terraces with soil of metamorphic origin. It tolerates alkalinity to a certain degree, growing well in soils with of pH of 7.5–8.5 and with carbonate content up to 29–30% without showing much iron chlorosis (Farré *et al.*, 1999). Slightly acidic (pH 6.5) to slightly alkaline (pH 7.5) soils are considered the most suitable for cherimoya cultivation (Pinto *et al.*, 2005; Duchi Albarracín, 2017). *Annona reticulata* (custard apple) trees adapt to various types of soil, but prefer well-drained sandy clay with a good depth and a pH of 5.5–6.5 (Vieira *et al.*, 1994).

*Annona muricata* (soursop) prefers slightly acidic to neutral soil for optimal growth. Soils with a pH range of 5.5–6.5 are optimally suitable for soursop growth, while those with a pH of 5.0–5.5 are moderately suitable and very acidic soils with a pH <5.0 are marginally suitable. Soils with a light or sandy texture such as loam, sandy loam, loamy sand and sandy clay loam are optimal for soursop (Kome *et al.*, 2024).

* Email: sisirm55@gmail.com



DOI: 10.1079/9781800627628.0008

## 8.3 Windbreaks

Annonas are shallow-rooted plants and susceptible to wind damage. Planting of windbreak trees is an important orchard practice, particularly during the first 3 years of planting. The windbreak trees should be flexible, long lived and well adapted to the site where they are being considered for planting. Major windbreaks commonly used in Australia are cadaga (*Eucalyptus torelliana*) and Dunn's white gum (*Eucalyptus dunnii*) (Sanewski, 1988). Casuarina (*Casuarina equisetifolia*) has been used in some regions of Brazil as a windbreak, although this species is very competitive for water and is susceptible to natural combustion, thus offering a fire risk (Pinto *et al.*, 2005).

## 8.4 Pruning

Cherimoya (*Annona cherimola*) and atemoya (*A. cherimola* × *A. squamosa*) trees are typically trained and pruned annually in some production areas in Australia, New Zealand, California, Chile and Spain (Crane *et al.*, 2004). Cherimoya and atemoya are generally trained to open-centre, open goblet or modified central-leader forms, thus avoiding many of the problems with limb breakage, fruit scarring and loss of the lower canopy that may occur with non-trained trees (George *et al.*, 1987; Sanewski, 1988). Agustin and Alviter (1996) described different pruning methodologies for cherimoya involving the maintenance of two or three principal branches. Pruning using the two-branch system should begin between month 4 and 5 after transplanting. Two principal branches should be selected according to their vigour and others should be removed establishing a V-shaped training system (resembling two fingers in an open position). The three-branch system requires the complete removal of the top part of the tree leaving 90 cm of the main trunk, which encourages the production of lateral branches (Nakasone and Paull, 1998). The lateral branches should be at angles of approximately 120° from each other (Agustin and Alviter, 1996). The new branches should be spaced at 15–25 cm above each other in different directions to develop a good scaffold by year 5 after planting, when the plant is about 2 m tall (Nakasone and Paull, 1998).

In Spain, cherimoya trees were traditionally trained in an open-vase system with three or four main branches maintained at a height of 1.5–2 m above the ground. This height allowed cultivation to be performed in between the plants using draught animals, but since the no-tillage approach was introduced, the branches have been allowed to grow at a lower height. Every year, at the end of winter and before new growth begins, the fruit branches and the growing points of the main branches are removed. During the vegetative period, suckers are clipped or removed (Farré *et al.*, 1999).

For trees with a single trunk where the canopy is too tall for easy tree management and productivity is low, total or drastic pruning is required. Cutting of the tree back to a height of 40–50 cm above ground is recommended. Subsequent regrowth should be trained. Shoot selection must be done leaving four well-spaced branches, which can be grafted with scions from a superior cultivar when the shoots become woody or reach a diameter of 1.5 cm. It is not necessary for new shoots of pruned trees of good cultivars to be grafted. The cut-back plants will fruit again in year 2 after pruning (Anon., 2008).

The main harvest season of cherimoya in Spain is autumn (October–November), but over-supply in the market causes a dramatic price drop. Soler and Cuevas (2008) developed a technique to produce cherimoya in winter. The trees were pruned back severely in January, followed by tipping the newly emerged shoots together with removal of leaves on either 1 or 15 June. Both dates were highly successful because all tipped shoots produced a second bloom in July. The second bloom lasted for about 2 weeks. Late-forming flowers were hand pollinated, which produced high-quality fruit about 50 days later than the ordinary crop. The late fruit also suffered less preharvest drop than an ordinary crop. The yield of the treated tree did not differ from that of the ordinary crop (~13 t $ha^{-1}$). The tip pruning and defoliation delayed harvesting to December and January when the market price is almost double compared with the main season. However, Gonzalez Vega (2013) stated that the lowest quality and largest fruit of cherimoya are produced on branches that are 2 or more years old but in smaller numbers.

Pruning is an important tree management operation of sugar apple (*A. squamosa*), as bearing sugar apple trees seem to deteriorate if not pruned every year (Coronel, 1994). The heading back of the branches at the onset of dormancy to rejuvenate sugar apple trees will result in resumed production after 2 years (Coronel, 1994). Pruning of sugar apple should begin with 1-year-old branches, cutting them back to 10 cm. Pruning should leave 120–150 branches per tree; root pruning is not recommended (Nakasone and Paull, 1998). Flower initiation begins at the basal end of the growing branches (Lo, 1987). In Taiwan, normal pruning is done in January–February with fruit thinning in June–October of selected trees, which leads to harvesting of fruit from October to the following March (Yang, 1987). Higher winter fruiting occurs when summer shoots are pruned, compared with pruning unfruiting shoots or pruning in late May (Nakasone and Paull, 1998).

In Thailand, sugar apple trees are pruned every year to produce new fruiting shoots. This operation consists of removing of all small shoots and heading back larger shoots to about 10–15 cm long (Coronel, 1994). The operation is carried out during the dry season just prior to shedding of the leaves; the plants are then irrigated to promote new shoots and irrigation is continued until the rainy season starts. The same procedure is followed for sugar apple in north-eastern Brazil (Pinto *et al.*, 2005). In São Paulo, Brazil, summer pruning is done from January to March and the fruit are harvested from August to October, as late pruning (after March) can promote flowering during periods of low temperature, leading to reduced fruit set (Kavati and Piza, 1997). Cultivars of cherimoya and sugar apple, which have strong apical dominance, should be pruned by application of chemicals that cause defoliation, as light pruning may not be sufficient to release many of the buds from induced dormancy (George *et al.*, 1987). A mixture of 250 g urea and 1 g ethephon, plus wetting agent, per 100 l of water successfully defoliated and induced buds. Two defoliation periods are possible: (i) at 5–10% bud break, when the new shoots are about 3–5 cm long; and (ii) at mid-season (first week of January in Australia), applied to non-bearing trees, usually 2–3 years of age (George *et al.*, 1987).

In India, *A. squamosa* (sugar apple, but often referred to as custard apple in India) flowers are borne on the current season's growth as well as on the previous season's growth. Pruning facilitates a proper canopy structure for the plant and helps in early flowering and fruiting. The intensity and time of pruning in sugar apple depends on cultivar, crop age and region (Choudhary and Meena, 2020). It is important to start pruning and training sugar apple from planting. The trees should be pruned to an open-vase shape and this shape should be maintained for the first 4 years. The canopy is then kept open, so that light can penetrate and achieve fruit distribution throughout the canopy (Fig. 8.1). Working with sugar apple 'Balanagar' in Karnataka state, Sonika *et al.* (2023) suggested pruning of 30 cm from the tip of the branch in January to encourage good vegetative growth and maximum yield. Jadhav *et al.* (2022) suggested light (3–5 mm branch thickness) to medium (6–10 mm branch thickness) pruning in the last week of May in Maharashtra. Pruning caused early harvesting with less-stony fruit. Light pruning of shoots (20 cm from tip) after harvesting was reported to be beneficial for improving the yield and fruit quality of sugar apple 'Dharur-6' (Kadam *et al.*, 2018). If not pruned regularly, the trees become overcrowded and unproductive (Fig. 8.2A). The tree needs hard pruning to open up the canopy for penetration of light and air (Fig. 8.2B). These trees become productive again and bear a good crop in the following year (Fig. 8.2C). In Brazil, the recommended practice is to prune 3-month-old branches when they are at least two-thirds of their final length, leaving 8–12 buds. In the first 2 years after planting, pruning three times per year was suggested, with two prunings the following years 6 months apart (Cavalcanti, 1973). In India, sugar apple 'VNR Madhur' developed by VNR Seeds, Raipur, is commercially grown in the state of Chhattisgarh and adjoining areas over approximately 160 ha. The trees are usually trained using an espalier system (Fig. 8.3). The system provides the maximum utilization of land and provides easier access for pruning and harvesting. The fruit can be harvested in the first year after planting as the plants are supported by wires so do not collapse under the weight of the fruit. The cultivar 'VNR Madhur' produces around 11 t $ha^{-1}$ in year 3 after planting compared with 8 t $ha^{-1}$ using the open-centre system.

Soursop trees naturally produce a symmetrical that is canopy well adapted to a central-leader system; therefore, a good alternative for

**Fig. 8.1.** (A) *Annona squamosa* orchard trained as an open-centre system at Taiwan. Photograph courtesy of Dr Sisir Mitra. (B) *A. squamosa* orchard trained as an open-centre system at the Indian Institute of Horticultural Research (IIHR), India, with mulching and a drip irrigation system. Photograph courtesy of Dr Binu Mathew.

**Fig. 8.2.** (A) Overcrowded branches need hard pruning. (B) Tree after hard pruning. (C) Fruit developing in a pruned tree. Photographs courtesy of Dr N.K. Meena.

**Fig. 8.3.** (A) Three-tier espalier system. (B) Trees trained into an espalier system. (C) Bearing espalier trees. Photographs courtesy of VNR Seeds, India.

soursop tree formation is to develop a mushroom-shaped tree that is topped at a height of 2.0–2.6 m (Nakasone and Paull, 1998). When the tree is properly trained, little pruning is required, except to thin out poorly placed and weak branches, as well as natural pruning of the longest branches extending horizontally and vertically (Nakasone and Paull, 1998). In Mexico, top pruning for maintenance of bearing trees was suggested after each harvest. A single cut 10 cm below the tip of each shoot helps to generate a greater number of productive branches (Reyes-Montero *et al.*, 2018).

Pruning for rejuvenation is done in old soursop orchards, when the semi-abandoned tall trees have excessive branches with low yield capacity. All vertical branches are eliminated including the thick lateral and terminal ones in order to attain a plant height of 2.5–3.0 m, and then the new lateral branches are trained into horizontal positions at a later date (Pinto *et al.*, 2005).

Atemoya trees produce narrow crotch angles, which can lead to breaking of the limbs in untrained trees under high-wind or icy conditions, when overhead irrigation is used for cold protection or when there is a heavy fruit load (Marler and Crane, 1994). Therefore, some training of the canopy is necessary when trees are young to induce wide crotch angles (Campbell and Phillips, 1980). Marler and Crane (1994) investigated the effect of increasing the scaffold branch angles of atemoya by manipulation of the primary bud complex. Pruning the lateral branches to a stub (<1 cm) was found to be more effective in inducing regrowth and increasing branch angle of the regrowth of atemoya 'Gefner' than stripping the lateral branches by hand. The numerous meristems within the primary bud complex at each node of the shoots allow removal of the primary lateral branch to obtain regrowth at the same node. A mean 15° increase in crotch angle can be expected to accompany the subsequent regrowth from a node after pruning the previous lateral branch to a stub. The largest increase in attachment angle may be obtained from pruning off primary branches with narrow attachment angles. The removal procedure can be continued with the secondary lateral branch to obtain a further increase in crotch angle for the tertiary lateral branch (Marler and Crane, 1994).

The annona cultivar 'Arka Shan', an interspecific hybrid of atemoya × *A. squamosa* developed in India, comes to harvest during August–September in a mild tropical climate. However, the monsoon rain in this period causes poor fruit quality and high susceptibility to anthracnose and fruit fly infestation. Chander *et al.* (2022) successfully developed a technique for an advancing fruiting season by pruning. The effect of three pruning levels (25%, 50% and 75% of the previous season's growth) at five different times (60, 75, 90, 105 and 120 days after fruit harvest of the previous crop) was analysed. They concluded that fruiting in 'Arka Sahan' could be advanced by 8–9 weeks to harvest in June with comparable or better fruit quality by pruning 75% of the last season's growth during October.

In Australia, hedgerows and the Maroochy V-trellis were compared with vase-trained atemoya cultivar 'KJ Pinks'. 'KJ Pinks' sets 50% of its flowers compared with older cultivars, which set less than 3%. This variety is very suitable for a high-density training system such as the Maroochy V-trellis and hedgerow system (Broadley *et al.*, 2008). The V-trellis is a support system constructed at 30° from the vertical with an internal angle between the two supports of 60°. Each V-trellis supports two rows of trees 1.2 m apart. The rows run north–south within the rows, and the trees are 4.5 m apart. Adjacent V-trellises are 5.5 m apart. The system accommodates about 750 trees $ha^{-1}$. There are four wires spaced 50 cm apart on each arm of the V-trellis (Fig. 8.4). The trees are grown with a central leader at 30° from the vertical (i.e. parallel to the support) with the branches espaliered along the wires (Stacey *et al.*, 2008). Four-year-old 'KJ Pinks' trees on the Maroochy V-trellis system yielded 38.6 kg per tree under exclusion nets and 27.3 kg per tree without netting. The advantage of growing on the V-trellis are: (i) ease of harvesting, with a subsequent reduction in labour costs; (ii) ease of pruning; (iii) ease of spraying, with more effective spray coverage; and (iv) less damage to trees (e.g. limb breakage) and fruit from wind and movement of people during routine orchard maintenance and harvesting operations (Stacey *et al.*, 2008). Based on the initial observations, another experiment was set up in 2004 to compare three training systems: (i) vase (standard) with 300–400 trees $ha^{-1}$; (ii) hedgerow with 600 trees $ha^{-1}$ and mechanically pruned; and (iii) Maroochy V-trellis with 800 trees $ha^{-1}$ (Broadley *et al.*, 2008) using 'KJ Pinks'. The open-vase system is a free-standing tree form. Trees trained in the open-vase system are large and are planted at low densities (200–300 trees $ha^{-1}$) and yield 12–18 t $ha^{-1}$. Trees trained with this system usually have very low (<5%) foliar light interception, which may negatively affect fruit quality. The hedgerow system is a linear version of the vase system. Trees in this system are planted with 4.5 m between trees and 6 m between rows. The tree spreads out along the row until it shares space with adjacent trees, essentially forming a hedge. The trees can be mechanically pruned, which reduces harvesting and maintenance costs by 30–40%. The 3-year-old 'KJ Pinks' atemoya produced 21.5, 13.9 and 12.6 kg per tree using the vase, hedgerow and Maroochy V-trellis systems, respectively (Broadley *et al.*, 2008).

**Fig. 8.4.** The open V-trellis system trained on a four-wire trellis. The trees are 3 years old. Photograph courtesy of Dr Grant Bignell.

## 8.5 Planting Systems and Spacing

Annonas are usually planted in square or rectangular systems. The usual distances followed for different *Annona* spp. are shown in Table 8.1.

If the plantation is on sloping land, it is recommended to establish terraces, following the contour lines, to try to distribute the plants at the recommended distance.

Plant spacing in annona depends on species, cultivar, rootstock, pruning management and soil condition. In Australia, cherimoyas are planted at 6 × 6 to 8 × 6 m (George and Nissen, 1986). Soursop is planted at 4 × 4 to 8 × 8 m in Brazil (Pinto and Ramos, 1997). Soursop trials in Hawaii showed that the spacing could be 4.6 × 6.0 m without affecting growth or interfering with cultural practices (Nakasone and Paull, 1998). In Florida, narrow plant (4–6 m) and row (6–7 m) spacing is used for cherimoya (Campbell and Phillips, 1983). In India, soursop is planted at 4 × 4 to 6 × 6 m spacing (Maheswari, 2021) and sugar apple at 6 × 6 m (Rajadurai and Rajkumar, 2022). The traditional planting distance of cherimoya in Spain is 8 × 8 m, but with pruning the planting distance may be increased to

**Table 8.1.** Plant spacing followed in different countries for *Annona* spp.

| Species | Spacing (m) | Country | Reference |
|---|---|---|---|
| Cherimoya | 4 × 5 or 4 × 4 | Ecuador | Guirado *et al.* (2003) |
| Sugar apple | 8 × 8 or 6 × 8 | Colombia | Rodriguez (2013) |
| Soursop | 4.6 × 6 | Hawaii, USA | Nakasone and Paull (1998) |
| Atemoya | 5–10 × 7–12 | Australia | Sanewski (1988) |
| | 4–6 × 6–7 | Florida, USA | Campbell and Phillips (1983) |
| Soursop | 4–6 × 4–6 | India | Maheswari (2021) |
| | 6–8 × 6–8 | Brazil | Pinto and Silva (1996) |
| Sugar apple | 6 × 4 or 8 × 5 | Brazil | Kavati and Piza (1997) |
| Sugar apple | 6 × 6 | India | Rajadurai and Rajkumar (2022) |

625 trees ha$^{-1}$ (Faree *et al.*, 1999). In southern Ecuador and northern Peru, cherimoya trees are spaced at 4 × 4 or 5 × 5 m, but when intercropped with coffee, banana or citrus trees, they are spaced at 10 × 10 or 8 × 8 m (Anon., 2008). The recommended plant spacing in Ecuador is 4 × 5 or 4 × 4 m (500–625 plants ha$^{-1}$); however, a few high-density orchards (with 1000, 1256 and 2500 plants ha$^{-1}$) have been planted for evaluation (Guirado *et al.*, 2003; Encalada, 2011).

In Mexico, Reyes-Montero *et al.* (2018) established a soursop orchard with a high density of 2222 plants ha$^{-1}$ (spacing of 3 × 1.5 m). At 10 years of age, the plants produced 28 t ha$^{-1}$. With very highly intensive pruning, the trees were maintained at a low height, making orchard management and harvesting easier. Feican *et al.* (2019) evaluated three planting densities (625, 1250 and 2500 plants ha$^{-1}$) of cherimoya 'Cumbe' in Ecuador. The two lowest densities had double the yield compared with the highest density, and also obtained the highest percentage of fruit in the best size categories (super extra and extra) and the harvested fruit weight was >400 g. Working with sugar apple 'APK-1', Rajadurai and Rajkumar (2022) showed that decreasing the plant spacing from 6 × 6 to 6 × 3 m increased the yield from 6.79 to 11.10 t ha$^{-1}$; however, the average fruit weight decreased from 156.0 to 139.1 g and the total soluble solids (TSS) content of fruit from 22.0°Brix to 19.5°Brix. Ramteke *et al.* (2023) suggested the use of 4 × 2.5 m spacing for sugar apple. Fruit from trees planted at 4 × 2.5 m produced higher fruit weight and better graded fruit yield compared with other spacings (4 × 4 and 3 × 3 m). In Australia, different systems of training with different spacing are used for atemoya cultivation. In the traditional vase system, 100–150 trees ha$^{-1}$ are accommodated, while the hedgerow system 400–500 trees ha$^{-1}$, vertical trellis system 600–800 trees ha$^{-1}$ and the open V-trellis system 700–1300 trees ha$^{-1}$ (Bignell and Bruun, 2014).

## 8.6 Fertilization

Most orchard soils growing annonas do not have enough of the right nutrients for the plants to reach their full potential. Therefore, applying fertilizers is important to meet the nutritional requirements of the plants and to promote the growth, development and yield as well as fruit quality of annonas. It is necessary to perform soil analysis to determine the amounts of available macro- and micronutrients. Based on this information, fertilizer recommendations for the initial years after planting and subsequent rates of fertilizer for tree maintenance can be prepared (Revelo *et al.*, 2010). The recommended quantity and time of application, however, depend on a variety of factors including soil type, rootstock, cultivar, plant density and weather conditions.

### 8.6.1 Organic manures and biofertilizers

Applying organic manure helps to improve the soil structure and nutrients, as well as the ability of the soil to take up and retain water. Organic manure also serves as a reservoir of nutrients, releasing significant amounts of nitrogen (N), phosphate and sulfur (S) as it decomposes.

Organic manure can introduce microorganisms, including beneficial ones, into the soil. Several organisms found in compost are important for mineralization of organic N, converting organic N into inorganic forms. Additionally, organic manure used as a compost mulch can help reduce erosion and increase the water-holding capacity of the soil. Integrated use of organic manure along with chemical fertilizers may be a sustainable approach to improving soil health and increasing the efficiency of fertilizer, while reducing both nutrient losses and environmental impact (van Tri *et al.*, 2024). Where organic manures are used, extra N and potassium (K) will be needed. Organic manures should be spread evenly in a broad band starting 20 cm from the trunk and extending to 1 m outside the canopy drip line and covered with mulching material (Sanewski, 1988).

Application of 60–80 kg of organic manure per tree year$^{-1}$ has been suggested (Morton, 1987). In soils with less than 3% organic matter, it is recommended to apply 5 t ha$^{-1}$ of well-decomposed organic manure in the planting holes 1 month before planting and then every 4 months. Young sugar apple 'NMK-1 Golden' plants showed increased growth when 75% of the requirement amount was supplied by vermicompost and 25% by inorganic fertilizer (Naik *et al.*, 2021).

Application of native mycorrhizal fungi and beneficial strains of bacteria and fungi ensures better adaptability and survival of plants under prevailing environmental conditions, and is an extremely important in terms of their long-term effects on plants (Regvar *et al.*, 2003). Enriching fertilizer with beneficial strains of bacteria and fungi can increase their effectiveness in crop production (Chen, 2006) by enhancing the physiology of the plants, and stimulating their growth and yield, as well as by increasing their resistance to environmental and biotic stresses (Corte *et al.*, 2014). There are, however, very few studies on the effect of beneficial bacteria and fungi on annonas. The response to biofertilizers by sugar apple was studied by Rajadurai and Rajkumar *al.* (2022) in the dry vertisols of Tamil Nadu, India. Their findings showed that combined application of *Azotobacter* ssp., *Azospirillum* spp., vesicular arbuscular mycorrhizal fungi and phosphobacteria each applied at 250 g per tree year$^{-1}$ along with 50% recommended dose (not specified) of NPK mixture, mixed with farmyard manure when applied to the tree basin, increased plant growth, fruit weight and TSS content of the fruit. The average fruit weight was 133.6 g versus 81.6 g for the control, while the TSS content of the fruit was 19.0°Brix versus 15.1°Brix in the control.

Application of arbuscular mycorrhizal fungi (AMF) to soursop seedlings has been reported to increase growth (da Silva *et al.*, 2008) and increase plant tolerance to the root-lesion nematode *Pratylenchus coffeae* (Angelica, 2003). Da Silva *et al.* (2008) reported that soursop seedling growth was improved with 10% organic manure with AMF (*Acaulospora longula* and *Gigaspora albida*). AMF inoculation increased phosphorus (P) uptake by roots in the nursery stage of soursop seedlings (Nadiah *et al.*, 2020).

### 8.6.2 Nutrient uptake

Cherimoya removed 95 kg N, 4.6 kg P or 11.3 kg $P_2O_5$, 38 kg K or 45 kg of $K_2O$, 9 kg of calcium (Ca) and 27 kg of magnesium (Mg) ha$^{-1}$ for a yield of 14 t ha$^{-1}$ (Guirado, 1988). The cultivation of soursop requires high levels of N, P and K. Each bearing tree of soursop extracts 19 kg K, 19 kg N, 8 kg P, 5 kg Ca and 0.9 kg Mg from soil (García-Soto *et al.*, 2012). Venkataratnam (1959) estimated that sugar apple growing in India needs 7.17 kg N and 5.19 kg of K to produce 1 t of fruit.

### 8.6.3 Role of nutrients

#### *8.6.3.1 Nitrogen*

N deficiency in cherimoya is indicated by leaves that are pale green to light yellow during summer and by poor shoot growth. In soursop, the seedling shows a reduction in height, and leaf yellowing and abscission occur more quickly than in cherimoya. The leaf blade also shows a green-yellow chlorosis and changes texture. Symptoms of N deficiency in sugar apple seedlings are irregular bud development and lack of branches. The leaves are small and show dusty spots; abscission occurs later (Pinto *et al.*, 2005).

#### *8.6.3.2 Phosphorus*

Cherimoya seedlings show P deficiency symptoms at 60 days after sowing. Irregular chlorosis in the basal leaves and a darker green colour of matured leaves are observed. As the deficiency progresses, the leaves became small and irregular in shape. Coffee-coloured spots develop on the blade and then leaf abscission occurs. Necrosis of the leaf tip occurs due to P deficiency in soursop and the plants become stunted. Sugar apple seedlings also show reduced growth with thin shoots, and the leaves develop abnormal brown spots at their tip and on the edges of the blade (Pinto *et al.*, 2005).

#### *8.6.3.3 Potassium*

Cherimoya has a high K requirement so a deficiency is likely, especially if organic fertilizer has been used exclusively. The first symptom of K deficiency is a slight yellowing of the leaf margins. This is followed by a marginal and interveinal scorch as the deficiency becomes more acute (Sanewski, 1988). In soursop seedlings, a reduction in growth and size of the leaves is followed by yellowing, and leaf abscission occurs. Sugar apple trees show retarded growth and dryness of the apical leaves due to K deficiency (Pinto *et al.*, 2005).

#### *8.6.3.4 Calcium*

Deficiency of Ca causes hard brown lumps in cherimoya fruit flesh, particularly around the seeds (Sanewski, 1988). In young seedlings, new leaves show interveinal chlorosis. In 70-day-old seedlings, the apical meristems of the leaves dies and the leaves stop growing and become twisted. The symptoms are similar in soursop seedlings (Pinto *et al.*, 2005).

#### *8.6.3.5 Magnesium*

Mg deficiency in cherimoya occurs first in new leaves. The first symptom is a yellow band on either side of the midrib. The whole leaf progressively turns yellow, but the veins retain some green. With acute deficiency, dead patches develop between the veins but close to the midrib (Sanewski, 1988). In soursop seedlings, chlorosis is progressive, turning the leaves necrotic (Pinto *et al.*, 2005).

#### *8.6.3.6 Sulfur*

The first symptoms of S deficiency occur in the young leaves – the new leaves of cherimoya and soursop seedlings are notably smaller than the old ones and the new leaves present an overall yellowing (Pinto *et al.*, 2005).

#### *8.6.3.7 Boron*

In cherimoya seedlings, leaf symptoms of boron (B) deficiency appear after 70 days. The upper leaves turn intense green, with some chlorosis. At 140 days, leaf chlorosis becomes more intense (Pinto *et al.*, 2005). Like Ca deficiency, hard brown lumps appear in the fruit flesh, particularly around the seeds (Sanewski, 1988).

#### *8.6.3.8 Zinc*

The deficiency symptoms of zinc (Zn) appear first on young growth. Mild deficiency results in yellowing of the leaves with green veins. The leaves are smaller than usual. There is no marginal or interveinal scorching. Severe deficiency causes a dramatic reduction in leaf size as well as mottling. If left untreated, limb dieback can occur. Zn deficiency is often referred to as 'little leaf' because of its dramatic effect on leaf size (Sanewaski, 1988).

#### *8.6.3.9 Iron*

New leaves are affected first by iron (Fe) deficiency. Deficient leaves are light green with green veins. As the deficiency worsens, the leaf veins can lose their green colour and the whole leaf will appear quite pale. The leaf size remains normal.

### 8.6.4 Leaf analysis

Leaf analysis can be used to check for nutrient deficiency or toxicity, particularly where suspected nutrient problems exist, or to adjust the annual fertilizer programme. Leaf nutrient element composition varies with different factors such as age of the leaf, fruiting or non-fruiting shoots, cultivar and time of sampling. The effects are large enough to influence nutrient diagnosis in many orchards, suggesting that the protocol for sampling leaf nutrient concentrations needs to be standardized across different growing regions.

In Australia, a sample of the youngest mature leaves (third or fourth leaf back from the growing point) is taken for atemoya at the end of

February or early March after the second major flush. Leaves are selected from non-fruiting shoots only (Sanewski, 1988). The seasonal pattern of leaf nutrient concentration was studied by George *et al.* (1989) for atemoya 'Pink's Mammoth' in subtropical Queensland. One group of nutrients showed a general decline in concentration in the leaves with time (K, 1.4–0.8%; P, 0.3–0.14%) while another group showed an overall increase (Ca, 0.4–1.4%; Mg, 0.3–0.54%; B, 24–66 μg $g^{-1}$). Leaf N levels decreased throughout the growing season (3.5–2.5%), with the pattern of decline strongly influenced by vegetative flushing. Leaf concentrations of Ca, Mg, manganese (Mn) and B generally increased, while those of N, P, K, copper (Cu) and Zn decreased with leaf age. Movement of more mobile elements (i.e. N, P, K, Cu and Zn) from older into the younger leaves was more rapid when trees were flushing compared with trees that had completed flushing. Tentative leaf nutrient standards for atemoya (Sanewski, 1988), cherimoya (Guirado *et al.*, 2003), sugar apple (Oliveira Dias *et al.*, 2003) and soursop (Gazel *et al.*, 1994) have been suggested (Table 8.2).

Cherimoya leaves should be collected from intermediate branches of the canopy during the fruiting period or late in the harvest season (Gonzalez and Esteban, 1974). The third and fourth leaf pairs from the intermediate branches of the canopy in the four cardinal points was suggested for foliar analysis of soursop (Laprode, 1991). Soursop leaves should be 8–9 months old (Pinto and Silva 1996). Sugar apple leaves should be collected from 2-month-old shoots of new growth, and the fifth leaf from the apex was suggested for leaf analysis (Bhargava and Raghupathi, 1993).

Among the tools available to estimate the nutritional status of plants that can be applied to annonas, the Diagnosis and Recommendation Integrated System (DRIS) appears suitable. This method uses bivariate relationships between nutrients, enabling the categorization of nutrients and identification of elements that are present in limitation or excess; thus, the indices can reveal the relative balance of the nutrients. Dos Santos and Rozane (2017) established the DRIS standard and range of nutrient sufficiency available in plant dry matter, particularly in leaf samples of atemoya 'Thompson'. They observed that highly productive populations showed the presence of the lack of nutrients (LF) or higher potential response to fertilization (PRA) for atemoya in the order of Zn > Mn > Cu = Fe > N = P > Ca = Mg = S > B > K, while for excess nutrient limitation (LE), the order was Zn = Mn > Fe = S > Cu = N = P = K > Ca = Mg + B. In the low-productivity population, the order in which the LF nutrients for atemoya appeared was Cu > Mn > Fe = Ca > B > Zn > N > P = S > Mg = S, whereas for LE the order was P > Mn > K > S > Ca = Cu > Mg = Fe > N = Zn > B. The DRIS indices of the nutrients, estimated for P, Ca, Mg, S, Cu, Fe, Mn and Zn from this study, were adequate to establish the deficiency, excess or balance of these nutrients (dos Santos and Rozane, 2017).

**Table 8.2.** Tentative leaf nutrient standards for *Annona* spp.

| | Atemoya 'Pink's Mammoth' (Sanewski, 1988) | Cherimoya (Guirado *et al.*, 2003) | Sugar apple (Oliveira Dias *et al.*, 2003) | Soursop (Gazel *et al.*, 1994) |
|---|---|---|---|---|
| Nitrogen (%) | 2.5–3.0 | 2.1–2.5 | 2.5 | 1.99–2.04 |
| Phosphorus (%) | 0.16–0.20 | 0.11–0.16 | 0.15–0.17 | 0.12–0.14 |
| Potassium (%) | 1.0–1.5 | 0.85–1.9 | 1.17–1.5 | 1.49–1.52 |
| Calcium (%) | 0.6–1.0 | 1.8–2.5 | 1.7–2.12 | 1.20–1.52 |
| Magnesium (%) | 0.35–0.5 | 0.3–0.6 | 0.35–0.5 | 0.19–0.22 |
| Sodium (%) | <0.02 | – | – | – |
| Chloride (%) | <0.3 | – | – | – |
| Manganese (p.p.m.) | 30–90 | 25–200 | – | – |
| Copper (p.p.m.) | 10–20 | 5–9 | – | – |
| Zinc (p.p.m.) | 15–30 | 30–100 | – | – |
| Iron (p.p.m.) | 50–70 | 60–150 | – | – |
| Boron (p.p.m.) | 15–40 | 50–100 | – | – |

### 8.6.5 Fertilizer rates and time of application

Prior soil analysis and constant foliar analysis are of fundamental importance for application of the correct amount of fertilizer as well as for monitoring soil fertility and plant nutrition and for making corrections. Annonas have an indeterminate growth habit (axillary flowering) and applying N at a somewhat excessive amount does not interfere greatly with floral initiation (Nakasone and Paull, 1998). However, excessive tree vigour is usually associated with reduced flowering and yields in atemoya (George *et al.*, 1989).

N fertilization increase the vegetative and reproductive stimulus of sugar apple plants (São José *et al.*, 2014). Sugar apple producers in Brazil modify the N:K ratio in the fertilization programme as a function of time of the year, from two to four parts N to one part K in autumn/winter pruning (São José *et al.*, 2014). It was observed that addition of humic substances increased the nutrient content in leaves by increasing absorption of the applied fertilizer (Cunha *et al.*, 2014). The 8-year-old sugar apple 'Balanagar' produced the highest yield (8.10 t $ha^{-1}$) by application of N, P and K at 250 g, 125 g and 125 g per tree $year^{-1}$. The best N source has been reported as ammonium sulfate (Solanke *et al.*, 2019). Cherimoya 'Concha Lisa' fruit in Colombia showed a decrease in TSS content and fruit pulp firmness and an increase in fruit acidity by increasing the rates of N from 4.14 to 6.72 kg per tree (as urea) (Undurraga *et al.*, 1995).

Annona fruits have a high demand for K, and the K content in leaves should be maintained above the suggested critical leaves for different *Annona* spp.

Araujo-Son *et al.* (1998) suggested the fertilizer schedule for sugar apple in Brazil where soil analysis information is not available (Table 8.3).

A fertilization guide for N, P and K based on P and K contents in soil for the establishment and maintenance of cherimoya orchards has been suggested, as shown in Table 8.4 (INIAP, 2014).

The primary sink for K in the atemoya is the fruit rather than the leaves, and thus there is a requirement for high levels of K. About 60% of this requirement is applied during the fruit development period. Atemoyas also have a fairly high requirement for Mg and Ca. A guide to annual application of NPK fertilizer for atemoya 'Pink's Mammoth' grown in Australia for trees of different ages and the percentage distribution of application during the annual growth and fruiting cycles was suggested by Sanewski (1988) (Table 8.5).

**Table 8.3.** Fertilizer schedule for sugar apple in Brazil. From Araujo-Son *et al.* (1998).

| | Nutrients (g per plant) | | |
|---|---|---|---|
| Age (years) | N | $P_2O_5$ | $K_2O$ |
| 1 | 60–90 | – | 30–45 |
| 2 | 45–75 | 210–270 | 30–45 |
| 3 | 120–150 | 210–270 | 60–75 |
| 4 and above | 150–210 | 270–360 | 75–105 |

**Table 8.4.** Fertilization guide for the establishment and maintenance of cherimoya orchards. From INIAP (2014).

| | | P in soil (p.p.m.) | | | K in soil (p.p.m.) | | |
|---|---|---|---|---|---|---|---|
| | | 0–10 | 11–20 | >20 | 0–45 | 46–90 | >90 |
| Plant age (years) | N g per plant | $P_2O_5$ per plant (g) | | | $K_2O$ per plant (g) | | |
| 0–1 | 50 | 60 | 40 | 20 | 70 | 40 | 20 |
| 1–2 | 100 | 80 | 60 | 40 | 60 | 40 | 20 |
| 3–4 | 120 | 120 | 80 | 60 | 120 | 80 | 60 |
| >4 | 180 | 160 | 120 | 80 | 200 | 140 | 80 |

**Table 8.5.** Fertilizer programme for atemoya trees using straight fertilizers.

| | Straight fertilizer rate (g per tree) | | | | | | |
|---|---|---|---|---|---|---|---|
| | Late September | | | Late November | | Late March | |
| Year of growth | Urea | Superphosphate | Potassium chloride | Urea | Potassium chloride | Urea | Potassium chloride |
| 2 | 100 | 560 | 80 | 120 | 100 | 40 | 60 |
| 3 | 150 | 840 | 120 | 180 | 150 | 60 | 90 |
| 4 | 200 | 1120 | 160 | 240 | 200 | 80 | 120 |
| 5 | 250 | 1500 | 250 | 300 | 400 | 100 | 150 |
| 6 | 300 | 1800 | 300 | 360 | 480 | 120 | 180 |
| 7 | 350 | 2100 | 350 | 420 | 560 | 140 | 210 |
| 8 | 400 | 2400 | 400 | 480 | 640 | 160 | 240 |
| 9 | 450 | 2700 | 450 | 540 | 720 | 180 | 270 |
| 10 | 500 | 3000 | 500 | 6000 | 800 | 200 | 300 |

## 8.7 Irrigation

*Annona* spp. are usually grown in tropical and subtropical regions. They are a water-demanding crop, and have been shown to grow and produce poorly without irrigation (Pinto *et al.*, 2005). However, annonas are grown in many areas without irrigation when rainfall is well distributed (Nakasone and Paull, 1998). There must be adequate soil moisture to encourage vegetative growth, as flowering occurs on new growth. Bearing atemoya trees may need up to 1440 l of water per tree every 4 weeks during the low-growth phase and 500–750 l per tree every 3–5 weeks during flowering, fruit set and fruit growth (Sanewski, 1988). Reducing irrigation in late winter to force atemoya and cherimoya trees into dormancy for 1–2 months in spring is recommended in Australia and California, respectively (Nakasone and Paull, 1998). A shortage of water adversely affects fruit growth of cherimoya, mainly in the first and last phases of development, when the process of cell division and elongation occur, respectively. The water balance maintained by the cherimoya tree affects the quality of the fruit, such as fruit size, firmness and TSS content (Pinto *et al.*, 2005).

Determining how much and when to water is critical, as it is related to the amount of water used by the tree in different stages. The crop coefficient ($K_c$) values reflect the water requirements of the plant in relation to climatic conditions. In Chile, Gardizabal and Rosenberg (1993) estimated $K_c$ values, which were modified for the growing conditions in Ecuador for cherimoya (Table 8.6) (INIAP, 2013). In Spain, modern cherimoya plantations use tensiometers to calculate actual water needs. The tensiometer readings are kept between −10 and −15 kPa during growth and between −30 and −35 kPa during winter. From mid-February/beginning of March (coinciding with the start of leaf fall) until mid-April (start of leaf development), cherimoya is not usually watered. Water consumption per hectare and per year is, however, very variable, depending on where the plantation is situated and which irrigation system is used. Indicative values for a hectare of cherimoya in patches reasonably protected from the wind and with a system of fixed irrigation (drip or spray irrigation) range from 5500 to 6500 $m^3$ $ha^{-1}$ $year^{-1}$ (Farré *et al.*, 1999). Franciosi Tijero (1992), however, stated that a producing cherimoya plantation needs 8000–10,000 $m^3$ water $ha^{-1}$ equivalent to 800–1000 mm rainfall $year^{-1}$. $K_c$ values of cherimoya were estimated by Durán-Zuazo *et al.* (2019). In the subtropical Mediterranean environment of Granada and Malaga (south-east Spain), the suggested values were 0.60–0.66, 0.64–0.71 and 0.48–0.62 at flowering, fruit set and fruit growth, respectively.

In India, generally sugar apple is grown under rainfed areas and irrigation is not required. However, uniform soil moisture throughout the fruit development stages ensures a high-quality yield (more edible pulp per segment) and also prevents fruit splitting. Irrigation is usually provided at 10–15-day intervals depending on the

**Table 8.6.** Crop coefficient ($K_c$) values of cherimoya at different phenological stages in the subtropical valleys of Ecuador. Modified from Gardizabal and Rosenberg (1993) and INIAP (2013).

| Month | Growth stage | $K_c$ |
|---|---|---|
| September | Bud sprouting and flowering | 0.30–0.35 |
| October | Flowering and growth initiation | 0.35–0.40 |
| November | Flowering and growth | 0.45–0.50 |
| December | Flowering and growth | 0.60 |
| January | Fruit growth and vegetative growth | 06–0.7 |
| February | Fruit growth and vegetative growth | 0.6–0.7 |
| March | Fruit growth and vegetative growth, beginning of harvest | 0.6–0.7 |
| April | Harvesting and to start restriction of vegetative growth | 0.60 |
| May | Harvesting and to induce vegetative dormancy | 0.45–0.50 |
| June | Leaf maturation and rest | 0.30–0.35 |
| July | Natural fall of leaves | 0.20 |
| August | Fall of leaves and rest | 0.20 |

soil moisture content in the orchard during the flowering and fruit growth stages (Yadav, 2018). More recently, some commercial orchards have developed a drip irrigation system (Fig. 8.5).

Conventionally, annona trees are irrigated by flooding in the field using furrows (small channels that carry water down the land slope between tree rows; water infiltrates the soil as it moves along the slope and the trees are usually grown on the ridges between the furrows) and ring basins (for each tree, a separate basin in made, which is usually circular in shape; water to the basin is supplied through small field channels). The conventional irrigation systems have several disadvantages in areas where water is a limiting resource, rains are uneven, drought is a recurring factor and the topography is undulating; moreover, these systems require a high volume of water for each irrigation. The quality of irrigation water is also an important issue. The presence of heavy metals and salt and the pH of the water may limit absorption of water and nutrients by the plants. Guirado *et al.* (2003) suggested the following analytical values for irrigating cherimoya:

- Electrical conductivity <1.7 mm h $cm^{-1}$ $h^{-1}$
- Sodium absorption ratio (RAS) <2
- Boron concentration <1 p.p.m.
- Chlorides <140 p.p.m.

Sugar apple is sensitive to saltwater so it should be avoided for irrigation (Yadav, 2018). The semi-arid region of north-eastern Brazil has water limitations in terms of both quantity and quality, with salt stress as a limiting factor for increasing yield. High water salinity of 4.0 ds $m^{-1}$ negatively affected the growth of sugar apple (Fernandes *et al.*, 2021). Plants that grow under saline conditions are exposed to various unfavourable conditions, such as low soil water potential and nutrient imbalance, with excessive accumulation of sodium ($Na^+$) and/or deficiency of $Ca^{2+}$, $Mg^{2+}$ and $K^+$ in plants (Lima *et al.*, 2020). Leaf water potential and gas exchange of sugar apple are affected by water stress whereas photosynthesis and water-use efficiency are the most sensitive to the excess of water (Moreira *et al.*, 2019).

**Fig. 8.5.** An orchard under drip irrigation. Photograph courtesy of Dr Binu Mathew.

Most of the high-input commercial annona orchards around the world are now using drip irrigation/fertigation systems due to their higher water-use efficiency. Drip irrigation can save 40–60% water compared with a conventional basin irrigation system. Drip irrigation is superior to flood and ring basin methods of irrigation in terms of water saving and water-use efficiency. Micro-sprinkler irrigation is also recommended for areas where water is limited. Like drip irrigation, this system allows more efficient water utilization.

In micro-sprinkler irrigation systems, there must be a coverage range between 30% and 40% of the soil surface; approximately 25 l $h^{-1}$ of water is used to humidify a 12 m radius area (Farré *et al.*, 1999). In drip irrigation, 1800–2000 drippers are used per hectare for a plantation with 315 trees $ha^{-1}$ (8 × 4 m and six drippers per tree). For trees planted at 8 × 4 m with an irrigation schedule of 24–25 l per tree, six drippers of 4 l $h^{-1}$ result in 20% of the area wetted (over a 25 cm depth), while one micro-sprinkler of 25 l $ha^{-1}$ wets 12 $m^2$ resulting in approximately 40% wetted area. In general, better results are obtained with spray irrigation. Because of high evaporation losses during the day, it is advisable to water at night (Farré *et al.*, 1999). El-Beltagi *et al.* (2019) compared the effect of mini-sprinkler and drip irrigation systems on sugar apple. Fruit numbers and fruit weight were found to be significantly improved by drip irrigation system by ten fruits and 77 g compared with the control. In semi-arid regions of Brazil where rainfall is highly irregular, Cavalcante *et al.* (2021) evaluated the production of soursop 'Morada' under rainfed and irrigated conditions with and without mulch. The irrigated plants were provided weekly with 17.2 mm water depth without mulching. Irrigation increased the yield from 5.3 to 34.7 t $ha^{-1}$ in the first harvest and from 9.4 to 57.7 t $ha^{-1}$ in the second harvest. Mulching with sisal residue further increased the yield to 45.6 and 70.3 t $ha^{-1}$ in the first and second harvest, respectively.

Experiments conducted with 5-year-old sugar apple 'Balanagar' revealed that application of the recommended dose of soluble fertilizers through the fertilization system resulted in increased fruit weight (184.93 g), fruit yield (13.03 kg per tree) and TSS:acid ratio of the fruit (84.35%) compared with 167.97 g average fruit weight, 11.87 kg per tree yield and 76.61% TSS:acid ratio when solid fertilizers were applied in the soil (Priya *et al.*, 2023). In an evaluation trial with different intensities of pruning and methods of irrigation (drip and double furrow), Anupama and Panchbhai (2020) reported maximum fruit weight (233.6 g) in sugar apple 'Balanagar' compared with 227.59 g when using double furrow irrigation and the lowest (202.2 g) in the control (no irrigation).

Commercial annona varieties are generally grown on seedling rootstocks and are susceptible to soil flooding (Popenoe, 1974). Except for *A. glabra*, which is extremely flood tolerant, all commercial *Annona* spp. are susceptible to flooding. Flooding of commercial *Annona* spp., even for short periods, reduces net carbon dioxide assimilation and vegetative growth (Núñez-Elisea *et al.*, 1999), may cause defoliation, and severely reduces flowering and fruit set (Marler *et al.*, 1994). Atemoya 'Gefner' can be successfully grafted on to *A. glabra* using '49-11' as an interstock. The resulting scion–interstock–rootstock combination is extremely flood tolerant or flood adapted (Núñez-Elisea *et al.*, 1999).

## 8.8 Intercrops, Cover Crops and Mulching

Annonas can be grown as a monocrop in high-density orchards or as an intercrop among larger fruit trees such as mango (Ochse *et al.*, 1991). In the initial years after planting, the space between the trees may be planted with fruit crops, such as papaya, which is a common practice in north-eastern Brazil (Pinto *et al.*, 2005). Short-duration vegetables such as tomato, onion, chilli, okra, brinjal, radish, cowpea, green gram, horse gram or any green manure crop can be intercropped, if irrigation facilities are available. These crops should not be raised too near the tree (Satish *et al.*, 2023). Growing intercrops such as cowpea (*Vigna unguiculata*), sorghum (*Sorghum bicolor*), horse gram (*Macrotyloma uniflorum*), stylo grass (*Stylosanthes guianensis*) and buffel grass (*Cenchrus ciliaris*) along with the application of recommended doses of manures and fertilizers in sugar apple orchards maintained the nutritional requirements of intercrops, suppressed weed growth, added more biomass, maintained soil fertility and gave additional income to the growers (Munde *et al.*, 2011). By sowing seeds of dhaincha (*Sesbania bispinosa*), sunhemp (*Crotalaria juncea*) and cowpea in the tree basin of sugar apple as a cover crop and applying the recommended doses of fertilizer and irrigation for the cover crops, Lokesha *et al.* (2022) observed that the cover crops helped to create a microclimate by reducing the canopy air temperature and increasing the canopy relative humidity, favouring better pollination and fruit set. Short grass can also be used as a cover crop in the first 12 months after transplanting (Nakasone and Paull, 1998).

Mulching in the tree basin is a useful practice particularly in areas where the crop is rainfed. Mulches are important in conserving soil moisture, maintaining root-zone temperature, controlling weeds and mitigating soil erosion. Mulching materials used are straw (Fig. 8.6), hay, wood chips and bark chips, which are incompletely degraded organic matter. A thin layer of mulch 5–10 cm thick applied from the tree drip line to within 15 cm of the trunk is recommended. Mulch materials should be spread away from the trunk of the tree to prevent the trunk area from becoming too moist, which can lead to bark disease problem (Crane *et al.*, 2005). Atemoya trees require reasonable light levels and temperatures for commercial maturation, and these levels are often significantly reduced in the lower canopy when trees are trained on an open V-trellis (Bignell, 2014). An experiment carried out in Australia with atemoya 'KJ Pinks' shows that using reflective mulch (Fig. 8.7) increased fruit set by at least 35%, increased average fruit weight by 6–12% and overall yield per tree by more than 40% (Bignell, 2014).

**Fig. 8.6.** Trees mulched with straw. Photograph courtesy of Dr Grant Bignell.

## 8.9 Weed Control

Annonas are shallow-rooted trees and thus limit the possibility of deep cultivation under the tree. Control of grasses and weeds must begin before planting by cultivating or applying translocated herbicides (Sanewski, 1988). The young trees should be protected from weeds by manual

**Fig. 8.7.** (A) Reflective mulch strips, 1 m wide. (B) Reflective mulch strips placed under the outer and centre bands of an open V-trellis. Photographs courtesy of Dr Grant Bignell.

weeding, mulching or use of contact herbicides (Nakasone and Paull, 1998).

In bearing trees, it is suggested that a bare strip is maintained along the row under the trees. The strip width varies from 1.5 to 3.0 m depending on tree size. The herbicide used to maintain the bare strip depends on the weeds, the cost of herbicides and their method of application. Best results are achieved if translocated herbicides are applied on a bright, sunny day to fast-growing weeds. In Australia, the herbicides paraquat, diquat, glyphosate, and a combination of paraquat and diquat have been registered for use in annona orchards (Sanewski, 1988). In Spain, oxyflurophen is applied as a pre-emergence herbicide. Perennial weeds are controlled with glyphosate using two or three applications per year. Manual weeding is also suggested three times a year (Farré *et al.*, 1999). In India, Kumar *et al.* (2017) suggested pre-emergence application of imazethapyr (0.03 kg $ha^{-1}$) or pendimethalin (0.75 kg $ha^{-1}$) with one hand weeding for weed management in sugar apple–sesame-based agrihorticultural systems. Araujo *et al.* (1999) suggested use of glyphosate (post-emergence), diuron (pre-emergence) or diuron plus paraquat (post-emergence to obtain a residual effect).

## 8.10 Conclusions

Orchard profitability and sustainability are largely dependent on the proportion of crop that can be marketed as first quality. Managing foliar light interception is one of the key factors creating the most productive fruit orchards. New training systems and the method and time of pruning have been standardized for commercial *Annona* spp. such as cherimoya, atemoya, sugar apple and soursop. The trellis system of training enhances precocity of bearing; however, not all species and cultivars respond equally, so this needs evaluation across all commercially grown *Annona* spp.

In this chapter, efforts have been made to discuss the various aspects of fertilization and irrigation scheduling for higher productivity and fruit quality of cherimoya, sugar apple and atemoya through better nutrient- and water-use efficiency. Detailed studies are required on plant–water relationships for fine-tuning of fertilizer and irrigation rates and their application timings. Such studies should take into consideration the nutrient availability in the plant body, residual nutrients in the soil from the application, local soil and climatic conditions, and the long-term impact of these inputs on soil and environmental health.

## References

Agustin, J.A. and Alviter, A.R. (1996) *El Cultivo de la Cherimoya (Annona cherimola Mill.) en el Estado de Michoacan*. Universidad Autonoma Chapingo (UAC), Texcoco, Mexico.

Angelica, C.B.J. (2003) Arbuscular mycorrhizal symbiosis of soursop (*Annona muricata*) in soil infested with *Pratylenchus coffeae*. Thesis, Federal University of Pernambuco, Recife, Brazil.

Anon. (2008) Promotion of Sustainable Cherimoya Production Systems in Latin America through Characterization, Conservation and Use of Local Germplasm Diversity. Reports of the Cherla Project No. 015100. Project co-funded by the European Commission within the Sixth Framework Programme (2002–2006).

Anupama, K.C. and Panchbhai, D.M. (2020) Effect of pruning severity and irrigation methods on quality production of custard apple. *International Journal of Current Microbiology and Applied Sciences* 9, 2272–2280. DOI: 10.20546/ijcmas.2020.912.269

Araujo-Son, G.C., Andrade, T.M.S., Castro, F.A. and Sa, F.T. (1998) *Technical Instructions for the Cultivation of Ateira*. Embrapa Tropical Agro-industry, Brazil.

Araujo, J.F., Araujo, J.F. and Alives, A.A.C. (1999) Technical instruction for growing pinecone (*Annona squamosa* L). *EBDA Technical Circular* 77.

Bhargava, B.S. and Raghupathi, H.B. (1993) Analysis of plant materials for macro and micronutrients. In: Tandon, H.L.S. (ed.) *Methods of Analysis of Soils, Plants, Water and Fertilizer*. Fertilizer Development and Consultation Organization, New Delhi, pp. 49–82.

Bignell, G. and Bruun, D. (2014) *Accelerating the Development of the Australian Custard Apple Industry*. Final Report for HAL Project CU 11000. Department of Agriculture, Fisheries and Forestry, Queensland, Australia.

Broadley, R., George, A., Price, S., Redpath, S., Olsen, T. *et al.* (2008) Queensland and NSW regional tree training trials – 2004–2007. In: Broadley, B. (ed.) *New Tree Training Systems for Custard Apple*. Horticultural Australia, Sydney, Australia, pp. 35–45. Available at: www.horticulture.com.au/globalassets/hort-innovation/historic-reports/new-tree-training-systems-for-custard-apple-cu04004.pdf (accessed 19 August 2025).

Campbell, C.W. and Phillips, R.L. (1980) *The atemoya*. Fruit Crop Fact Sheet FC-64. Florida Coop Extension Service. University of Florida, Gainesville, Florida.

Campbell, C.W. and Phillips, R.L. (1983) *The Sugar Apple*. Fruit Crops Factsheet. Florida Extension Service, University of Florida, Gainesville, Florida.

Cavalcante, L.F., Silva-Matos, R.R.S., Souto, A.G. de L., de Lima Neto, A.J., Nunes, J.C. and Mesquita, F. (2021) Soursop production under supplementary irrigation and mulching in the semiarid region of Brazil. *Revista Brasileira de Engenharia Agrícola e Ambiental* 25, 522–528. DOI: 10.1590/1807-1929/agriambi.v25n8p522-528

Cavalcanti, R. (1973) The culture of the pinha (*Annona squamosa* L.). In: *En-Contro Estadual de Fruticultural 1 Cruz Das Almas*. EMBRAPA/CNPMF, Cruz das Almas, Brazil, p. 159.

Chander, S., Kurian, A.M., Satisha, J., Upreti, K.K. and Kaxman, R.H. (2022) Advancing fruiting season in Annona cv. Arka Sahan through pruning. *Journal of Horticultural Sciences* 17, 316–324. DOI: 10.24154/jhs.v17i2.1434

Chen, J. (2006) The combined use of chemical and organic fertilizers and/or biofertilizers for crop growth and soil fertility. In: *International Workshop on Sustainable Management of the Soil – Rhizospheres System for Efficient Crop Production and Fertilizer Use, 16–20 October, Bangkok*. Land Development Department, Bangkok, pp. 1–11.

Choudhary, K. and Meena, N.K. (2020) Benefits of pruning in custard apple. *Indian Horticulture* 65, 38–40.

Coronel, R.E. (1994) Atis. In: Coronel, R.E., Zuno, J.C. and Sotto, R.C. (eds) *Promising Fruits of Philippines*. College of Agriculture, University of the Philippines at Los Banos, Laguna, Philippines, pp. 1–18.

Corte, L., Dell'Abate, M.T., Magini, A., Migliore, M., Felici, B. *et al.* (2014) Assessment of safety and efficiency of nitrogen organic fertilizers from animal-based protein hydrolysates-a laboratory multidisciplinary approach. *Journal of the Science of Food and Agriculture* 94, 235–245. DOI: 10.1002/jsfa.6239

Crane, J.H., Balerdi, C.F. and Kohout, M. (2004) Effect of four years of tree maintenance pruning on yields of Red Sugar apple (*Annona squamosa* L.). *Proceedings of the Interamerican Society for Tropical Horticulture* 48, 152–154.

Crane, J.H., Balerdi, C.F. and Maguire, I. (2005) *Sugar Apple Growing in the Florida Home Landscape*. IFAS Extension Fact Sheet HS38. Horticultural Sciences Department, UF/IFAS Extension, University of Florida, Gainesville, Florida. Available at: https://edis.ifas.ufl.edu/publication/MG330 (accessed 19 August 2025).

Cunha, M.S., Silva Júnior, G.B., Cavalcante, I.H.L., Santos, E.M., Albano, F.G. and Rocha, L.F. (2014) Nutritional status of custard apple (*Annona squamosa*) as a function of nitrogen fertilizing and humic substances. *Journal of the Faculty of Agronomy* 31, 493–509.

da Silva, D.K.A., Silva, da F.S.B., Yano-Melo, A.M. and Maia, L.C. (2008) Use of earthworm manure improves growth of soursop seedlings (*Annona muricata* L. 'Morada') associated with arbuscular mycorrhizal fungi. *Acta Botanica Brasilica* 22, 863–869. DOI: 10.1590/S0102-33062008000300021

dos Santos, E.M.H. and Rozane, D.E. (2017) DRIS standard and normal ranges of foliar nutrients for the culture of 'Thompson' atemoya. *Ciência Rural* 47: e20160613. DOI: 10.1590/0103-8478cr20160613

Duchi Albarracín, M.I. (2017) [Pomological and agromorphological characterization of cherimoya (*Annona cherimoya* la Mill.) existing at the INIAP and University of Cuenca]. Thesis, University of Cuenca, Cuenca, Ecuador. Available at: https://rest-dspace.ucuenca.edu.ec/server/api/core/bitstreams/043d9565-4cef-4108-8578-77f24ddf8294/content (accessed 19 August 2025).

Durán-Zuazo, V.H., Franco Tarifa, D., García-Tejero, I.F., Gutiérrez Gordillo, S., Cermeño Sacristan, P. and Roldan, J.J.P. (2019) Water use and leaf nutrient status for terraced cherimoya trees in a subtropical Mediterranean environment. *Horticulturae* 5: 11. DOI: 10.3390/horticulturae5020046

El-Beltagi, H.S., Sassine, Y.N., Hammoud., M. and El-Sebaaly, Z. (2019) Effect of different localized irrigation systems on growth and production of sugar apple (*Annona squamosa* L.). *Fresenius Environmental Bulletin* 28, 10206–10213.

Encalada, C. (2011) Planting densities in cherimoya. In: *III International Cherimoya Seminar*. Faculty of Agronomy, University of Cuenca, Cuenca, Ecuador.

Farré, J.M., Hermoso, J.M., Guirado, E. and García-Tapia, J. (1999) Techniques of cherimoya cultivation in Spain. *Acta Horticulturae* 497, 91–118. DOI: 10.17660/ActaHortic.1999.497.5

Feican, C., Vanegas, I., Encalada, C., Brito, B., Moreira, R. and Viera, W. (2019) Effect of planting density on the fruit quality of cherimoya (*Annona cherimola* Mill.). *Tropical and Subtropical Agroecosystenus* 22, 795–801. DOI: 10.56369/tsaes.2395

Fernandes, E.A., Soares, L.A. dos A., Lima, G.S. de, Silva Neta, A.M. de S., Roque, I.A. *et al.* (2021) Cell damage, gas exchange, and growth of *Annona squamosa* L. under saline water irrigation and potassium fertilization. *Semina* 42, 999–1018. DOI: 10.5433/1679-0359.2021v42n3p999

Franciosi Tijero, R. (1992) *El cultivo del chirimoyo en el Peru*. Ediciones Fundeagro, Lima, Peru, p. 108.

García-Soto, A., Ettiene, E.G., Pérez-Pérez, L., Sandoval, L., Montilla, L. and Solo, E. (2012) Propagation and fertilization of soursop cultivation. II: Chemical characteristics of fruits. *Revista de la Facultad de Agronomía* 29, 20–36.

Gardizabal, F. and Rosenberg, G. (1993) *Cultivation of Cherimoya*. Catholic University of Valparaiso, Faculty of Agronomy, Chile.

Gazel, F.A.B., Carvalho, A.C.A. and Menezes, A.J.E.A. (1994) Teores de Macronutrients em Folhas de Graviola. *Revista Brasileria de Fruticultura* 16, 121–124.

George, A.P. and Nissen, R.J. (1986) Edible fruits and nuts: *Annona cherimola* Mill. and *Annona squamosa*.L. In: Verherij, E.W.M. and Coronel, R.E. (eds) *Plant Resources of South-East Asia*, Vol. 2. PUDOC, Bogor, Indonesia, pp. 71–75.

George, A.P., Nissen, R.J. and Brown, B.I. (1987) The custard apple. *Queensland Agriculture Journal* 113, 287–297.

George, A.P., Nissen, R.J. and Carseldine, M.L. (1989) Effect of season (vegetative flushing) and leaf position on the leaf nutrient composition of *Annona* spp. hybrid cv. Pink's Mammoth in south eastern Queensland. *Australian Journal of Experimental Agriculture* 29, 587–595. DOI: 10.1071/EA9890587

Gonzalez, C. and Esteban, E. (1974) Nutricion del Cherimoya: Ciclo Annual. *Anales de Edafologia e Agrobiologia* 33, 371–380.

Gonzalez Vega, M.R. (2013) Chirimoya (*Annona cherimola* Miller), tropical and subtropical fruit tree with promises. *Tropical Crops* 34, 52–63.

Guirado, E. (1988) *Chirimoyo. Comparacion de Metodos de Polinizalion Artificial. Analis de Componetes del Fruto. Effecto del Acido Giberelico en la Dessarrolo del Fruto y Sus Components*. UITA, La Rabida, Huelva, Spain.

Guirado, E., Hermoso, J., Perez, M. and Faree, M. (2003) *Introduction to Cherminoya Cultivation*. Experimental Farm, The Nacia, Granada Rural Bank.

INIAP (National Institute of Agricultural Research) (2013) *Annual Program Report*. Fruit Growing Experimental Station of the Austro, INIAP, Cuenca, Ecuador.

INIAP (National Institute of Agricultural Research) (2014) *Annual Program Report*. INIAP, Cuenca, Ecuador.

Jadhav, K.P., Panchbhai, D.M., Bahadure, A., Patil, S.R. and Ramteke, N.H. (2022) Effect of severity of pruning and fruit retention on growth and harvesting of custard apple. *Pharma Innovation Journal* 11, 1122–1126.

Kadam, S.R., Dheware, R.M. and Urade, P.S. (2018) Effect of different levels of pruning on quality of custard apple (*Annona squamosa* L.). *International Journal of Bio-resource and Stress Management* 9, 573–575. DOI: 10.23910/IJBSM/2018.9.5.1865c

Kavati, R. and Piza, C.T. Jr (1997) Formação e Manejo do Pomar de Fruta-do-Conde, Atemoia e Cherimoia. In: São José, A.R., Souza, I.V.B., Morais, O.M. and Rebouças, T.N.H. (eds) *Anonaceas Producao e Mercado*. Universidade Estadual do Sudoesta da Bahia, Vitoria de Conquista, Bahia, Brazil, pp. 76–83.

Kome, G.K., Enang, R.K., Silatsa, F.B.T., Yerima, B.P.K. and van Ranst, E. (2024) Baseline edaphic requirements of soursop (*Annona muricata* L.). *Tropical Plant* 3: e022. DOI: 10.48130/tp-0024-0023

Kumar, M.K.M., Deepak, D. and Singh, R.K. (2017) Weed control in custard apple–sesame based agrihorticulture system. *Agricultural Research Journal* 54, 139–140. DOI: 10.5958/2395-146X.2017.00028.X

Laprode, S.C. (1991) Variacion Estacional de Nutrimentos Folares em Guanabana (*Annona muricata* L.). *Corbana (Costa Rica)* 15, 6–10.

Lima, G.S.D., Fernandes, C.G.J., Soares, L.A.D.A., Gheyi, H.R. and Fernandes, P.D. (2020) Gas exchange, chloroplast pigments and growth of passion fruit cultivated with saline water and potassium fertilization. *Revista Caatinga* 33, 184–194. DOI: 10.1590/1983-21252020v33n120rc

Lo, S.S. (1987) Pruning technique, use of sprouting chemical and flower initiation in sugar apple (*Annona squamosa*). In: Chang, L.R. (ed.) *Forcing Culture of Horticultural Crops*. Special Publication No. 10. Taichung District Agricultural Improvement Station, Taichung, Taiwan, pp. 147–150.

Lokesha, R., Athani, S.I., Gollagi, S.G., Reddy, S.G., Nadaf, A.M. *et al.* (2022) Influence of different cover crops on flowering and fruit set of custard apple (*Annona squamosa* L.) cv. Balanagar under northern dry zone of Karnataka. *Pharma Innovation Journal* 11, 1557–1562.

Maheswari, T.U. (2021) Soursop – a wonder fruit. *Indian Horticulture* 32–35.
Marler, T.E. and Crane, J.H. (1994) Increasing scaffold branch angle of atemoya by manipulation of the primary bud complex. *HortScience* 29, 818–820. DOI: 10.21273/HORTSCI.29.7.818
Marler, T.E., George, A.P., Nissen, R.J. and Anderson, P.C. (1994) Miscellaneous tropical fruit. In: Schaffer, B. and Andersen, P.C. (eds) *Handbook of Environmental Physiology and Fruit Crops, Vol. 2. Tropical and Subtropical Crops*. CRC Press, Boca Raton, Florida, pp. 199–224.
Mitra, S. and Pathak, P.K. (2021) Orchard management. In: Sisir, M. (ed.) *Guava: Botany, Production and Uses*. CAB International, Wallingford, UK, pp. 172–185.
Moreira, R.C.L., Brito, M.E.B., Fernandes, P.D., Sá, F.V. da S., Silva, L. de A. *et al.* (2019) Growth and physiology of *Annona squamosa* L. under different irrigation depths and phosphate fertilization. *Bioscience Journal* 35, 389–397. DOI: 10.14393/BJ-v35n2a20198-41931
Morton, J.F. (1987) *Fruits of Warm Climates*. Creative Resource Systems, Winterville, North Carolina.
Munde, G.R., Hiwale, B.G., Nainwad, R.V. and Dheware, R.M. (2011) Effect of intercrops on growth and yield of custard apple. *Asian Journal of Horticulture* 6, 29–31.
Nadiah, N.S.H., Nursyahidah, R., Jaafar, N.M., Zaharah, S.S. and Muharam, F.M. (2020) Arbuscular mycorrhizal Fungi (AMF) and NPK fertilisation rate on the growth of soursop (*Annona muricata* L.). *Malaysian Journal of Soil Science* 24, 147–159.
Naik, S., Bahadur, V., Prasad, V.P., Saket, M., Paul, A. and John, S.. (2021) Effect of different combinations of organic manures and inorganic fertilizers on growth and establishment of custard apple (*Annona squamosa*) cv. NMK-1 Golden. *Biological Forum* 13, 206–211.
Nakasone, N.Y. and Paull, R.E. (1998) Annonas. In: Nakasone, N.Y. and Paull, R.E. (eds) *Tropical Fruits*. CAB International, Wallingford, UK, pp. 45–75.
Núñez-Elisea, R., Schaffer, B., Fisher, J.B., Colls, A.M. and Crane, J.H. (1999) Influence of flooding on net $CO_2$ assimilation, growth and stem anatomy of *Annona* species. *Annals of Botany* 84, 771–780. DOI: 10.1006/anbo.1999.0977
Ochse, J.J., Soule, Jr., M.J., Dijkman, M.J. and Wehlburg, C. (1991) Otros cultivos frutales. In: *Cultivo y Mejoramiento de Plantas Tropicales y Subtropicales*. Editorial Limusa, Mexico, pp. 587–818.
Oliveira Dias, N., Matsumoto, S.N., Rebouças, T.N.H., Viana, A.E.S., São José, A.R. and Souza, I.V.B. (2003) The influence of pruning on branches of different diameter in the vegetative and reproductive growth of sugar apple (*Annona squamosa* L.). *Revista Brasileira de Fruticultura* 25, 100–103. DOI: 10.1590/S0100-29452003000100029
Pinto, A.C. de Q. and Ramos, V.H.V. (1997) Graviola: Formação do pomar e tratos culturais. In: São José, A.R., Souza, I.V.B., Morais, O.M. and Reboucas, T.N.H. (eds) *Anonáceas, Produção e Mercado*. Universidade Estadual Do Sudoesta Da Bahia, Vitoria de Conquista. Bahia, Brazil, pp. 94–104.
Pinto, A.C. de Q. and Silva, E.M. (1996) *Graviola para Exportacao, Aspectos Tecnicas da Producao*. EMBRAPA/SPI, Brasilia.
Pinto, A.C. de Q., Cordeiro, M.C.R., de Andrade, S.R.M., Ferreira, F.R., Filgueiras, H.A. de C. *et al.* (2005) *Annona species*. International Centre for Underutilized Crops, University of Southampton, Southampton, UK.
Popenoe, W. (1974) *Manual of Tropical and Subtropical Fruits*. Hafner Press, New York.
Priya, B., Kurubar, A.R., Ashok, H., Ramesh, G., Undaykumar, N. *et al.* (2023) Influence of flower, fruit yield and quality of custard apple (*Annona squamosa* l.) by drip fertigation. *Annals of Plant and Soil Research* 25, 466–472. DOI: 10.47815/apsr.2023.10293
Rajadurai, K.R. and Rajkumar, J. (2022) Standardization of plant density through canopy management in custard apple (*Annona squamosa*). *Pharma Innovation Journal* 11, 1538–1540.
Ramteke, N.H., Bharad, S.G., Panchbhai, D.M., Nagre, P.K. and Patil, S.R. (2023) Effect of planting density and fruit load on flowering, fruit set and yield of custard apple. *Pharma Innovation Journal* 12, 1894–1903.
Regvar, M., Vogel-Mikuš, K. and Ševerkar, T. (2003) Effect of AMF inoculum from field isolates on the yield of green pepper, parsley, carrot, and tomato. *Folia Geobotanica* 38, 223–234. DOI: 10.1007/BF02803154
Revelo, J., Viteri, P., Vasquez, W., Valverde, F., Leon, J. *et al.* (2010) *Manual del Cultivo Ecológico de la Naranjilla.* Technical Manual No. 77. INIAP, Quito, Ecuador.
Reyes-Montero, J.A., Aceves-Navarro, E., Camal-Velazquez, J.H. and Alamilla-Magana, J.C. (2018) Producción de guanábana (*Annona muricata* L.) en alta densidad de plantación, como alternativa para productores con superficies reducidas. *Agroproductivity* 11, 37–42. DOI: 10.32854/agrop.v11i9.1212
Rodriguez, M. (2013) Study of the custard apple and signature pastry proposal. Thesis, Universidad Internacional del Ecuador, Quito, Ecuador
Sanewski, G.M. (1988) *Growing Custard Apples*. Information Series Q187014. Queensland Department of Primary Industry, Brisbane, Australia.

São José, A.R., Barreto do Prado, N., Bomfim, M.P., Rebouças, T. and Mendes, H.T.A. (2014) March of nutrient absorption in anonaceous. *Revista Brasileria de Frutisultura* 36, 176–183. DOI: 10.1590/S0100-29452014000500021

Satish, J., Kurian, R.M., Sujatha, S., Linta, V. and Anushma, P. (2023) *Production Technology of Tropical and Subtropical Fruits: A Hand Book*. Indian Institute of Horticultural Research, Bangalore, India.

Solanke, A.A., Kadam, A.S. and Padekar, V.D. (2019) Effect of different nitrogenous fertilizers on quality parameters of custard apple (*Annona squamosa* L.). *International Journal of Current Microbiology and Applied Sciences* 8, 2073–2079. DOI: 10.20546/ijcmas.2019.8DOI: 10.241

Soler, L. and Cuevas, J. (2008) Development of a new technique to produce winter cherimoyas. *HortTechnology* 18, 24–28. DOI: 10.21273/HORTTECH.18.1.24

Sonika, A.S., Sabrad, A.I., Jalawadi, S., Gorabal, K. and Prashantha, A. (2023) Influence of time and pruning intensity on growth and yield attributes in custard apple (*Annona squamosa* L.) cv. Balanagar. *Biological Forum* 15, 357–360.

Stacey, P., Broadley, R., George, A., Nissen, R. *et al.* (2008) Maroochy V-trellis trial – Alstonville 2003-7. In: Broadley, R., *et al* (ed.) *New Tree Training Systems for Custard Apple*. Horticultural Australia, Sydney, pp. 8–18. Available at: www.horticulture.com.au/globalassets/hort-innovation/historic-reports/new-tree-training-systems-for-custard-apple-cu04004.pdf (accessed 19 August 2025).

Undurraga, P.L., Olaeta, J.A. and Marnich, M.L. (1995) Effect of increasing nitrogen fertilizer rates on cherimoya (*Annona cherimola* Mill.) cv. Concha Lisa fruits in refrigerated storage. Harvest and postharvest technologies for fresh fruits and vegetables. In: *Proceedings of the American Society of Agricultural Engineers, Guanajuanto, Mexico*, pp. 405–413.

van Tri, M., Phuong, D.T.Y. and Mitra, S. (2024) Orchard management. In: Sisir, M. (ed.) *The Dragon Fruit: Botany, Production and Uses*. CAB International, Wallingford, UK, pp. 102–146.

Venkataratnam, L. (1959) Sitaphal and other *Annona* fruits in India. In: *Farm Information Newsletter*, Vol. 54. Department of Extension, New Delhi, p. 36.

Vieira, V.J.S., Mello, J.J. de L., Rosa, J.C.M. and Silva, F. (1994) *Pinheiro (Annona squamosa) Cultivation under Irrigated Condition*. Embrapa, Recife, Bazil.

Yadav, V. (2018) Cultivation of custard apple fetches more. *Indian Horticulture* 63, 42–46.

Yang, C.S. (1987) Production of sugar apple fruit in winter in. In: Chang, L.R. (ed.) *Forcing Culture of Horticultural Crops*. Special Publication No. 10. Taichung District Agricultural Improvement Station, Taichung, Taiwan, pp. 129–140.

# 9 Flowering, Pollination and Fruit Set

**Hiroshi Matsuda[1], Hirokazu Higuchi[2] and Sisir Mitra[3]***

*[1]Tropical Agriculture Research Front, Japan International Research Center for Agricultural Sciences, Ishigaki, Okinawa, Japan; [2]Graduate School of Agriculture, Kyoto University, Kyoto, Japan; [3]Former Professor, Faculty of Horticulture, Bidhan Chandra Krishi Viswavidyalaya (State Agricultural University), Mohanpur, West Bengal, India*

## 9.1 Introduction

The family Annonaceae consists of 109 recognized genera (Leal and Paull, 2023) and is distributed mainly in the subtropics and tropics. It is an early-divergent angiosperm family that originated more than 100 million years ago, with the genus *Annona* having diverged around 30–40 million years ago, between the end of the Eocene and the Oligocene (Couvreur *et al.*, 2011). Annonaceae have been studied as one of the key groups providing important information on the evolutionary trends of floral organs in the reproductive biology of angiosperms (e.g. Tsou and Fu, 2002; Lora *et al.*, 2009a,b; Saunders, 2020). The genus *Annona* includes about 100 species indigenous to Central and South America, with entomophilous hermaphroditic flowers that show protogynous dichogamy (e.g. Wester, 1910). They are characterized by edible aggregate fruit containing tens to hundreds of seeds. The mesocarp covering the seeds is the edible portion and has a sweet or sour taste with an aromatic flavour for fresh consumption and processing. The climacteric fruits are eaten after softening for several days after harvest.

*Annona* spp. are indigenous to regions ranging from tropical highlands to tropical lowlands, so the climates suitable for fruit production also vary. Cherimoya (*Annona cherimoya*) is a subtropical species indigenous to the Andean tropical highlands (~1500 m elevation), and a subtropical to temperate climate of 18–25°C in summer and 5–18°C in winter is suitable for fruit production (Morton, 1987). Sugar apple (*A. squamosa*) is also grown in some temperate regions such as Israel and northern India, but is indigenous to tropical lowlands in Central and South America; fruit production is more prevalent in tropical lowlands such as Thailand, the Philippines and Taiwan, and climates with summer temperatures ranging from 25°C to 41°C and average winter temperatures above 15°C are suitable for fruit production (Lim, 2012). Atemoya (*A. cherimola* × *A. squamosa*; synonym, *A. atemoya*) is intermediate between these, with high yields at temperatures of 22–32°C (Brandão and Santos, 2016). Soursop (*A. muricata*) is the most tropical-climate-preferring *Annona* spp. and grows well under temperature conditions ranging from 21°C to 30°C (Kitamura, 2002), with suitable average annual temperatures of 24–27°C (Calzavara and Müller, 1987). Mountain soursop (*A. montana*) and bullock's heart (*A. reticulata*) prefer tropical climates for cultivation but are more tolerant of low temperatures than soursop, and average annual temperatures of 21–26°C are suitable for cultivation of mountain soursop

* Email: sisirm55@gmail.com



DOI: 10.1079/9781800627628.0009

(Lim, 2012). Ilama (*A. diversifolia*), soncoya (*Annona purpurea*), pond apple (*A. glabra*), and biriba (*A. mucosa*, syn. *Rollinia mucosa*; *Rollinia* have been reported to be considered a clade within the genus *Annona* (Chatrou *et al.*, 2009)) are well adapted to tropical lowland climates, similar to soursop (Morton, 1987). This chapter covers various fruit crops of *Annona* spp. cultivated in various climates, from phenological and physiological insights on flowering and fruit set, to cultivation management techniques from flowering to harvest.

## 9.2 Flowering

Most of *Annona* spp. are semi-deciduous and do not require dormancy for flower bud differentiation and bud formation, although axillary buds may become dormant due to low temperatures or drought. Flower bud differentiation and bud formation occur with the growth of new shoots, and if environmental conditions are favourable, flowering occurs year-round (Higuchi and Utsunomiya, 1999); the name *Annona* is derived from the Latin *annona* meaning 'annual harvest'. Flower buds are formed at the base of new shoots emerging from axillary buds and at the middle of the new shoots (Fig. 9.1A). The flower buds that develop in the middle of the shoot are the result of floral bud differentiation in the apical bud as the shoot elongates and are formed on the opposite side of the node from the new leaves. If the nutritional conditions of the shoot and the climatic conditions are favourable, multiple flower buds will form in clusters at the same site (Fig. 9.1B, C).

**Fig. 9.1.** (A) Formation of cherimoya flower buds on the opposite side of the node. (B) and (C) Formation of multiple flower buds in a cluster, cherimoya and sugar apple respectively.

### 9.2.1 Floral morphology

The flower is an aggregate of several hundred pistils, and the conical ovary is surrounded by numerous stamens (Fig. 9.2B–J). The petals are succulent and develop between the sepals and stamens, surrounding the stamens in a valvate pattern. The outer surface of the petals is a darker colour than the inner surface. Cherimoya, sugar apple, atemoya, bullock's heart, biriba and ilama have three petals, while soursop, mountain soursop, pond apple and soncoya have three outer petals alternately surrounding three inner petals. The outer petals are thicker and have a prominent midrib-like protrusion that develops in the centre of the petal. The size of the petals, pistils and stamens varies among species, and the flower organs of species with six petals are larger in size (Fig. 9.2A). Some species have red or purple coloration at the base of or inside the petals, especially the petals of pond apple, which have a beautiful red pattern on the inside (Fig. 9.2H). Soncoya flowers have a purple coloration over the entire inner petal and the inner surface of the outer petal, from where its specific name *A. purpurea* derives. Ilama flowers are completely maroon in colour (Fig. 9.2K), and the stigmatic papillae develop prominently (Fig. 9.2J). *Annona* flowers have papillae structures on the stigma surface (Vithanage, 1984). *Rollinia* spp. such as biriba has three petals, as does cherimoya, but the petal shape differs markedly from that of

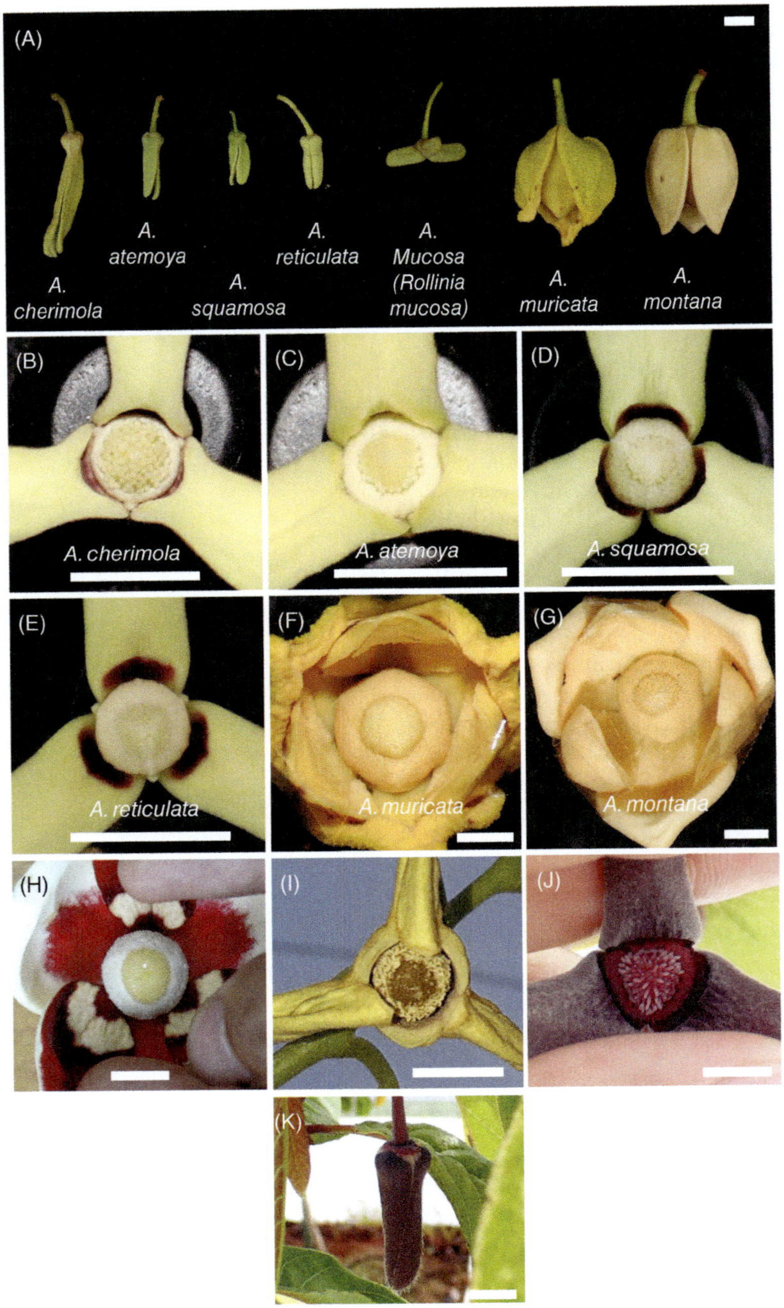


**Fig. 9.2.** (A) Flowers of different *Annona* spp. (B–J) Annona flowers showing pollen and ovary surrounded by numerous stamens. (H) Petals of pond apple. (I) Propeller-shaped horizontal petals in biriba. (J) Stigmatic papilla in ilama. (K) Maroon-coloured flower in ilama. Bars indicate 1 cm.

other *Annona* spp., with the petals developing in a propeller shape horizontally rather than downwards (Fig. 9.2I). Cherimoya and ilama have minute, soft, short hairs on the surface of the petals and peduncles. The embryo sac of cherimoya was reported to exhibit *Polygonum*-type morphogenesis (Matsuda and Higuchi, 2012). The ovule is anatropous, bitegmic, and crassinucellate with an inner integument forming a micropyle (Lora *et al.*, 2009a, 2011a). In receptive pistils, the surface of the stigma is covered with a viscous secretion.

### 9.2.2 Flower induction

Multiple bud complexes consisting of several floral bud and vegetative meristems can already be differentiated in the leaf axil of 1-week-old cherimoya (Higuchi and Utsunomiya, 1999). In cherimoya, sugar apple, atemoya, bullock's heart, ilama and pond apple, the axillary buds of new shoots are completely covered by the petiole (Fig. 9.3A–F); the axillary buds require petiole dropout to break dormancy, and thus axillary branches are not often formed from the axils of elongating new shoots. In contrast, for soursop, mountain soursop and biriba, the axillary buds are not completely covered by the petiole (Fig. 9.3G–I), and axillary branches often grow simultaneously with the elongation of new shoots, and flower buds are also formed on these branches (Fig. 9.3J). Despite their tropical origin, some *Annona* spp. show day-length sensitivity in the growth of shoots, and long-day conditions promote shoot growth (Inoue *et al.*, 2009), which induces floral bud differentiation and flowering. In atemoya, mild to moderate soil drought during the flowering period promotes flower bud formation, although photosynthesis is reduced, resulting in lower fruit quality (George and Nissen, 2002).

### 9.2.3 Flowering hours

The flowers show protogynous dichogamy where the pistils become receptive before anther dehiscence. A few days before anther dehiscence, gaps between petals begin to open and release an aromatic fragrance, and the pistil is receptive to pollen. Thereafter, the petals completely open and the anthers dehisce to shed pollen in the late afternoon to around sunset in cherimoya (Wester, 1910; Gazit *et al.*, 1982; Rosell *et al.*, 1999; Matsuda *et al.*, 2016) and bullock's heart (Wester, 1910; Venkataratnam, 1959). Petal opening and anther dehiscence occur early to late in the morning in sugar apple (Wester, 1910; Kumar *et al.*, 1977; Gazit and Eisenstein, 1985; Sahoo *et al.*, 2000) and soursop (Venkataratnam, 1959), and late at night to early in the morning in pond apple (Venkataratnam, 1959). In atemoya, the hours of anther dehiscence vary among cultivars, but the anthers dehisce a little later than in cherimoya at around sunset in the major commercial

**Fig. 9.3.** (A–F) Axillary buds completely covered by the petiole in cherimoya (A), sugar apple (B), atemoya (C), bullock's heart (D), ilama (E) and pond apple (F). (G–I) Axillary buds not completely covered by petiole in soursop (G), mountain soursop (H) and biriba (I). (J) Axillar buds growing simultaneously with the elongation of new shoots and flower buds in soursop.

cultivars 'Gefner', 'Pink's Mammoth' and 'African Pride' (Morton, 1987; Suzaki *et al.*, 2011). The pistil loses receptivity coinciding with the time of anther dehiscence, so that the time overlap between the female and male stages in the same flower is slight, if it occurs at all. In cherimoya, the female and male stages barely overlap, except in some cultivars such as 'Big Sister' (Morton, 1987).

### 9.2.4 Gene regulation

Analysis of flowering-related genes has progressed in recent years in sugar apple. Quantitative real-time polymerase chain reaction analysis has shown that a reduction in expression of two genes, *LYF* and *AP2*, is associated to dormancy breaking and flowering induction by defoliation of sugar apple (Mo *et al.*, 2015). Liu *et al.* (2015) used transcriptome analysis in sugar apple to create a cDNA library and identified the unigenes involved in plant hormone signalling events and circadian rhythms, as well as those similar to flowering-related genes. Comparative transcriptome analysis has identified a number of differentially expressed genes associated with malformed flower development that impair fruit quality, and the expression of these genes related to flowering and hormones was found to decrease in malformed flowers (Liu *et al.*, 2016). A comprehensive analysis of seven MADS-box genes involved in flowering and other reproductive development in sugar apple showed that many of the MADS-box genes were responsive to phytohormone treatment (Liu *et al.*, 2017a,b). Liu *et al.* (2018) performed proteomic analysis in different developmental stages of flower buds and identified 36 differentially expressed proteins, the majority of which were related to respiration and energy metabolism.

Genome sequencing has been completed in cherimoya (Talavera *et al.*, 2023) and is now available for omics analysis such as using RNA sequencing. Gene analysis related to flowering and fruiting in *Annona* spp. is expected to accelerate.

### 9.2.5 Anther dehiscence

When cherimoya petals open and the anthers dehisce, the water content of the petals is at a maximum, and at the same time viscous secretion on the stigma surface is reduced (Koura *et al.*, 2001). During cherimoya anther dehiscence, the pollen swells and the internal starch is hydrolysed to sugars (Lora *et al.*, 2009a; 2012). Thus, water transfer from the stigma to the petals and stamens is considered to occur during anther dehiscence. In the early spring, which is the early flowering period of cherimoya, the pollen is not fully developed, and when stained with iodine solution, the entire pollen is stained, showing that it contains a lot of starch; the germination percentage of such pollen is low, making it difficult to produce fruit if used for pollination (Saavedra, 1977). The timing of starch hydrolysis seems to vary depending on the cherimoya cultivar and growing environment: in 'Fino de Jete' in Spain it was reported to occur just before anther dehiscence (Lora *et al.*, 2012), while in 'Big Sister' under greenhouse cultivation in Japan it was reported to occur after anther dehiscence (Matsuda *et al.*, 2016). *Annona* spp. pollen remains as a tetrad even after anther dehiscence (Fig. 9.4) (Saavedra, 1977; Tsou and Fu, 2002; Lora *et al.*, 2009b), which is a feature that has arisen independently in various taxa during the evolution of animal-pollinated angiosperms (Harder and Johnson, 2008). Most angiosperms have bicellular pollen at anther dehiscence, and it is rare to find a mixture of bicellular and tricellular pollen species in the same genus (Brewbaker, 1967), but in cherimoya, pollen from a single species contains both bicellular and tricellular pollen at anther dehiscence (Rosell *et al.*, 1999). This phenomenon is quite unusual and is observed mainly in early-divergent angiosperm species (Sampson, 1969; Gardner, 1974). High temperatures near anther dehiscence time in cherimoya

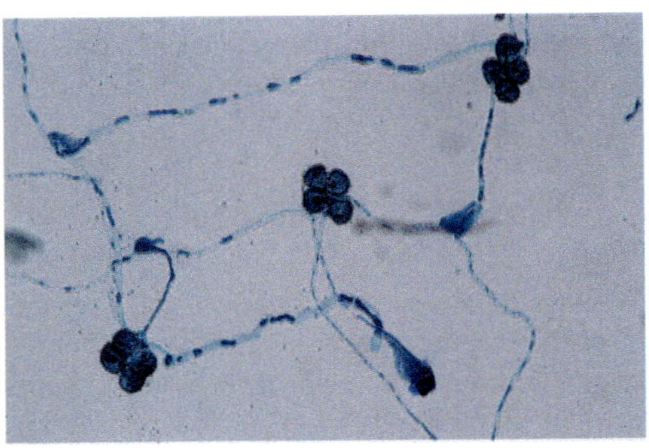

**Fig. 9.4.** Tetrad pollen of cherimoya.

promote mitosis II and increase tricellular pollen (Lora *et al.*, 2009a). As cell division ceases when pollen dehydration occurs, tricellular pollen is an advantageous trait in aquatic taxa (Brewbaker, 1967), which is consistent with the characteristic of cherimoya that it prefers humid conditions during anther dehiscence and quickly loses germination ability under dry conditions (Matsuda and Higuchi, 2015). The time of anther dehiscence varies depending on the environmental conditions of the day. In cherimoya, the temperature, not the humidity, has a strong influence, with uniform anther dehiscence at an early time when the night-time temperature just before anther dehiscence was 18–22°C and the daytime temperature was 20°C, while anther dehiscence was delayed at both lower and higher day temperatures and varied at daytime temperatures above 30°C (Matsuda and Higuchi, 2015; Matsuda *et al.*, 2016). Cherimoya anther dehiscence was delayed by about a day when the flowers encountered low night temperatures below 8°C in the early morning of the expected anther dehiscence day, and shrivelled pollen grains were observed after anther dehiscence when the night temperature was below 6°C (Matsuda *et al.*, 2016). Similarly in other *Annona* spp., anther dehiscence time is considered to be affected by environmental conditions, but there are no reports yet of detailed investigations.

## 9.3 Pollination

The main cultivated Annonaceae species are allogamous plants, with hermaphroditic flowers; however, they present the phenomenon of protogynous dichogamy, that is, when pollen grains are released, the female organs are no longer receptive, generally preventing self-pollination in the same flower (Pereira *et al.*, 2021). It was also observed that flowers of the same genotype usually open synchronously and, consequently, pollen transfer between different flowers of the same genotype is difficult (Lora *et al.*, 2010). This phenomenon was confirmed by phenological studies in sugar apple (de Moura *et al.*, 2019), cherimoya (Cautín and Agustí, 2005) and soursop (Jiménez-Zurita *et al.*, 2023).

The flowers of the majority of Annonaceae species commonly exhibit a trimerous perianth more or less distinctly differentiated into sepals and petals. Sepals are arranged in one whorl and petals are all arranged in two whorls; stamens and carpels have a helical arrangement. Valvate and imbricate aestivations are most common. Petals in sexually mature flowers may be reflexed, spreading, erect or connivent. The majority of Annonaceae have many stamens, which are frequently tightly packed and shielded by a broadened apical prolongation of the connective. Most Annonaceae are apocarpous, both in flower and fruit, and carpel number varies from one to many within and among genera (van Heusden, 1992). The numerous, centrally located carpels have sessile stigmas of various shapes. In the pistillate phase, stigmas in many species are covered with a sticky secretion, and usually become detached at the end of the pistillate phase. The sticky mass of stigmas function as a compitum (van Heusden, 1992).

In cherimoya, the flower is hermaphroditic and has stamens and pistils grouped in a three-sided pyramid. Surrounding the base is the mass of stamens, which are white during the so-called female state and pale cream during the male state (Soria *et al.*, 1991). From the time the flower bud appears, the flower develops but remains closed for approximately 30 days. Once it has reached its final size, opening takes place in three consecutive phases. In the first, the so-called pre-female phase, the petals begin to separate from the end. In Spain, this phase takes 6–15 h, finishing just after noon (~13:00–15:00). In this state, the flower is receptive, but insects cannot pollinate it. During the second phase, the female phase, the petals continue to open, allowing small pollinating insect to make contact with the stigmatic mass. The duration of this phase is approximately 26–27 h occurring from around 13:00–15:00 on day 1 to 16:00–18:00 the following day. The stigmas are receptive for the whole period, except for the last 3 h. Finally, during the third or male phase, the petals separate totally over 20–30 min, coinciding with the ripening of the stamens. The step from the female to the male phase always takes place in the afternoon at around 16:00–18:00. When temperatures are high, the cycle shortens, and when the temperature is low, it takes longer (Farré *et al.*, 1999).

Some researchers have observed that the rhythm of change from female flower to the male stage occurs synchronously and simultaneously at the tree level and even at the plot level (Rosell and Galan, 1995), preventing the simultaneous presence of female and male flowers and impeding insect pollination, while other researchers have described the presence of overlap of the two flowering stages (Garcia del Corral, 1989). The presence of natural pollination is considered the reason for natural fruit set; however, Gazit *et al.* (1982) and George *et al.* (1989) stated that certain environmental conditions of cool temperature and high humidity (>70% relative humidity) can prevent the loss of stigma receptivity before anther dehiscence, thus allowing natural self-pollination.

Soursop has solitary and hermaphroditic flowers, with flower buds of globose or pyramidal shape. The inflorescence has two to 13 flowers (rarely solitary), and the corolla has three to six petals; when it has six, the internal ones are rudimentary (Romero-Soler and Cetzat-Ix, 2015). The flowers open in the early hours of the day, when the anthers are beginning to release pollen. The outer petals come off after a few hours of opening; the inner ones can take days to fall or they may come off together (Hernandez *et al.*, 2013). To have a good fruit set, it is very important that the flowers are properly pollinated. The best way to ensure this is by pollinating manually: this is done with an already developed flower from which the pollen is extracted, and by means of a small brush, each flower is opened and the pollen is inserted.

### 9.3.1 Pollen germination

A number of studies have tested *in vitro* pollen germination in *Annona* spp. (Saavedra, 1977; Rosell *et al.*, 1999; Cavalcante *et al.*, 2009; Nietsche *et al.*, 2009; Mendes *et al.*, 2012; Pereira *et al.*, 2014; Maita *et al.*, 2022). Cherimoya releases mixed (bicellular and tricellular) pollen in tetrads (Lora *et al.*, 2009a). The bicellular pollen is characterized by a longer life and a slower germination, while the tricellular pollen has a faster germination but a shorter life with greater sensitivity to dehydration, which makes its utilization more problematic (Lora *et al.*, 2012). In cherimoya, the optimum *in vitro* pollen germination was at 20–25°C, and pollen required the provision of calcium and boron as well as sucrose at a concentration of 5–10% (Rosell *et al.*, 1999). In sugar apple, pollen viability in seeded and seedless accessions was 38.5 and 52.5%, respectively. Pereira *et al.* (2014) reported that the percentage *in vitro* germination for the sugar apple cultivars 'Red' and 'Lessard Thai' was significantly greater for pollen collected from female-stage flowers and used 18 h after collection; in contrast, the percentage *in vitro* germination was significantly lower from pollen collected from male-stage flowers (2.8%). The percentage of *in vitro* atemoya pollen germination in Brazil ranged from 5.9% (Mendes *et al.*, 2012) to 19.7% (Bettiol Neto *et al.*, 2009). Atemoya 'Gefner' pollen did not germinate when pollen was collected from male-stage flowers and when using female-stage flowers collected in the late afternoon and stored for 72 h. The highest pollen germination (37.2%) was noted in atemoya 'Gefner' when pollen was collected from the female flowering stage in the late afternoon (Pereira *et al.*, 2014). Maita *et al.* (2022), working with cherimoya ecotypes 'Fabulosa' and 'Austro' and cultivar 'Fino de Jete' observed the highest pollen germination in the male phenological stage, at 2 and 24 h after incubation at 21°C. 'Fabulosa' had the highest percentage of germination in the female stage. The ideal phenological stage for pollen collection was found at anther dehiscence. 'Austro' and 'Fabulosa' had the longest pollen tubes after 2 and 24 h in both phenological stages. Both the ecotypes and 'Fino de Jete' showed promising pollen characteristics for hand pollination of cherimoya.

Rodrigues *et al.* (2016) studied different temperatures for the *in vitro* pollen grain germination of sugar apple and found that the temperature of 25 ± 1°C provided the best *in vitro* germination (48.13%); as the temperature increased, the germination percentage decreased significantly. Working with sugar apple at Roraima state, Brazil, Araújo *et al.* (2021) observed beginning of anthesis at 00:00 (second assessment day) in 53% of marked flowers and extending until 12:00 of the same day, when 100% of flowers reached anthesis. The transition from the female to the male stage occurred between 06:00 and 12:00 of the following day (second assessment day). The transient phase of female to male flowers is characterized by full opening of flowers and anther dehiscence. The female reproductive structure of sugar apple matures first and can remain receptive for up to 24 h. The sugar apple pollen grains stored under ambient (27 ± 2°C) and refrigerated (5 ±2°C) conditions showed an excellent germination rate at the beginning (storage time 0) of 65.75% and 50.25%, respectively. In the first 4 h of storage, the germination percentage decreased to 51.25% and 45.0%, respectively, with a significant decrease in later hours (Araújo *et al.*, 2021). The sugar apple cultivar 'Brazilian Seedless' showed 52.5% *in vitro* pollen grain germination at 27.2°C and 43% humidity (Mendes *et al.*, 2012). Rodrigues *et al.* (2016) demonstrated that pollen-tube growth was significantly affected by different temperatures. Low temperatures (10°C and 15°C) reduced the elongation of pollen tubes, which produced mean lengths of 74.30 and 107.37 μm, respectively. The increase in pollen-tube length observed started at 20°C (255.42 μm) and at 25°C it further increased (536.45 μm), however above 30°C, the pollen-tube length started to be affected, showing a significant decrease to 276.41 μm at 30°C, 239.85 μm at 35°C and 204.00 μm at 40°C.

### 9.3.2 Pollinators

Members of the Annonaceae family with large flowers are mostly pollinated by large scarab beetles, but those with small flowers such as cherimoya, sweetsop and atemoya attract small beetles such as those in the Nitidulidae and Staphylinidae families (Gottsberger, 1989). Nitidulid beetles that breed and feed on decaying fruit or sap flows are the main pollinators of commercial *Annona* spp. (Peña *et al.*, 2002). Research in Spain has shown that Hemiptera insects of the genus *Orius* could also be effective as pollinating vectors in cherimoya (Sánchez *et al.*, 2001). Other beetles such as weevils (Curculionidae), cucujids (Cucujidae), scarab beetles (Scarabaeidae), rove beetles (Staphylinidae) and anthicids (Anthicidae) can also participate in pollination of annona fruit crops (Peña *et al.*, 2002).

A study in Israel found that four species of nitidulids (*Carpophilus hemipterus*, *Urophorus humeralis*, *Carpophilus mutilatus* and *Haptoncus luteolus*) were common visitors throughout the flowering season (Gazit *et al.*, 1982). These authors also observed that fruit set depended on the presence of nitidulid beetles, but there was no difference between the four species of beetles in their efficiency to pollinate atemoyas. Nitidulid beetles were found to be the most common visitors (51.4%) to atemoya flowers in Florida. Of the nine species of nititulid beetles found to visit atemoya flowers, *Carpophilus fumatus* was the most common (Nagel *et al.*, 1989). In Japan, the cherimoya insect visitors included *Phloeonomus* sp., *Cortinicara gibbosa*, *Mimemodes monstrosus*, and *Haptoncus ocularis* as its major consituents (Tsukada *et al.*, 2005). Among these, *H. ocularis* and *M. monstrosus* were identified as most effective pollinators for cherimoya (Higuchi *et al.*, 2014a).

Nitidulid beetles were found to be associated with fruit set of atemoyas and cherimoyas in Australia, Chilli and the USA (Florida) (George *et al.*, 1989; Laport and Dent, 1992; Nagel *et al.*, 1989). In Australia, fruit set has been shown to increase linearly with increasing numbers of nitidulid beetles (*C. hemipterus*) per flower (George *et al.*, 1989).

Several *Annona* spp. have large flowers (3–4 cm diameter) with thick petals such as *A. montana*, *A. glabra*, *A. paludosa*, *A. densicoma*, *A. coriacea*, *A. aurantiaca*, *A. warmingiana*, *A. monticola*, *A. crassiflora*, *A. dioica*, *A. tomentosa* and *A. cornifolia*, and attract large (17–21 mm long) dynastid scarab beetles of the genus *Cyclocephala* (Gottsberger, 1989).

The pollination syndrome of *Annona senegalensis* was investigated by Dao *et al.* (2023) in the Sudanian and Sudano–Sahelian climatic zone of Burkina Faso during the flowering season, from May to June. A total of nine species of insects including Coleoptera (Curculionidae, Scarabeidae), Hymenoptera (Formicidae) and Diptera (Tephritidae) were observed visiting flowers. Two sympatric species of weevil (Curculioninae, Ochyromerini, *Endaeus castus* and *E. floralis*) were found to be the most abundant. Both species were observed on the fleshy petals of the cantharophilous flowers of *A. senegalensis* that they used as substrate for the development of larval stages. This study provides the evidence of brood-site mutualism between an *Annona* sp. and its specialized weevil pollinators.

The beetles *Cyclocephala atricapilla*, *C. undata*, *C. ohausiana*, *C. octopunctata* and *Arriguttia brevissima* are pollinators of *Annona coriacea* found in Cerrado of Mato Grosso State, Brazil (Costa *et al.*, 2017). However, *C. atricapilla* was considered the main pollinator due to its abundance in relation to the other species. Floral odours given off by several species of Annonaceae act as chemical signals to attract insects to feeding sites and for breeding (Jürgens and Dötterl, 2004; Maia *et al.*, 2012; Costa *et al.*, 2017). Maia *et al.* (2012) identified 4-methyl-5-vinylthiazole as a floral signal utilized by several *Annona* spp. that convergently attracts their *Cyclocephala* beetle pollinators.

### 9.3.3 Hand pollination

Hand pollination is suggested to ensure pollination and good fruit shape. Hand pollination can be performed in two different ways: (i) pollen is collected in the morning from fully open flowers (male stage) and open flowers (female stage) and then hand pollinated immediately; or (ii) open flowers are collected in the evening at the female stage, held in a paper bag at room temperature, and the next morning flowers are shaken over a shallow tray or paper to collect the pollen, which is transferred to a small container and hand pollination is performed using a small brush, hand blower or pollination gun (Fig. 9.5). Pollen may be held in the refrigerator for use the next morning (Nakasone and Paull, 1998; Pereira *et al.*, 2011). Sometimes the crude pollen (anthers and pollen) collected is cleaned by passing it through a 2–3 mm mesh sieve; gentle hand pressure is sufficient to separate the pollen grains from the anthers. The pollen can be stored in a refrigerator for a day at 3–7°C and applied the following day (Pereira *et al.*, 2011). Viable pollen is yellow-orange and pollen that is black or dark yellow is discarded as being over-mature. Hand pollination is normally carried out in the morning from 07:00 to 12:00; the pollen is taken from the refrigerator and rapidly applied using a camel hair brush or atomizer to recently opened flowers when the stigma is most receptive (Leal and Paull, 2023).

Hand pollination has several advantages: (i) fruit set is guaranteed; (ii) increased potential yield; (iii) increased fruit size; (iv) more symmetrical fruit shape; and (v) greater harvest efficiency (fruit accessible from the ground). There are also a few disadvantages: (i) specialized hand labour is needed for pollen collection, preparation and pollination; (ii) pollinated fruit is located only on the lower branches; (iii) a good tree form required and therefore specialized pruning is needed, (iv) the tree size needs to be limited; and (v) the fruit may have more seeds (Fioravanço and Caleffi Paiva, 1994).

Three methods of hand pollination were tested by Richardson and Anderson (1990) on cherimoya in New Zealand. They compared pure pollen or a pollen/lycopodium mixture using a brush and the pollen/lycopodium mixture using a hand puffer. Lycopodium powder is the spores of a primitive non-vascular plant in the genus *Lycopodium* and acts as an inert diluent for pollen. Puffer application of the pollen/lycopodium mixture produced significantly higher fruit set than the other two methods, but the fruit shape did not differ significantly among the three methods. Using the brush application alone, the authors found that the pollen/lycopodium mixture caused improved fruit set over pure

**Fig. 9.5.** (A) Pollination gun. (B) Pollinating the female flower. Photographs courtesy of Dr Gary Bender.

pollen. Puffer application was found to be more efficient than the brush method, producing a fivefold increase in number of flowers pollinated per unit time.

### 9.3.4 Effect of pollen sources on fruit quality

Pollen sources and pollen viability from related species can affect fruit set and fruit quality (Chander and Kurian, 2019; Matsuda and Higuchi, 2019; Meena *et al.*, 2023).

Meena *et al.* (2023) investigated the effect of different pollen sources and time of pollination for hand pollination on biochemical and organoleptic properties in the sugar apple cultivar 'Arka Sahn'. Three pollen sources ('Balanagar', 'Raydurg' and 'APK Ca-1') and three time durations for pollination (first pollination, 18–15 July; second pollination, 15–22 July, and third pollination, 22–29 July) were tested. Pollinating 'Arka Sahn' with 'Balanagar' between 8 and 15 July showed the maximum pulp weight (249 g), seed ratio (24.24), rind weight (65.33 g), number of seeds per fruit (28.0), total soluble solids content (32.33°Brix), total sugar (25.41%) and ascorbic acid (27.87 mg 100 $g^{-1}$) in the fruit. Sensory evaluation also showed the highest rating for texture, primarily due to the presence of grittiness in the pulp. The authors concluded that 'Balanagar' is a good pollen source for hand pollination of 'Arka Sahan'. In sugar apple 'Sindhan', the highest pulp weight of fruit (191.33 g) was obtained by hand pollination using a paint brush compared with 179.67 g by pollinating with a sprayer. Using a paint brush for pollination produced a yield of 23.05 t $ha^{-1}$ compared with 20.21 t $ha^{-1}$ when a hand gun was used (Sanghani and Varu, 2022).

Working with cherimoya 'Big Sister' that was pollinated using pollen from 'Big Sister' and cherimoya 'Fino de Jete' and from atemoya 'Gefner' and 'Pink's Mammoth', Matsuda and Higuchi (2019) reported a decrease in seed set, fruit weight and fruit symmetry when atemoya pollen was used. 'Gefner' pollen produced globose fruit, while 'Fino de Jete' pollen produced fruit of higher eating quality. Atemoya pollen decrease the eating quality, although the pollen parent did not affect the soluble solids content of the fruit (Matsuda and Higuchi, 2019). Pollen dilution with *Lycopodium* spores reduced the fruit set and size and favoured the appearance of misshapen fruit in cherimoya (González *et al.*, 2006).

## 9.5 Cultivation Management for Sustainable Production

Cultivation management is becoming more sophisticated in *Annona* spp. For example, the BBCH scale, a code for identifying the seasonal developmental stages of plants, is being extended to sugar apple (Liu *et al.*, 2015) to help develop protocols for more efficient agronomic management, such as cultivation and breeding. (Phenological stages of different *Annona* spp. are discussed in Chapter 2, this volume.) Almost 100% fruit set can be achieved if hand pollination is performed properly. Reliable fruit set requires vigorous elongation of the pollen tube, supported by high receptivity of the pistil and subsequent fertilization. The detrimental effects of high and low temperatures in cherimoya, which are a species particularly sensitive to environmental conditions such as temperature and humidity, have been studied extensively in both male and female organs.

### 9.5.1 Effects of temperature, humidity and photoperiod

In cherimoya, day/night temperatures of 30/25°C negatively affect reproductive organ development and result in lower fruit set (Higuchi *et al.*, 1998). Furthermore, pollen germination is suppressed (Rosell *et al.*, 1999) and stigma receptivity is shortened (Lora *et al.*, 2011a) under high temperature conditions (>30°C). High temperatures in midsummer reduce flower number and inhibit flower bud development (George *et al.*, 1987; Higuchi and Utsunomiya, 1999; Matsuda and Higuchi, 2012). Cherimoya pollen germination is encouraged by temperatures of 22–25°C (Rosell *et al.*, 1999; Yonemoto *et al.*, 1999), and night temperatures of 17–22°C are suitable for a reliable fruit set (Matsuda *et al.*, 2015). In the early spring, low night temperatures below 10°C inhibit reproductive organ development (Matsuda and Higuchi, 2012), while low night temperatures below 14°C reduce pollen vigour (Matsuda *et al.*,

2016), in addition to nutrient competition with new shoot growth (Schroeder, 1943; Matsuda and Higuchi, 2012), which can cause low fruit set and malformed and smaller fruit. Later in the warm season, flowers tend to produce larger fruit (Richardson and Anderson, 1996; Soler and Cuevas, 2008).

The reproductive organs of *Annona* spp. prefer high humidity (George and Nissen, 1988; Lora *et al.*, 2011a), and thus avoiding a dry environment during pollination is a key factor for stable fruit production. Cherimoya maintains a high germination percentage of pollen for a longer time under relative humidity conditions higher than 80% (Matsuda and Higuchi, 2015), and pollen-tube elongation is also superior at lower vapor pressure deficit (VPD) with moderate temperatures (Murai *et al.*, 2021), while pollen is susceptible to wetting and fruit set is reduced on rainy or excessively humid days (George *et al.*, 1987). George *et al.* (1987) also noted that, in Queensland, Australia, fruit set by artificial pollination decreased on hot days with low humidity (<70% relative humidity), as well as on wet and excessively humid days. Similarly, in sugar apple, production is difficult during the dry season due to a decrease in pollen vigour and poor pollination in Thailand (Thanarut and Ko, 2017) and Brazil (Rodrigues *et al.*, 2016). In areas where temperatures tend to be colder and drier in winter and early spring, and warmer in the summer, high-quality fruit production can be achieved for a longer period under greenhouse cultivation with appropriate heating, cooling and humidifying (e.g. using boilers, heat pumps, fog cooling).

Under favourable environmental conditions, shoot growth and flower development occur throughout the year, but in temperate and subtropical areas, axillary buds become dormant due to low temperatures in winter. For species with axillary buds that are completely covered by the petiole, such as cherimoya, atemoya and sugar apple, defoliation is necessary to artificially induce flowering. Pruning and defoliation in early spring will promote the axillary buds to break dormancy and accelerate flowering. Meanwhile, in cooler springs in warm temperate areas, the pollen become immature (Saavedra, 1977), and fruit set tends to decrease during cooler springs (Schroeder, 1941; Yonemoto and Nakao, 1993; Matsuda and Higuchi, 2012). Cherimoya flower development tends to be impaired when temperatures frequently drop below 10°C (Matsuda and Higuchi, 2012). Insufficient leaf numbers results in a lack of assimilates, which delays flower bud development and the flowering period (Olesen and Muldoon, 2012). Thus, in areas where such temperatures occur in early spring, it is desirable to promote growth of shoots and flower buds by appropriate temperature management in early spring under greenhouse cultivation for high-quality fruit production.

New axillary shoot growth and flower bud formation can be induced by pinching the growing shoot tips and removing leaves near the tips under suitable temperature and sufficient day-length conditions; in Spain, a winter fruit production technique for cherimoya has been established in which leaves are removed in early summer (June) to ensure sufficient flowering in July, just before midsummer, and harvested in December or later (Soler and Cuevas, 2008, 2009). In Taiwan and Okinawa, Japan, there are cultivation systems in which sugar apple and atemoya are given supplementary illumination during winter to create long-day conditions for the growth of shoots, to increase the number of flowers and to expand the harvest period (Inoue *et al.*, 2009; Lin *et al.*, 2014).

### 9.5.2 Training and defoliation

*Annona* spp. trees are generally trained in an open-vase system (Fig. 9.6A) for convenience for artificial pollination and harvesting. In addition to the open-vase system, cherimoya and atemoya are also trained using the horizontal trellis system in Taiwan, Australia, India and Japan (Fig. 9.6B). The horizontal trellis system is based on the Japanese pear cultivation technique, in which the fruit-bearing branches are placed horizontally about 10 cm higher than the grower's height. In addition, cherimoya and atemoya are susceptible to strong sunlight, which causes the branches and fruit to be sunburned, and the horizontal trellis system has the advantage of protecting the fruit, branches and trunks from the summer sunlight with the bushy and upright new shoots while improving pollination, fruit thinning and harvesting efficiency. Sugar apple trees are less vigorous than cherimoya and atemoya, making them unsuitable for the horizontal trellis system. In Taiwan, the open-vase

**Fig. 9.6.** (A) Trees trained in an open-vase system. (B) Trees trained using the horizontal trellis system. (C) High-density sugar apple trees trained in an open-vase system.

system with high planting density and low tree height is the most popular for sugar apple in order to protect against strong typhoon winds (Fig. 9.6C), similar to the planting system for apple trees in temperate zones. Soursop trees are vigorous enough to be trained using the horizontal trellis system, as are cherimoya and atemoya. Cautín *et al.* (2010) developed a high-density tunnel training system in cherimoya that maintained high humidity, low temperatures and low light in the canopy interior and reported that this system advanced the hours of anther dehiscence at which time the pistils tended to remain receptive.

*Annona* spp. trees often weaken due to the fruit-bearing load, and recovering tree vigour by pruning is important to ensure stable production. In particular, sugar apple trees are very susceptible to fruit load, so thinning out small branches and cutting back large branches during the dry season is performed to recover the vigour of the tree (Coronel *et al.*, 1994; Wongs-Aree and Noichinda, 2011). Depending on the cultivar and climate, in sugar apple, the proper fruit load is 50–60 fruit year$^{-1}$ on a 5–6-year-old trees (Hoa *et al.*, 2017), and a leaf:fruit ratio of 50 or more is appropriate (Rathod, 2019). Similarly, the proper fruit set is about 60 fruit per tree in hybrid cultivars from backcrossing of atemoya with sugar apple (Chander and Kurian, 2019). The proper leaf:fruit ratio for cherimoya is also 50 or higher (Matsuda *et al.*, 2015). In bullock's heart, a branch bearing fruit will die back to the base after harvesting the fruit. In cherimoya, sugar apple and other *Annona* spp., they do not die back, but even if the branches that did produce fruit are cut back and pruned, the following year's productivity is not as good. Defoliation techniques using urea and ethrel spray treatments on cherimoya (González *et al.*, 2013) and sugar apple (Chander *et al.*, 2019a) have been developed to reduce the cost of labour for manual defoliation and to control flowering and harvesting times.

### 9.5.3 Hand pollination and other fruit set managements

*Annona* spp. tend to produce malformed fruits with few seeds when pollination is insufficient. Thus, hand pollination is generally performed to produce large, symmetrical fruits with higher commercial value. Pollen for hand pollination is usually collected by gathering excess flowers before anther dehiscence and allowing them to open indoors; cherimoya pollen with remarkably higher germinability (>80%) can be obtained by having detached flowers attain anther dehiscence at a constant temperature and humidity environment in an incubator rather than collecting pollen from flowers just after anther dehiscence in the orchards (Matsuda and Higuchi, 2015). Hand pollination is necessary for stable fruit production, but pollen storability is poor. Although pollen can be stored for 2–3 months using liquid nitrogen without losing germinability in cherimoya (Lora *et al.*, 2006) and sugar apple (Chander *et al.*, 2019b), development of simpler and more effective storage techniques is still needed.

Due to the poor storability of pollen and its characteristic of rapidly losing germination vigour, hand pollination immediately after anther dehiscence is recommended to obtain high fruit set (Schroeder, 1943), although pollen can be stored in a refrigerator overnight. Therefore, various pollen handling methods have been developed, mainly in cherimoya, that maintain high pollen vigour and facilitate hand-pollination operations. For example, small glass or plastic vials are used to carry pollen, eliminating the influence of the worker's body temperature. The pollination is then performed using a small paintbrush. In addition, Japanese persimmon pollination guns combined with pollen-diluting powders such as *Lycopodium* spores have been used successfully in California (George *et al.*, 1987; Fig. 9.5). Because pollen is sensitive to high temperatures and dryness (Rosell *et al.*, 1999; Matsuda and Higuchi, 2015), keeping the pollen in cool and humid surroundings is recommended when transporting it during hand pollination.

Performing hand pollination is a very time-consuming and labour-intensive operation. To mitigate this problem, insect pollinators are used such as *Orius laevigatus* (Caleca *et al.*, 1998), *Carpophilus tersus* (Caleca *et al.*, 2002), *Carpophilus marginellus* and *Mimemodes monstrosus* (Higuchi *et al.*, 2014a,b) for pollination of cherimoya, as well as efficient propagation methods using pollinators in laboratory conditions (Tsukada *et al.*, 2008) and by intercropping with maize (Farré *et al.*, 1999). However, the productivity and fruit quality are not as good as from hand pollination.

Self-incompatibility or intersterility between cultivars is not observed in *Annona* spp. (e.g. Schroeder, 1943); however, in cherimoya, using the cultivar 'Bronceada' pollen to pollinate 'Cumbe' can produce higher fruit set than self-pollination (Duarte and Escobar, 1997). Metaxenia is observed in *Annona* spp. fruit. Cherimoya fruit quality and seed traits vary with pollen-parent cultivars (Kahn *et al.*, 1994); the eating quality of 'Big Sister' cherimoya is more stable when pollinated with 'Fino de Jete' pollen, as well as varying seed traits (Matsuda and Higuchi, 2019); in atemoya, 'Hillary White' pollinated by 'African Pride' results in large, symmetrical fruit and improved productivity (Pritchard and Edwards, 2006). Selecting suitable cultivar combinations of the fruit-bearing and pollen-source trees is important for high-quality fruit production. In addition, many interspecific cross-compatible combinations are available among *Annona* spp. (e.g. Kumar and Jalikop, 2000; Goenaga and Jenkins, 2016). Using pollen of sugar apple or *A. senegalensis* to pollinate cherimoya is effective in stable fruit setting (Schroeder, 1943). However, pollinating cherimoya with atemoya pollen can have negative effects on fruit texture and taste (Matsuda and Higuchi, 2019). Thus, the use of pollen from other *Annona* spp. is recommended after thorough investigation of the pollen-parent effects on fruit quality.

*Annona* spp. fruit are less convenient to process and sometimes unpleasant to eat because each fruit contains dozens to hundreds of seeds, but several cherimoya cultivars can produce seedless fruit when treated with gibberellic acid ($GA_3$) at high concentrations of 1500–1600 p.p.m. on unpollinated stigmas (Saavedra, 1979; Yonemoto *et al.*, 2000; Koura *et al.*, 2004). In atemoya 'Gefner', a concentration of 1000 p.p.m., lower than in cherimoya, is suitable for seedless fruit production (Pinto *et al.*, 2018). Sugar apple has several seedless strains (Lora *et al.*, 2011b; Mendes *et al.*, 2012), and breeding for seedless atemoya that does not require $GA_3$ treatment is anticipated.

## 9.6 Conclusions

*Annona* spp. are characterized by an indeterminate growth habit and associated axillary flowering. Low humidity reduces flowering, fruit set and fruit size in cherimoya and atemoya. Sugar apple, cherimoya and atemoya have hermaphroditic protogynous dichogamous flowers, which drastically reduces the chance of self-pollination. Flowers first undergo a female phase where stigmas are receptive to pollen, followed by a male phase when the anthers dehisce the pollen but the stigmas are no longer receptive. Natural pollination of cherimoya, sugar apple and atemoya is affected by various nitidulid beetle species. Hand pollination is commonly practised to increase the potential for commercially viable *Annona* spp. fruit production. Plant growth regulators (e.g. $GA_3$) increase fruit set and can induce seedless fruit.

## References


Araújo, D.C.B., Chagas, P.C., Chagas, E.A., Moura, E.A., Oliveira, R.R. *et al.* (2021) Flower stages, germination and viability of pollen grains of *Annona squamosa* L. in tropical conditions. *Acta Scientiarum – Technology* 43: e51013. DOI: 10.4025/actascitechnol.v43i1.51013

Bettiol Neto, J.E., Del Nero, M., Kavati, R. and Pinto-Maglio, C.A.F. (2009) Viability and conservation of pollen from tree commercial annonas. *Bragantia* 68, 825–837. DOI: 10.1590/S0006-87052009000400002

Brandão, A.P.M.E. and Santos, D.Y.A.C. (2016) Nutritional value of the pulp of different sugar apple cultivars (*Annona squamosa* L.). In: Simmonds, M.S.J. and Preedy, V.R. (eds) *Nutritional Composition of Fruit Cultivars*. Elsevier, Amsterdam, Netherlands, pp. 195–214. DOI: 10.1016/B978-0-12-408117-8.00009-X

Brewbaker, J.L. (1967) The distribution and phylogenetic significance of binucleate and trinucleate pollen grains in the angiosperms. *American Journal of Botany* 54, 1069–1083. DOI: 10.1002/j.1537-2197.1967.tb10735.x

Caleca, V., Lo Verde, G., di Chiara, S.R. and Tsolakis, H. (1998) Effetti di lanci di *Orius laevigatus* (Fieber) (Rhynchota Anthocoridae) e dell'impollinazione manuale sulla fruttificazione dell'Anona (*Annona cherimola* Mill.). *Bollettino Di Zoologia Agraria e Di Bachicoltura* 30, 207–212. Available at: https://riviste.unimi.it/index.php/bzab/article/view/27448

Caleca, V., Lo Verde, G., Ragusa, S. and Tsolakis, H. (2002) Insect and hand pollination of *Annona* spp. in Sicily. *Phytophaga* 12, 117–127.

Calzavara, B.B.G. and Müller, C.H. (1987) *Fruticultra tropical: A gravioleira (Annona muricata L.)*. EMBRAPA-CPATU, Belém, Pará, Brazil. Available at: www.infoteca.cnptia.embrapa.br/infoteca/handle/doc/382561

Cautín, R. and Agustí, M. (2005) Phenological growth stages of the cherimoya tree (*Annona cherimola* Mill.). *Scientia Horticulturae* 105, 491–497. DOI: 10.1016/j.scienta.2005.01.035

Cautín, R., Reig, C., Mesejo, C., Martinez-Fuentes, A. and Agustí, M. (2010) Flowering and fruit set in cherimoya (*Annona cherimola* Mill.) as affected by the tree-training system. *Journal of Horticultural Science and Biotechnology* 85, 511–515. DOI: 10.1080/14620316.2010.11512706

Cavalcante, T.R.M., Naves, R.V., Franceschinelli, E.V. and da Silva, R.P. (2009) Pollination and fruit development in araticum. *Bragantia* 68, 13–21. DOI: 10.1590/S0006-87052009000100002

Chander, S. and Kurian, R.M. (2019) Effect of crop load, fruit position and shoot vigour on yield and quality of *Annona atemoya* × *Annona squamosa* in India. *Journal of Horticultural Science and Biotechnology* 94, 507–512. DOI: 10.1080/14620316.2019.1592712

Chander, S., Kurian, R.M., Satisha, J., Upreti, K.K. and Laxman, R.H. (2019a) Chemical interventions for advancing the fruiting season of sugar apple (*Annona squamosa* L.) cv. Balanagar. *International Journal of Chemical Studies* 7, 774–781.

Chander, S., Rajasekharan, P.E. and Kurian, R.M. (2019b) Pollen storage studies in sugar apple (*Annona squamosa* L.) cv. Balanagar. *Israel Journal of Plant Sciences* 66, 196–202. DOI: 10.1163/22238980-20191080

Chatrou, L.W., Escribano, M.P., Viruel, M.A., Maas, J.W., Richardson, J.E. and Hormaza, J.I. (2009) Flanking regions of monomorphic microsatellite loci provide a new source of data for plant species-level phylogenetics. *Molecular Phylogenetics and Evolution* 53, 726–733. DOI: 10.1016/j.ympev.2009.07.024

Coronel, R.E., Zuno, J.C. and Sotto, R.C. (eds) (1994) *Promising Fruits of the Philippines*. College of Agriculture, University of the Philippines at Los Baños, Laguna, Philippines.

Costa, M.S., Silva, R.J., Paulino-Neto, H.F. and Pereira, M.J. (2017) Beetle pollination and flowering rhythm of *Annona coriacea* Mart. (Annonaceae) in Brazilian cerrado: behavioral features of its principal pollinators. *PLOS ONE* 12: e0171092. DOI: 10.1371/journal.pone.0171092

Couvreur, T.L.P., Pirie, M.D., Chatrou, L.W., Saunders, R.M.K., Su, Y.C.F. *et al.* (2011) Early evolutionary history of the flowering plant family Annonaceae: steady diversification and boreotropical geodispersal. *Journal of Biogeography* 38, 664–680. DOI: 10.1111/j.1365-2699.2010.02434.x

Dao, Z.A., Romba, R., PEÑALVER-CRUZ, A., Satour, P., Jaloux, B. *et al.* (2023) Green fruit visitor insects of *Annona senegalensis* Pers. (Magnoliales: Annonaceae) and their potential role in the plant fruiting. Preprint. Available at: https://assets-eu.researchsquare.com/files/rs-2470333/v1/3a647de1-3d2b-4b4a-857d-4ed1892728db.pdf?c=1673970268 (accessed 2 September 2025).

de Moura, E.A., Chagas, P.C., Chagas, E.A., de Oliveira, R.R., Araújo, W.F. *et al.* (2019) Phenological study of sugar apple (*Annona squamosa* L.) in dystrophic yellow latosol under the savanna conditions of Roraima. *Australian Journal of Crop Science* 13, 1467–1472. DOI: 10.21475/ajcs.19.13.09.p1557

Duarte, O. and Escobar, O. (1997) Improving fruit set of cherimoya (*Annona cherimola* Mill.) cv. Cumbe, by autogamous and allogamous hand pollination. *Proceedings of the International Society for Tropical Agriculture* 41, 162–165.

Farré, J.M., Hermoso, J.M., Guirado, E. and García-Tapia, J. (1999) Techniques of cherimoya cultivation in Spain. *Acta Horticulturae* 497, 91–118. DOI: 10.17660/ActaHortic.1999.497.5

Fioravanço, J.C. and Paiva, M.C. (1994) Tratos culturais. In: Manica I. (ed.) *Fruticultura – Cultivo das Anonáceas*. Evangraf, Porto Alegre, Brazil, pp. 62–77.

Garcia del Corral, F. (1989) Polinización natural y artificial del chirimoyo. Estudio de polen y de la receptividad de la flor. MSc thesis, Escuela Universitaria de Ingeniería Técnica Agrícola, Seville, Spain.

Gardner, R.O. (1974) Trinucleate pollen in *Beilschmiedia nees* (Lauraceae). *New Zealand Journal of Botany* 12, 243–244. DOI: 10.1080/0028825X.1974.10428867

Gazit, S. and Eisenstein, D. (1985) Floral biology of *Annona squamosa* and *Annona cherimola* in relation to the spontaneous appearance of atemoya in Israel. In: *Proceedings of the 21st Annual Meeting of Caribbean Food Crops Society, 8–13 September, Port of Spain, Trinidad and Tobago*, pp. 66–67. DOI: 10.22004/ag.econ.261374

Gazit, S., Galon, I. and Podoler, H. (1982) The role of nitidulid beetles in natural pollination of *Annona* in Israel. *Journal of the American Society for Horticultural Science* 107, 849–852. DOI: 10.21273/JASHS.107.5.849

George, A.P. and Nissen, R.J. (1988) The effects of temperature, vapour pressure deficit and soil moisture stress on growth, flowering and fruit set of custard apple (*Annona cherimola* × *Annona squamosa*) 'African pride'. *Scientia Horticulturae* 34, 183–191. DOI: 10.1016/0304-4238(88)90091-X

George, A.P. and Nissen, R.J. (2002) Effects of drought on fruit set, yield and quality of custard apple (*Annona* spp. hybrid) 'African pride' plants. *Journal of Horticultural Science and Biotechnology* 77, 418–427. DOI: 10.1080/14620316.2002.11511515

George, A.P., Nissen, R.J. and Brown, B.I. (1987) The custard apple. *Queensland Agricultural Journal* 113, 287–286.

George, A.P., Nissen, R.J., Ironside, D.A. and Anderson, P. (1989) Effects of nitidulid beetles on pollination and fruit set of *Annona* spp. hybrids. *Scientia Horticulturae* 39, 289–299. DOI: 10.1016/0304-4238(89)90122-2

Goenaga, R. and Jenkins, D. (2016) Yield and fruit quality traits of atemoya hybrids grown in Puerto Rico. *HortTechnology* 26, 547–551. DOI: 10.21273/HORTTECH.26.4.547

González, M., Baeza, E., Lao, J.L. and Cuevas, J. (2006) Pollen load affects fruit set, size, and shape in cherimoya. *Scientia Horticulturae* 110, 51–66. DOI: 10.1016/j.scienta.2006.06.015

González, M., Hueso, J.J., Alonso, F. and (2013) Foliar application of urea advances bud break, bloom and harvest in cherimoya (*Annona cherimola* Mill.). *Acta Horticulturae* 975, 269–274. DOI: 10.17660/ActaHortic.2013.975.30

Gottsberger, G. (1989) Comments on flower evolution and beetle pollination in the genera Annona and Rollinia (Annonaceae). Plant Systematics and Evolution 167, 189–197.

Harder, L.D. and Johnson, S.D. (2008) Function and evolution of aggregated pollen in angiosperms. *International Journal of Plant Sciences* 169, 59–78. DOI: 10.1086/523364

Hernandez, F.L.M., Gomez, J.R. and Andres, A.J. (2013) *Important Insect Pests and Fungal Diseases of Soursop Cultivation*. Technical Book No.1, Santiago Iccuintla Experimental Field, Nayarit, Mexico.

Higuchi, H. and Utsunomiya, N. (1999) Floral differentiation and development in cherimoya (*Annona cherimola* Mill.) under warm (30/25°C) and cool (20/15°C) day/night temperatures. *Journal of the Japanese Society for Horticultural Science* 68, 707–716. DOI: 10.2503/jjshs.68.707

Higuchi, H., Utsunomiya, N. and Sakuratani, T. (1998) High temperature effects on cherimoya fruit set, growth and development under greenhouse conditions. *Scientia Horticulturae* 77, 23–31. DOI: 10.1016/S0304-4238(98)00160-5

Higuchi, H., Tsukada, M., Ninomiya, T., Furukawa, T., Yoshida, A. and Kuriyama, M. (2014a) Pollination of cherimoya (*Annona cherimola* Mill.) by the mass-propagated beetles *Haptoncus ocularis*, *Mimemodes monstrosus*, and *Carpophilus marginellus*. *Tropical Agriculture and Development* 58, 18–25. DOI: 10.11248/jsta.58.18

Higuchi, H., Tsukada, M., Yoshida, A. and Furukawa, T. (2014b) Effective pollinators among Japanese fauna of the insect visitors of cherimoya (*Annona cherimola* Mill.). *Tropical Agriculture and Development* 58, 33–36. DOI: 10.11248/jsta.58.33

Hoa, N.V., Dien, L.Q., Uyen, D.T.K. and Hang, N.T.N. (2017) Overview on guava, pineapple, wax apple, sugar apple research and production in Vietnam. *Acta Horticulturae* 1166, 15–24. DOI: 10.17660/ActaHortic.2017.1166.3

Inoue, H., Yonemoto, Y., Ogata, T. and Higashikayama, H. (2009) Effect of day length on shoot growth of atemoya cv Jeffner. *Research for Tropical Agriculture* 2, 15–17.

Jiménez-Zurita, J.O., Alia-Tejacal, I., Balois-Morales, R., Villarrcal-Funtes, J.M., Nunez-Colin, C.A. *et al.* (2023) Phenological growth stages of soursop trees (*Annona muricata* L.) based on the extended BBCH-scale. *Revista Chapingo Serie Horticulturae* 29, 5–18. DOI: 10.5154/r.rchsh.2022.03.006

Jürgens, A. and Dötterl, S. (2004) Chemical composition of anther volatiles in Ranunculaceae: genera-specific profiles in Anemone, Aquilegia, Caltha, Pulsatilla, Ranunculus, and Trollius species. *American Journal of Botany* 91, 1969–1980. DOI: 10.3732/ajb.91.12.1969. PMID: 21652345.

Kahn, T.L., Adams, C.J. and Arpaia, M.L. (1994) Paternal and maternal effects on fruit and seed characteristics in cherimoya (*Annona cherimola* Mill.). *Scientia Horticulturae* 59, 11–25. DOI: 10.1016/0304-4238(94)90087-6

Kitamura, M.C. (2002) Propagação precoce da gravioleira (*Annona muricata* L.) por enxertia. PhD thesis, Universidade Federal de Lavras, Lavras, Brasil.

Koura, S., Hasegawa, K., Suzuki, H., Yamamoto, Y. and Yonemoto, Y. (2001) Influence of high temperature and amount of viscous fluid on the surface of pistils on the fruit set of cherimoya (*Annona cherimola* Mill.). *Japanese Journal of Tropical Agriculture* 45, 275–280. DOI: 10.11248/jsta1957.45.275

Koura, S., Hasegawa, K., Yamamoto, Y. and Yonemoto, Y. (2004) Fruit set and fruit growth of seedless cherimoya (*Annona cherimola* Mill.) induced by $GA_3$ under greenhouse cultivation in Japan. *Acta Horticulturae* 653, 63–66. DOI: 10.17660/ActaHortic.2004.653.7

Kumar, P.S. and Jalikop, S.H. (2000) Cross-compatibility among *Annona* species. *Indian Journal of Horticulture* 57, 309–313.

Kumar, R., Hoda, M.N. and Singh, D.K. (1977) Studies on the floral biology of custard apple (*Annona squamosa* Linn.). *Indian Journal of Horticulture* 34, 252–256.

Laport, E.L. and Dent, R.R. (1992) Arthropods associated with cherimoya (*Annona cherimola* Mill.) flowering in Quillota, Chile. *Acta Entomológica Chilena* 17, 101–106.

Leal, F. and Paull, R.E. (2023) The genus *Annona*: botanical characteristics, horticultural requirements and uses. *Crop Science* 63, 1030–1049. DOI: 10.1002/csc2.20833

Lim, T.K. (2012) *Edible Medicinal and Non-Medicinal Plants: Vol. 1, Fruits*. Springer, Berlin.

Lin, H.S., Lu, P.S. and Jiang, S.W. (2014) Regulation of sugar apple and atemoya production season and its impact on industry in Taiwan. *Journal of the Taiwan Society for Horticultural Science* 60, 149–166.

Liu, K., Li, H., Yuan, C., Huang, Y., Chen, Y. and Liu, J. (2015) Identification of phenological growth stages of sugar apple (*Annona squamosa* L.) using the extended BBCH-scale. *Scientia Horticulturae* 181, 76–80. DOI: 10.1016/j.scienta.2014.10.046

Liu, K., Feng, S., Pan, Y., Zhong, J., Chen, Y. *et al.* (2016) Transcriptome analysis and identification of genes associated with floral transition and flower development in sugar apple (*Annona squamosa* L.). *Frontiers in Plant Science* 7: 1695. DOI: 10.3389/fpls.2016.01695

Liu, K., Feng, S., Jiang, Y., Li, H., Huang, S. *et al.* (2017a) Identification and expression analysis of seven MADS-box genes from *Annona squamosa*. *Biologia Plantarum* 61, 24–34. DOI: 10.1007/s10535-016-0688-1

Liu, K., Li, H., Li, W., Zhong, J., Chen, Y. *et al.* (2017b) Comparative transcriptomic analyses of normal and malformed flowers in sugar apple (*Annona squamosa* L.) to identify the differential expressed genes between normal and malformed flowers. *BMC Plant Biology* 17: 170. DOI: 10.1186/s12870-017-1135-y

Liu, K.D., Mo, Y.W., Feng, S.X., We, W.Y., Li, H.L. *et al.* (2018) Comparative proteomic analysis in different developmental stages of sugar-apple (*Annona squamosa* L.) flowers. *Scientia Agricultura Sinica* 51, 149–159.

Lora, J., de Oteyza, M.A.P., Fuentetaja, P. and Hormaza, J.I. (2006) Low temperature storage and *in vitro* germination of cherimoya (*Annona cherimola* Mill.) pollen. *Scientia Horticulturae* 108, 91–94. DOI: 10.1016/j.scienta.2005.12.003

Lora, J., Herrero, M. and Hormaza, J.I. (2009a) The coexistence of bicellular and tricellular pollen in *Annona cherimola* (Annonaceae): implications for pollen evolution. *American Journal of Botany* 96, 802–808. DOI: 10.3732/ajb.0800167

Lora, J., Testillano, P.S., Risueño, M.C., Hormaza, J.I. and Herrero, M. (2009b) Pollen development in *Annona cherimola* Mill. (Annonaceae). Implications for the evolution of aggregated pollen. *BMC Plant Biology* 9: 129. DOI: 10.1186/1471-2229-9-129

Lora, J., Hormaza, J.I. and Herrero, M. (2010) The progamic phase of an early-divergent angiosperm, *Annona cherimola* (Annonaceae). *Annals of Botany* 105, 221–231. doi: 10.1093/aob/mcp276

Lora, J., Herrero, M. and Hormaza, J.I. (2011a) Stigmatic receptivity in a dichogamous early-divergent angiosperm species, *Annona cherimola* (Annonaceae): influence of temperature and humidity. *American Journal of Botany* 98, 265–274. DOI: 10.3732/ajb.1000185

Lora, J., Hormaza, J.I., Herrero, M. and Gasser, C.S. (2011b) Seedless fruits and the disruption of a conserved genetic pathway in angiosperm ovule development. *Proceedings of the National Academy of Sciences USA* 108, 5461–5465. DOI: 10.1073/pnas.1014514108

Lora, J., Herrero, M. and Hormaza, J.I. (2012) Pollen performance, cell number, and physiological state in the early-divergent angiosperm *Annona cherimola* Mill. (Annonaceae) are related to environmental conditions during the final stages of pollen development. *Sexual Plant Reproduction* 25, 157–167. DOI: 10.1007/s00497-012-0187-2

Maia, A.C., Dötterl, S., Kaiser, R., Silberbauer-Gottsberger, I., Teichert, H. *et al.* (2012) The key role of 4-methyl-5-vinylthiazole in the attraction of scarab beetle pollinators: a unique olfactory floral signal shared by Annonaceae and Araceae. *Journal of Chemical Ecology* 38, 1072–1080. DOI: 10.1007/s10886-012-0173-z

Maita, S., Minchala, N. and Orellana, R. (2022). Assessment of *in vitro* pollen germination and pollen tube growth of *Annona cherimola* Mill. *International Journal of Fruit Science* 22, 57–63. DOI: 10.1080/15538362.2021.1988810

Matsuda, H. and Higuchi, H. (2012) Anatomical study on seasonal changes in pistil receptivity of cherimoya (*Annona cherimola* Mill.). *Tropical Agriculture and Development* 56, 95–103. DOI: 10.11248/jsta.56.95

Matsuda, H. and Higuchi, H. (2015) Effects of temperature and humidity conditions on anthesis and pollen germinability of cherimoya (*Annona cherimola* Mill.). *Tropical Agriculture and Development* 59, 57–62. DOI: 10.11248/jsta.59.57

Matsuda, H. and Higuchi, H. (2019) Effects of the pollen parent on cherimoya fruit set and quality. *Tropical Agriculture and Development* 63, 87–92. DOI: 10.11248/jsta.63.87

Matsuda, H., Higuchi, H., Kozai, N. and Ogata, T. (2011) Effect of temperature on the time requirement of pollen tubes to penetrate into the embryo sac after pollination in cherimoya (*Annona cherimola* Mill.). *Tropical Agriculture and Development* 55, 157–161. DOI: 10.11248/jsta.60.155

Matsuda, H., Higuchi, H. and Ogata, T. (2015) Sectional temperature control and anatomical study for critical post-pollination night temperatures to prevent fruit set failure in cherimoya (*Annona cherimola* Mill.). *Tropical Agriculture and Development* 59, 28–34. DOI: 10.11248/jsta.59.28

Matsuda, H., Higuchi, H. and Ogata, T. (2016) Anatomical observations of pollen starch accumulation and pollen germinability as affected by pre-anthesis night temperatures in cherimoya (*Annona cherimola* Mill.). *Tropical Agriculture and Development* 60, 155–161. DOI: 10.11248/jsta.60.155

Meena, H., Meena, N.K., Singh, J., Jain, S.K., Gupta, A.K. *et al.* (2023) Pollen sources and time of pollination for hand pollination affect biochemical and organoleptic attributes of sugar apple (*Annona squamosa* L.) cv. Arka Sahan. *Journal of Applied Horticulture* 25, 179–183. DOI: 10.37855/jah.2023.v25i02.31

Mendes, H.T.A., Costa, M.R., Nietsche, S., Oliveira, J.A.A. and Pereira, M.C.T. (2012) Pollen grain germination and fruit set in 'Brazilian seedless' sugar apple (*Annona squamosa* L.). *Crop Breeding and Applied Biotechnology* 12, 277–280. DOI: 10.1590/S1984-70332012000400007

Mo, Y.W., Liu, K.D., Song, H.W., Deng, X. and Chen, J.J. (2015) Effect of pruning and illumination on the flowering and gene expression of sugar apple (*Annona squamosa* L.). *Journal of Agricultural Biotechnology* 23, 991–1001. DOI: 10.3969/j.issn.1674-7968.2015.08.002

Morton, J.F. (1987) Annoaceae. In: Dowling, C.F. (ed.) *Fruits of Warm Climates*. J.F. Publishing, Morton, pp. 65–90.

Murai, K., Higuchi, H. and Matsuda, H. (2021) Effects of vapor pressure deficit on ovary morphogenesis and pollen-tube growth in pistils of cherimoya (*Annona cherimola* Mill.). *Tropical Agriculture and Development* 65, 185–194. DOI: 10.11248/jsta.65.185

Nagel, J., Peña, J.E. and Habeck, D. (1989) Insect pollination of atemoya in Florida. *Florida Entomologist*. 72, 207–211.

Nakasone, H.Y. and Paull, R.E. (1998) *Tropical Fruits*. CAB International, Wallingford, UK.

Nietsche, S., Pereira, M.C.T., Oliveira, C., Dias, M.M. and dos Reis, S.T. (2009) Viability of the sugar apple (*Annona squamosa*) pollen grains at different hours of the day. *Ciência e Agrotecnologia* 33, 527–531. DOI: 10.1590/S1413-70542009000200026

Olesen, T. and Muldoon, S.J. (2012) Effects of defoliation on flower development in atemoya custard apple (*Annona cherimola* Mill. × *A. squamosa* L.) and implications for flower-development modelling. *Australian Journal of Botany* 60: 160. DOI: 10.1071/BT11299

Peña, J.E., Nadel, H.M., Pereira, B. and Smith, D. (2002) Pollinators and pests of *Annona* species. In: Peña, D.E., Sharp, J.L. and Wysoki, M. (eds) *Tropical Fruit Pests and Pollinators: Biology, Economics, Natural Enemies and Control*. CABI, Wallingford, UK. pp. 197–221.

Pereira, M.C.T., Nietsche, S., Costa, M.R., Crane, J.H., Corsato, C.D.A. and Mizobutsi E.H. (2011) Anonáceas: pinha, atemóia e graviola. *Informe Agropeuario* 32, 26–34.

Pereira, M.C.T., Crane, J.H., Montas, W., Nietsche, S. and Vendrame, W.A. (2014) Effects of storage length and flowering stage of pollen influence its viability, fruit set and fruit quality in 'Red' and 'Lessard Thai' sugar apple (*Annona squamosa*) and 'Gefner' atemoya (*A. cherimola* × *A. squamosa*). *Scientia Horticulturae* 178, 55–60. DOI: 10.1016/j.scienta.2014.08.004

Pereira, M.C.T., São José, A.R., Botelho, A.P. and Nietsche, S. (2021) Tratos culturais, In: Pereira, M.C.T. and Borém, A. (eds) *Anonáceas: do Plantio a Colheita*. Editora UFV, Viçosa, Brazil, pp. 120–142.

Pinto, V.O., Pereira, M.C.T., Nietsche, S., Mendes, D.S., Mota, M.F.C. *et al.* (2018) Characterization of 'Gefner' atemoya seedless fruits with $GA_3$ application. *Revista Brasileira Ciências Agrárias* 13, 1–9. DOI: 10.5039/agraria.v13i3a5562

Pritchard, K.D. and Edwards, W. (2006) supplementary pollination in the production of custard apple (*Annona* sp.) – the effect of pollen source. *Journal of Horticultural Science and Biotechnology* 81, 78–83. DOI: 10.1080/14620316.2006.11512032

Rathod, A.R. (2019) Studies on leaf to fruit ratio on growth, yield and quality of custard apple. MSc thesis, Dr Panjabrao Deshmukh Krishi Vidyapeeth, Akola, India.

Richardson, A C. and Anderson, P.A. (1990) Is hand pollination of cherimoya necessary? *The Orchardist* November, 21–25.

Richardson, A.C. and Anderson, P.A. (1996) Hand pollination effects on the set and development of cherimoya (*Annona cherimola*) fruit in a humid climate. *Scientia Horticulturae* 65, 273–281. DOI: 10.1016/0304-4238(96)00878-3

Rodrigues, B.R.A., dos Santos, R.C., Nietsche, S., Mercadante-Simões, M.O., da Cunha, I.R.G. and Pereira, M.C.T. (2016) Determination of cardinal temperatures for sugar apple (*Annona squamosa* L.). *Ciência e Agrotecnologia* 40, 145–154. DOI: 10.1590/1413-70542016402039115

Romero-Soler, K.J. and Cetzal-Ix, W. (2015) Cultivated species of the genus *Annona* (Annonaceae) from the Yucatan Peninsula. *Mexico CICY Herbarium* 7, 147–153.

Rosell, P. and Galan, V. (1995) Notes on rhythms observed in the duration of flower anthesis throughout flowering in cherimoya on the island of Tenerife. *Fruits* 50, 233–237.

Rosell, P., Herrero, M. and Saúco, V.G. (1999) Pollen germination of cherimoya (*Annona cherimola* Mill.). *In vivo* characterization and optimization of *in vitro* germination. *Scientia Horticulturae* 81, 251–265. DOI: 10.1016/S0304-4238(99)00012-6

Saavedra, E. (1977) Influence of pollen grain stage at the time of hand pollination as a factor on fruit set of cherimoya. *HortScience* 12, 117–118. DOI: 10.21273/HORTSCI.12.2.117

Saavedra, E. (1979) Set and growth of *Annona cherimola* Mill. fruit obtained by hand-pollination and chemical treatments. *Journal of American Horticultural Science* 104, 668–673.

Sahoo, S.C., Panda, J.M. and Mohanty, D. (2000) Studies on floral biology of custard apple (*Annona squamosa*) under Bhubaneswar condition. *Orissa Journal of Horticulture* 28, 43–45.

Sampson, F.B. (1969) Cytokinesis in pollen mother cells of angiosperms, with emphasis on *Laurelia novae-zelandiae* (Monimiaceae). *Cytologia* 34, 627–634. DOI: 10.1508/cytologia.34.627

Sánchez, E.G., González, J.M.H., Perez de Oteyza, A., Bello, J.G.-T. and Massip, J.M.F. (2001) *Polinización del Chirimoyo.* Consejería de Agricultura y Pesca, Granada, Spain. Available at: www.juntadeandalucia.es/export/drupaljda/1337166159Polinizacixn_del_Chirimoyo.pdf (accessed 2 September 2025).

Sanghani, J. and Varu, D.K. (2022) Effect of different pollination methods on fruit set and yield in custard apple cv. Sindhan (*Annona squamosa* L.). *Pharma Innovation Journal* 11, 2426–2430.

Saunders, R.M.K. (2020) The evolution of key functional floral traits in the early divergent angiosperm family Annonaceae. *Journal of Systematics and Evolution* 58, 369–392. DOI: 10.1111/jse.12645

Schroeder, C.A. (1941) Hand pollination effects in the cherimoya (*Annona cherimola*). *Yearbook of the California Avocado Society* 26, 94–98.

Schroeder, C.A. (1943) Hand pollination studies on the cherimoya. *Proceedings of the American Society of Horticultural Science* 43, 39–41.

Soler, L. and Cuevas, J. (2008) Development of a new technique to produce winter cherimoyas. *HortTechnology* 18, 24–28. DOI: 10.21273/HORTTECH.18.1.24

Soler, L. and Cuevas, J. (2009) Early flower initiation allows ample manipulation of flowering time in cherimoya (*Annona cherimola* Mill.). *Scientia Horticulturae* 121, 327–332. DOI: 10.1016/j.scienta.2009.02.005

Soria, J.T., Hermoso González, J.M. and Farré, J.M. (1991) Polinización del chirimoyo. In: *II Jornadas de la Asociación de Productores de Frutos Subtropicales.* Ayuntamiento de Almuñecar, Granada, Spain, pp. 11–25.

Suzaki, N., Ichtinokiyama, H. and Suzuki, K. (2011) Cultural adaptability of custard apple (atemoya) to the East-Kisyu District of Mie Prefecture. 3. Effects of flowering season and time of artificial pollination on the fruit setting of atemoya variety 'Pinks Mammoth' and 'Hillary White'. *Bulletin of the Mie Prefectural Agricultural Research Institute* 33, 5–11.

Talavera, A., Fernandez-Pozo, N., Matas, A.J., Hormaza, J.I. and Bombarely, A. (2023) Genomics in neglected and underutilized fruit crops: a chromosome-scale genome sequence of cherimoya (*Annona cherimola*). *Plants, People, Planet* 5, 408–423. DOI: 10.1002/ppp3.10366

Thanarut, C. and Ko, J.C. (2017) Guava, wax apple, sugar apple and pineapple productions in Thailand. *Acta Horticulturae* 1166, 7–14. DOI: 10.17660/ActaHortic.2017.1166.2

Tsou, C.H. and Fu, Y.L. (2002) Tetrad pollen formation in *Annona* (Annonaceae): proexine formation and binding mechanism. *American Journal of Botany* 89, 734–747. DOI: 10.3732/ajb.89.5.734

Tsudaka, M., Higuchi, H., Furukawa, T. and Yoshida, A. (2005) Flower visitors to cherimoya, *Annona cherimola* (Magnoliales: Annonaceae) in Japan. *Applied Entomology and Zooology* 40, 317–324. DOI: 10.1303/aez.2005.317

Tsukada, M., Tanaka, D. and Higuchi, H. (2008) Thermal requirement for development of *Carpophilus marginellus* (Coleoptera: Nitidulidae), a potential pollinator of cherimoya and atemoya trees (Magnoliales: Annonaceae). *Applied Entomology and Zoology* 43, 281–285. DOI: 10.1303/aez.2008.281

van Heusden, E.C.H. (1992) Flowers of Annonaceae: morphology, classification, and evolution. *Blumea* 7 (Suppl.), 1–218.

Venkataratnam, L. (1959) Floral morphology and blossom biology studies on some Annonaceae. *Indian Journal of Agricultural Science* 29, 69–76.

Vithanage, H.I.M.V. (1984) Pollen–stigma interactions: development and cytochemistry of stigma papillae and their secretions in *Annona squamosa* L. (Annonaceae). *Annals of Botany* 54, 153–168. DOI: 10.1093/oxfordjournals.aob.a086779

Wester, P.J. (1910) Pollination experiments with anonas. *Bulletin of the Torrey Botanical Club* 37: 529. DOI: 10.2307/2479045

Wongs-Aree, C. and Noichinda, S. (2011) Sugar apple (*Annona squamosa* L.) and atemoya (*A. cherimola* Mill. × *A. squamosa* L.). In: Yahia, E.M. (ed.) *Postharvest Biology and Technology of Tropical and Subtropical Fruits: Mangosteen to White Sapote*. Woodhead Publishing, Sawston, UK, pp. 399–427. DOI: 10.1533/9780857092618.399

Yonemoto, Y. and Nakao, S. (1993) Fruit growth curves of cherimoya (*Annona cherimola* Mill.) from different blooming time in a plastic house and fruit qualities. *Japanese Journal of Tropical Agriculture* 37, 7–8.

Yonemoto, Y., Higuchi, H., Nakanishi, T. and Tomita, E. (1999) Conditions of artificial media for pollen germination and tube growth of cherimoya (*Annona cherimola* Mill.). *Japanese Journal of Tropical Agriculture* 43, 260–264. DOI: 10.11248/jsta1957.43.260

Yonemoto, Y., Ikeda, M., Higuchi, H. and Tomita, E. (2000) Investigation on effects of the concentration, amount and number of gibberellic acid ($GA_3$) treatments on the production of seedless cherimoya (Annona cherimola Mill.) fruits grown under plastic house conditions. *Japanese Journal of Tropical Agriculture* 44, 44–50.

# 10 Pests and Nematodes

**Nagalapura Ramakrishnappa Prasannakumar*, Manikyahalli Chandrashekar Keerthi and Mansheppa Devindrappa**
*Division of Crop Protection, ICAR-IIHR, Hessaraghatta Lake, Bengaluru, India*

## 10.1 Introduction

The genus *Annona* (family Annonaceae) includes several economically significant fruit crops such as *Annona cherimola* (cherimoya), *A. muricata* (soursop), *A. reticulata* (custard apple), *A. squamosa* (sugar apple) and *A. senegalensis* (wild soursop). These species originate from South America, Mesoamerica and Eastern Africa, each displaying unique climatic adaptations. *A. cherimola* thrives in subtropical highlands, while *A. muricata* and other species favour tropical lowlands but are cultivated in subtropical regions as well. Cherimoya and soursop have received international recognition for their nutritional and commercial value, as highlighted by studies from the US National Academy of Sciences. Additionally, species like *A. purpurea* (soncoya), *A. scleroderma* (poshte), and *A. crassiflora* (araticum) show potential for expansion due to their distinct flavours, medicinal applications or regional popularity. However, challenges persist, such as unsuccessful introductions of *A. glabra* (pond apple) in South-east Asia. Plant-parasitic nematodes continue to threaten horticultural crops throughout the world, particularly in tropical and subtropical regions. These nematodes act as pathogens themselves, but also predispose easy entry of soil-borne fungi and bacteria into the host plants. Thus, nematodes cause huge economic losses quantitatively and qualitatively.

## 10.2 Economic Importance

Annona fruits hold significant economic value in tropical and subtropical agricultural markets due to their rich flavour, high nutritional content and diverse applications in food, beverages and medicinal products. Cherimoya and soursop, in particular, command high market prices in international trade, with increasing demand in Europe and North America for the fresh fruit, processed pulp and functional health products. Additionally, soursop is widely researched for its potential anticancer properties. Despite their high market value, annona cultivation faces significant economic threats from pests, leading to reduced yields, fruit deformities and postharvest losses. Farmers often suffer major financial setbacks due to high pest control costs and the need for integrated pest management strategies.

## 10.3 Pest Diversity and Losses

A recent review by Romero *et al.* (2025) listed 614 arthropod species associated with 29 *Annona*

* Email: prasannaent@gmail.com


DOI: 10.1079/9781800627628.0010

spp., with most association recorded on the cultivated species *A. muricata* L., *A. squamosa* L. and *A. cherimola* Mill. The most common orders involved are Hemiptera, followed by Lepidoptera and Coleoptera. Different phytophagous arthropod species can become serious pests, causing significant damage to annona plants by feeding on various parts (Hernández-Fuentes *et al.*, 2022). The major pests causing fruit losses are fruit and seed borers, fruit flies, mealybugs and scale insects.

The production of *Annona* spp. is constrained by significant pest-related challenges. In Mexico, Vidal Hernández (1982) reported up to 70% crop loss due to insect damage, which also facilitated fungal infections in trees. Insect damage is particularly severe in tropical and subtropical climates, where favourable conditions allow high reproductive rates of pests. The lack of effective pest-resistant cultivars and dependence on chemical control methods further exacerbate the issue. Various studies have reported yield losses ranging from 30% to 80%, depending on the region, pest intensity and management practices employed. Overall average annual yield loss of the world's major horticultural crops due to damage by plant-parasitic nematodes is estimated at 13.54% (Reddy 2012; Rao *et al.*, 2015a, 2016).

## 10.4 Pests of *Annona* Spp.

### 10.4.1 Trunk borers

Trunk borers are among the most destructive pests of *Annona* spp. These insects belong to the order Coleoptera and include species such as *Cratosomus bombina*, *Eurypages pennatus* and *Heilipus catagraphus* (Curculionidae: Coleoptera). Compared with other species of trunk borers, *Cratosomus* sp. is a serious pest. It is referred to as the soursop twig and stem weevil.

#### *10.4.1.1 Distribution*

*Cratosomus* sp. was first observed in Costa Rica, where its host is the soursop.*Cratosomus* sp. produces eggs on stems and branches.

#### *10.4.1.2 Life cycle and biology*

The larva is about 40 mm long, creamy and soft-textured, with a prominent brown mark on its back near the end of the abdomen; the head possesses powerful mandibles. The pupae are exaggerated. The adult is dark with several protuberances on its body and is 30–35 mm in length. The rostrum, or beak, is long. Initial damage is difficult to detect; trees with severe damage wither and die. Before becoming a pupa, a larva can construct a 40 cm-long tunnel. The larvae of these weevils bore into the trunks, forming tunnels that interfere with the tree's ability to transport nutrients and water, leading to plant decline and, in severe cases, death (Fig. 10.1). A key symptom of infestation is the exudation of black oxidized sap from small holes where the insects have entered the trunk (Oliveira *et al.*, 1992). Over time, infested trees become structurally weak and susceptible to environmental stress and secondary infections.

#### *10.4.1.3 Management and control*

Management of trunk borers requires both cultural and chemical control strategies. One effective approach is the injection of insecticides, such

**Fig. 10.1.** Annona tree damaged by *Heilipus catagraphus*.

as deltamethrin and D-tetramethrin, directly into the borer holes to eliminate larvae feeding inside the trunk (Oliveira *et al.*, 2001). Additionally, in Brazil, farmers have successfully used a 1% creolin solution mixed with 10% lime to protect tree trunks from infestations (Oliveira *et al.*, 2001). Regular monitoring of trees for sap exudation and borer entry points is crucial to detect infestations early and prevent widespread damage. Proper pruning and crop sanitation help with the effective control of these pests.

### 10.4.2 Fruit borers

Fruit borers are another major threat to *Annona* spp. crops, with species such as *Cerconota anonella* (annona fruit borer; Depressariidae: Lepidoptera) and *Anonaepestis bengalella* (atis borer moth; Pyralidae: Lepidoptera) being the most significant. These pests are notorious for their ability to damage the fruit, rendering it unmarketable.

#### *10.4.2.1 Damage*

Females of *C. anonella* typically oviposit in the epidermis of the fruit at different stages of maturation, but in years of high insect population density, oviposition may even occur on the flowers. On emergence, the larvae immediately move towards the fruit pulp and start feeding, thereby causing three types of damage: necrosis, partial maturation of fruit and the fall of unripe fruit. Moreover, the openings made by the larvae on the fruit surface allow the invasion of the fungal *Colletotrichum* spp., which result in fruit mummification (Grissel and Schauff, 1990; Bustillo and Peña, 1992).

#### *10.4.2.2 Life cycle and biology*

The forewings of the moths are white, with scattered fuscous specks on the anterior side and a straight hazy fuscous shade from one-fifth of the costa to the middle of the dorsum, as well as a small dark fuscous spot denoting the second discal stigma. The wingspan is around 22 mm. The hind wings are grey. Its larvae burrow into the fruit, consuming the pulp and seeds, leading to premature fruit drop and secondary infections. The life cycle of *C. anonella* lasts about 36 days, with adult moths being attracted to black-light traps, which can be used for pest monitoring (Bustillo and Peña, 1992). Similarly, the *A. bengalella* is considered one of the most destructive pests of sugar apple in the Philippines (George and Nissen, 1992). In atemoya, the borer prefers to lay eggs in small fruit (1.5–5.5 cm in diameter) and does not usually lay eggs on larger fruit (do Nascimento *et al.*, 2023). It is believed that seed hardness and pulp thickness of ripe fruit are important for the occurrence of oviposition (Peña *et al.*, 2002). Larger fruit are attacked when the population is very high (do Nascimento *et al.*, 2023).

#### *10.4.2.3 Management and control*

Effective management of fruit borers requires an integrated pest management approach. Cultural practices such as sanitation (removing and destroying fallen or infested fruit) help reduce pest populations. Fruit bagging with chlorpyrifos-treated bags has been shown to prevent up to 92% of infestations, making it a reliable control method (Bustillo and Peña, 1992). Leite *et al.* (2012) suggested bagging of fruit when they are less than 2 cm in diameter with a transparent plastic bag. In Colombia and Ecuador, biological control methods using braconid parasitoids have proven effective in reducing *C. anonella* populations (Bustillo and Peña, 1992). Release of *Bacillus thuringiensis*, which acts in the insect digestive tract, especially in caterpillars, was found to be effective (Oliveira *et al.*, 2017). Additionally, chemical control through the application of triclorphon or fenthion (0.1%) every 15 days, targeting young fruit, has been recommended for suppressing borer activity (Torres and Sánchez, 1992). The insecticides triflumuron (2.4–4.8 g active ingredient (a.i.)) and imidacloprid (4–16 g a.i.) were found to be effective against fruit borer; however, imidacloprid left residue in the fruit after 21 days of application (Oliveira *et al.*, 2017).

### 10.4.3 Seed borers

Seed borers pose a unique threat to *Annona* spp., as they directly impact seed viability and fruit quality. *Bephratelloides cubensis* (Ashmead) (annona seed borer; Hymenoptera: Eurytomidae) and *Talponia batesi* (Tortricidae: Lepidoptera) are the most significant pests in this category, with the former being particularly damaging to soursop, sugar apple and custard apple. These insects belong to the order Hymenoptera (wasps) and are known for their ability to lay eggs inside developing seeds. Economic damage occurs when the adults chew their way out of the fruit, creating a tunnel that provides entry for other insects and decay organisms (Carrillo *et al.*, 2017).

#### 10.4.3.1 Distribution

Hernández-Fuentes *et al.* (2010) classified the genus *Bephratelloides* as neotropical, with *B. cubensis* having the most widespread range. According to Castañeda-Vildózola *et al.* (2010), *B. cubensis* first reported from Cuba and can now be found in the Caribbean, Mexico, Central America, northern South America and the USA (Florida and Hawaii).

#### 10.4.3.2 Damage

The larvae consume the seed's internal contents, weakening the fruit and making it susceptible to secondary infection. Seed borer damage consists of holes 0.1–0.15 cm in diameter on green fruit. The emerging adult females tunnel a distance of 0.6–1.1 cm through the pulp. Exit holes are used frequently for the entrance of sap beetles. This factor contributes to increased necrosis in the whole fruit (Peña *et al.*, 1984).

#### 10.4.3.3 Life cycle and biology

The newly emerged wasps prefer to oviposit on fruit. In one study, only females ($n = 159$) were observed, suggesting the possibility of parthenogenic reproduction (Peña *et al.*, 1984). Female body length ranges between 0.6 and 0.8 cm. The characteristic coloration is reddish-brown or light brown in some individuals. Under laboratory conditions (27 ± 2°C, 75 ± 1% relative humidity), female lifespan ranged between 1 and 11 days. The duration of oviposition was 15–30 min. Eggs are oviposited in developing seeds. Egg incubation lasts 12–14 days and the larval stage lasts 42–45 days (Bruner and Acuña, 1967). The pupal stage is 12–13 days in summer and 15–18 days in winter. It is believed that at least four generations (66–87 days per generation) occur per year in south Florida (Peña *et al.*, 1984).

#### 10.4.3.4 Management and control

Managing seed borers requires a combination of cultural, biological and chemical control methods. Collecting and destroying infested fruit can significantly reduce the population of these pests. Additionally, bagging fruit can prevent female wasps from laying eggs inside the seeds. Chemical control options include applying deltamethrin at 0.05% every 15 days during the early fruit development stages, which has proven effective in controlling infestations (Torres and Sánchez, 1992). Furthermore, biological control using entomopathogenic fungi such as *Beauveria bassiana* has demonstrated up to 90% mortality in adult *B. cubensis* in laboratory settings (Peña *et al.*, 2002).

### 10.4.4 Fruit flies

Fruit flies from the genera *Anastrepha*, *Ceratitis* and *Bactrocera* (Tephritidae: Diptera) are frequently cited as major insect pests affecting *Annona* spp., particularly cherimoya, in many regions (Alvarez *et al.*, 1999; Farré *et al.*, 1999).

#### 10.4.4.1 Distribution

Several studies have reported severe economic losses due to fruit fly infestations, with the Mediterranean fruit fly (*Ceratitis capitata*) affecting up to 50% of cherimoya fruit in Spain (Farré *et al.*, 1999) and cherimoya fruit flies (*Anastrepha* spp.) infesting more than 94% of cherimoya fruit in Ecuador (Alvarez *et al.*, 1999). However, in Brazil, where cherimoya is cultivated on a smaller scale, fruit flies have not been reported as a major pest (Kavati *et al.*, 1997; Bonaventure, 1999). Similarly, infestations on other *Annona* spp. such as soursop, sugar apple and custard apple are generally insignificant, with only a few documented cases, such as the attack of *Anastrepha ludens* on soursop in Mexico (Ponce and Vidal, 1981). In Australia, the Queensland fruit fly (*Bactrocera tryoni*) causes damage to atemoya in coastal regions of Queensland and northern New South Wales, while Mediterranean fruit fly is present is Western Australia (Bull, 2009). The persistence of fruit flies in certain areas is attributed to favourable climatic conditions, high reproductive potential and the absence of natural enemies, making their control particularly challenging (Farré *et al.*, 1999).

#### 10.4.4.2 Life cycle and biology

The eggs of *Bactrocera dorsalis* are elliptical, smooth, elongated, slightly curved and tapering at one end. The adult males of *B. dorsalis* are easily distinguishable by the presence of a tapering abdomen extending into an ovipositor and are comparatively larger than the females. Each female lays a mean total of 230–315 eggs during its life cycle. The developmental period from immature to adult stages varies from 16 to 22 days (Ashoka *et al.*, 2020). Females of the Queensland fruit fly lay 200–300 eggs in a lifetime. In summer, the life cycle takes about 17–21 days but may be as long as 2 months during the winter. The female fly 'stings' the fruit with her ovipositor and deposits one to 16 eggs in the fruit. After 2 days, the eggs hatch into maggots. Over the next 4–7 days, they will go through two moults (first, second and third instars), the last of which burrows to the outside of the fruit and drops to the ground where it changes to the pupal or resting stage in the litter under the fruit tree. Depending on climatic conditions, it will stay there for 6–14+ days and will then hatch as an immature fly, completing the life cycle (George *et al.*, 2015).

#### 10.4.4.3 Damage

*B. dorsalis* infests semi-ripened fruit by laying eggs, and the maggots feeds on the flesh when they enter the fruit (Fig. 10.2), puncturing the skin and causing direct damage (Navale *et al.*, 2024). The Queensland fruit fly mainly infests atemoya fruit maturing from March to May in Australia. The fly's larvae feed inside the fruit, causing breakdown of the flesh and rendering it inedible. It is not easy to detect symptoms of fruit fly infestation in unripe fruit. In near-ripe fruit, the fruit softens around the egg-laying sites, sometimes with associated yellowing and exudates (George *et al.*, 2015).

**Fig 10.2.** The fruit fly *Bactrocera dorsalis*.

#### 10.4.4.4 Management and control

Due to the underground pupation behaviour of fruit fly larvae, most control measures target adult flies through insecticide applications (Farré *et al.*, 1999). Traditional methods involve the spraying of toxic baits containing 4% hydrolysed protein and 0.15% dimethoate over the upper portions of trees, a practice that has been used to control *C. capitata* on cherimoya (Fuentes, 1999). However, concerns over environmental safety have led researchers to explore integrated pest management strategies. Studies in Spain demonstrated that a combination of phosphate baits, pheromone traps and improved field hygiene (removal of fallen fruit) can reduce fruit fly infestations to 4–7% without insecticide applications (Farré *et al.*, 1999). Additionally, treating cherimoya fruit with hot water at 45–47°C for 60 min further decreased infestations to 0.5–2.0% (Farré *et al.*, 1999). In Mexico, simple cultural practices such as removing overripe and fallen fruit have reduced fruit fly populations by up to 80% (Rebollar-Alviter *et al.*, 1997). Fruit bagging has also proven to be an effective mechanical control against fruit fly oviposition (Nakasone and Paull, 1998). Bagging fruit with white wax-coated bags was suggested by Navale *et al.* (2024) against *B. dorsalis*. Another effective practice, commonly used in orchards in north-eastern Brazil, involves harrowing the soil beneath tree canopies to disrupt the pupation phase and interrupt the insect's life cycle. In Mexico, biological control methods have been successfully implemented to manage fruit fly populations affecting *Annona* spp., with the parasitoid wasps *Biosteres longicaudatus* and *Aceratoneuromyia indica* being utilized for their effectiveness in targeting fruit fly larvae (Rebollar-Alviter *et al.*, 1997). Cover sprays of insecticide are not desirable, as they are harmful to the natural enemies of other pests. However, when more than 30 fruit flies are caught in the traps, the broad-spectrum insecticide trichlorfon (Lepidex or Dipterex) can be used to control them (George *et al.*, 2015).

### 10.4.5 Mealybugs

Some species of mealybugs can cause significant crop loss. In Florida, the pink hibiscus mealybug (*Maconellicoccus hirsutus* (Green)) may cause severe damage to soursop (Carrillo *et al.*, 2017). The most commonly seen mealybug in California is the long-tailed mealybug (*Pseudococcus adonidum*) (Phillips *et al.*, 1987). The pink hibiscus mealybug was first detected in northern regions of Brazil in 2010. It causes high economic damage to fruit and trees. (Oliveira *et al.*, 2014). In Australia, citrus mealybug (*Planococcus citri*) affecting atemoyas occurs throughout Australia but is much more common in coastal districts and in areas north of Sydney (George *et al.*, 2015). In India, the mealybugs *Planococcus pacificus* and *M. hirsutus* have been reported to cause damage on sugar apple (Shukla and Tandon, 1984; Sahu *et al.*, 2024). The striped mealybug (*Ferrisia virgata*) has also been recorded in sugar apple, while the cotton mealybug (*Phenacoccus solenopsis*) is reported to cause severe damage in the production of soursop in Bangladesh (Ahmed *et al.*, 2022).

**Fig.10.3.** (A) Mealybug-infested cherimoya. Photograph courtesy of Dr Ben Faber. (B) Mealybug-infested sugar apple. Photograph courtesy of Dr Prawit Chumchu. (C) Argentine ant. Photograph courtesy of Dr Ben Faber.

#### *10.4.5.1 Damage*

Long-tailed mealybugs usually attack the stem end of the fruit first and eventually cover the entire surface of the fruit (Fig. 10.3A, B). Pink hibiscus mealybugs suck sap from young stems, flowers and fruit (Vitullo *et al.*, 2009). The toxic saliva of the mealybugs causes shortening of internodes, curling and crinkling of leaves, and deformation and early fall of fruit and flowers (Negrini *et al.*, 2018). Indirect damage is caused by development of sooty mould, a fungus that covers the plant structure, growing on the honeydew secreted by the mealybug. In some cases, the mealybugs covered the entire fruit surface (George *et al.*, 2015).

#### *10.4.5.2 Biology*

Adult females of citrus mealybugs are white, about 3 mm long and covered by a white mealy wax.

The males are short lived. During summer, the life cycle takes about 4–6 weeks. There are about four to six generations of citrus mealybugs per season (George *et al.*, 2015). The striped mealybug is named for the two dark, dorsal stripes that run longitudinally down its body. The length of the adult female body is approximately 2.0–4.5 mm. The striped mealybug has three nymphal instars, and at the end of the third instar, the female moults into the sexually mature adult form. There are three to five generations per year (Awadallah *et al.*, 1979). Development of the pink hibiscus mealybug occurs in 3–4 weeks and each female can produce about 500 eggs. Eggs hatch in 3–8 days and the nymphal stage lasts 10–22 days (Carrilo *et al.*, 2017).

#### 10.4.5.3 Control

Long-tailed mealybugs are generally under good biological control in Californian orchards, except where Argentine ants (*Iridomyrmex humilis*) (Fig. 10.3C) are foraging for honeydew as a food source. Use of sticky bands on the lower trunk was suggested to restrict movement of the ants (Phillips *et al.*, 1987). In Australia, sulfoxaflor (Transform), imidacloprid (Confidor) buprofezin (Applaud) and paraffin oils (Biopest, Bioclear) shows good control of citrus mealybug on atemoya (George *et al.*, 2015).

Parasitic wasps (*Leptomastix dactylopii*), mealybug ladybirds (*Cryptolaemus montrouzieri*) and lacewings (*Oligochrysa lutea*) are the preferred biological control agents (George *et al.*, 2015). Biological control with the parasitoid *Anagyrus kamali* and crop management, mainly fructification pruning, are effective strategies for reducing the population levels of *M. hirsutus* (Negrini *et al.*, 2018). Ahmed *et al.* (2022) suggested bagging of fruit against the mealybug *P. solenopsis* affecting soursop.

### 10.4.6 Mites

Mites such as spider mites are a serious problem in sugar apple and atemoya orchards in Taiwan (Tsai and Hsu, 2017). The two-spotted mite (*Tetranychus urticae*) is widely distributed through all the major atemoya-growing regions of Australia (George *et al.*, 2015).

#### 10.4.6.1 Damage

Spider mite-infested leaves initially show rust spots followed by yellowing, browning and, under severe infestation, leaf fall. They also damage the fruit and produce wounds visible as black dots on stem ends or scales at later stage, and impact on the appearance of the fruit (Tsai and Hsu, 2017). Two-spotted mites cause the leaves of trees to turn a dull greyish colour and in some cases, the leaves become chlorotic and fall (George *et al.*, 2015). The adult female of the two-spotted mite is 0.5 mm long and greenish-yellow, with a dark spot on either side of the body.

Leaves infested by *Aceria annonae* (Keifer) (Prostigmata: Eriophyidae) were observed in the group, with protuberances that, as the leaves develop, increase in size in the early stages and then shrink or bend; on the underside of leaves there is excessive formation of trichomes, known as erinea (Ochoa *et al.*, 1994).

#### 10.4.6.2 Distribution

Hsu (2014) reported four species of mites on *Annona* spp. in Taitung, Taiwan, between 2009 and 2012. *Oligonychus coffeae*, *Tetranychus kanzawai*, *T. piercei* and *T. urticae*. In north-eastern Brazil, *Amblyseius aerialis* (Muma), *Tetranychus mexicanus* (McGregor), *Paraponematus acaciae* Baker, *Pronematus ubiquitus* (McGregor) and *Agistemus floridanus* Gonzales were found to be associated with soursop, sugar apple and atemoya (de Sousa *et al.*, 2015). The eriophyid *Aceria annonae* (Keifer) infests soursop between February and May in Nayarit, Mexico (Flores-Canales *et al.*, 2019).

#### 10.4.6.3 Control

Mites are more likely to be a problem if there is excessive application of broad-spectrum pesticides. They are normally controlled by predatory phytoseiid mites and also devoured by *Stethorus* spp. ladybirds, lacewing larvae and predatory thrips (George *et al.*, 2015).

### 10.4.7 Fruit-piercing moths

The fruit-piercing moth *Gonodonta nutrix* Stoll, 1780 (Lepidoptera: Erebidae) is found from North to South America. It is present in the USA, Colombia, Barbados, Cuba, Mexico, Brazil, Ecuador, Paraguay, Surinam, Venezuela and Panama, as well as other countries.

#### *10.4.7.1 Damage*

The larvae of *G. nutrix* typically target plants in the Solanaceae family as well as species in the genus *Annona* (Kondo *et al.*, 2024). Both immature stages and adult moths of *Gonodonta* spp. are known to cause damage. The first- and second-instar larvae of *G. nutrix* feed superficially on the leaf epidermis of young leaves of *A. glabra* and the last three instars feed on the leaves and young shoots (King and Thompson, 1958).

Infestation of *Gonodonta* spp. larvae defoliating sugar apple and atemoya was reported from Minas Gerais, Brazil (Silva *et al.*, 2017). The species identified were *Gonodonta bidens* Geyer, *G. nutrix* Stoll and *G. pyrgo* Cramer. Eighty-five per cent of trees in the orchard was found to be affected by the larvae and the average defoliation percentage was 30% (Silva *et al.*, 2017). In Nayarit, Mexico, heavy damage to soursop foliage by *G. pyrgo* was observed during August and September (Hernández-Fuentes *et al.*, 2015). This insect lays eggs on leaves and young shoots. The larvae feed on leaves, damaging small fruit and flowers, causing the fall of leaves, flower and fruit. When it is about to pupate, the larvae come down to the ground, form a cocoon with small pieces of soursop leaves and hides under the leaves. The larvae were detected from August to December with infestations of up to 46 larvae per tree; in some areas, 40% of defoliation occurred. Two population peaks, one in August and one in November, were recorded, indicate the existence of two generations per year (Hernández-Fuentes *et al.*, 2015). In Florida, two species, *G. nutrix* and *G. unica*, were reported to infest *Annona* spp., but these species do not cause significant damage because they are heavily parasitized by a species of braconid wasp (Peña and Crane, 2006). Kondo *et al.* (2024) found that the larvae of *G. nutrix* were parasitized by a braconid parasitoid from the genus *Glyptapanteles* Ashmead 1904 (Hymenoptera: Braconidae: Microgastrinae).

## 10.5 Nematodes

Several plant-parasitic nematodes (PPNs) have been detected in association with *Annona* spp. in rhizosphere soil, including *Aphelenchus*, *Pratylenchus*, *Helicotylenchus*, *Rotylenchulus*, *Dorylaimida*, *Monochus* and *Meloidogyne* spp. These nematodes appear in studies on the diversity of the nematode community of which a few are correlated with damage to the annona crop.

Soursop was found to be a suitable host for several *Helicotylenchus* spp., including *H. cavenessi* (Caveness, 1967). *Pratylenchus coffeae* has been shown to be the causal agent of 'sudden death' of soursop in Brazil (de Moura *et al.*, 1999). The nematodes associated with soursop in south Bahia, Brazil, were identified as *Aphelenchoides*, *Aphelenchus*, *Helicotylenchus*, *Mesocriconema*, *Rotylenchulus* and *Tylenchus* spp. (Silveira *et al.*, 2008). In sugar apple, the presence of *Meloidogyne*, *Rotylenchulus* and *Criconemoides* spp. was detected in root and soil samples in the Philippines (Bumanglag *et al.*, 2020). Three nematode species (*Helicotylenchus*, *Dorylaimida* and *Monochus* spp.) were recorded in soursop orchards in Port Harcourt, Nigeria (Nengi-Benwari and Marcus, 2025). In south Florida, McSorley *et al.* (1982) recorded a number of PPNs in different *Annona* spp.: *Criconemella sphaerocephala* was recoded in sugar apple, custard apple and mountain soursop (*Annona montana*); *Rotylenchulus reniformis* in custard apple, mountain soursop and pond apple; *Xiphinema americanum* in sugar apple and mountain soursop; *Helicotylenchus dihystera* in pond apple and custard apple; and *Quinisulcius actus* in pond apple.

Although very little plant damage was evident in the different reports, soursop harboured high populations of several different nematode species and should be monitored for potential problems. In soursop, reniform nematode parasitism resulted in a very typically high level of isolate necrosis on younger roots and coalescent necrosis along the older roots. Secondary symptoms caused stunting, yellowing and defoliation (de Moura *et al.*, 2005). From a pot experiment, Tanimola and Ezeunara (2021) found that soursop seedlings were resistant to root-knot nematode (*Meloidogyne incognita*) and did not support reproduction and subsequent damage. Atemoya and soursop are non-hosts of *M. enterolobii* (Freitas *et al.*, 2017).

Simply identifying the nematodes and applying the appropriate nematicide is not a permanent/long-lasting solution for nematode management. Some of the chemical nematicides used are carcinogenic and toxic to humans, animals and the environment (Rao *et al.*, 2015b). They are also known to cause deterioration of soil quality and contaminate ground water (Pulavarty *et al.*, 2021). Conventional management practices include the use of clean planting materials, biofumigation of fields, heat treatment and organic amendments. Chemical fumigants such as Paladin (dimethyl disulfide), Telone II, Vapam (metam sodium) and chloropicrin were found to be successful to a certain extent (Hajihassani, 2018). Vapam demonstrated higher efficiency against root-knot and cyst nematodes when applied by shank injection (Li *et al.*, 2006). The fumigant Dazomet is also effective against root-knot and cyst nematodes. It is applied in the soil and the treatment is completed by irrigation to ensure chemical transformation of the product into methyl isothiocyanate (MITC), formaldehyde and carbonyl sulfide, which provides pest control. For application of the fumigants, it is necessary to cover the soil surface with an impermeable plastic film to avoid aerial dispersion of methyl isothiocyanate in the environment (Sasanelli *et al.*, 2021).

Cultivating cover crops increases soil fertility and reduces PPN populations. These crops can act as: (i) a non-host or low-susceptibility crop that does not allow PPNs to complete their life cycle; (ii) producers of secondary 'allelochemical' metabolites that affect nematode populations; and (iii) factors creating favourable conditions for the development of microflora and microfauna antagonists of PPNs (Westerdahl, 2020).

Biocontrol strategies that use nematode-trapping fungi, bacteria and other nematode-parasitic organisms are considered safe. Nematodes are preyed upon by these natural predators, lowering their populations and limiting crop damage (Zhang *et al.*, 2020). Nematophagous fungi – a small group of microfungi that can capture, kill and digest nematodes – are very effective against several nematodes. The advantages of using fungi for biocontrol include selectivity, environmental safety, long-term effectiveness, sustainability and compatibility with other management strategies (Rahman *et al.*, 2024).

## 10.6 Conclusion

Production of clean planting material is of the utmost importance in the current Indian scenario for annona crop production, as horticultural nurseries play a major role in dissemination of PPNs, especially in fruit crops. As most of nematodes feed on plant roots and underground plant parts, farmers, nurserymen and orchardists remain ignorant about their infestations and disseminations through planting material. This unawareness about nematode damage potential and its diagnosis by the crop producers is one of the main reasons behind the non-adoption of technologies available to manage nematodes. Hence, diagnosis and detection of nematode problems in the early stages of crop production is very important in terms of taking precaution measures for sustainable management practices. Production of *Annona* spp. is constrained by significant pest-related challenges. The important pests affecting the tree and fruit are: trunk borer, fruit borer, seed borer, fruit flies and mealy bugs. Given that the incidence of pest species and damage vary according to biogeographic conditions, each country should promote tailor-made pest management packages that are adapted to local scenarios.

## References

Ahmed, T., Hasan, M.M., Hassan, K., Ahmed, J.D., Ahmed, K.S.D. *et al.* (2022) Fruit bagging of custard apple (*Annona reticulata*) as an eco-friendly protection approach against mealybug (*Phenacoccus solenopsis*) infestation in the north-Eastern Bangladesh. *International Journal of Tropical Insect Science* 42, 723–732. DOI: 10.1007/s42690-021-00596-3

Alvarez, J.V.U., Motoche, J.P.R. and Scheldeman, X. (1999) Estudio del cultivo de la chirimoya (*Annona cherimola* Mill.) en las Provincia de Loja, Ecuador. In: *II Congresso International de Anonáceas Universad de Ciencias y Artes Del Estado de Chiapas*. Tuxtla Gutiérrez, Chiapas, Mexico, pp. 230–237.

Ashoka, K.S., Reddy, B.T., Bharathi, M.C. and Abhishek, B.M. (2020) Biology and morphometry of different life stages of oriental fruit fly, *Bactrocera dorsalis* on custard apple reared under laboratory condition. *Journal of Entomology and Zoology Studies* 8, 1522–1525.

Awadallah, K.T., Ammar, E.D., Tawfik, M.F.S. and Rashad, A. (1979) Life-history of the white mealy-bug *Ferrisia virgata* (CKLL.) (Homoptera, Pseudococcidae). *Deutsche Entomologische Zeitschrift* 26, 101–110. DOI: 10.1002/mmnd.19790260111

Bonaventure, L. (1999) The cultivation of the cherimoya and of its hybrid atemoya in Brazil. *Acta Horticulturae* 497, 143–152. DOI: 10.17660/ActaHortic.1999.497.7

Bull, R. (2009) *Control of Fruit Files*. Newsletter of the Subtropical Fruit Club of Queensland, Mount Samson, Queensland, Australia.

Bumanglag, J.D.M., Soliman, M.D.E., Alvarez, L.V., Mapanao, C.P., Rendon, A.O. and Zurbano, L.Y. (2020) Plant parasitic nematodes associated with *Annona squamosa* L. *Journal of Experimental Biology and Agricultural Sciences* 8, 818–827. DOI: 10.18006/2020.8(6).818.827

Bruner, S.C. and Acuña, J. (1967) Sobre la biologia de *Bephrata cubensis* Ashm., el insect perforador de las frutas anonáceas. *Academia de Ciencias de Cuba, Instituto de Agronomía, Serie Agrícola* 1: 14.

Bustillo, A.E. and Peña, J.E. (1992) Biology and control of the *Annona* fruit borer *Cerconota annonella* (Lepidoptera: Oecophoridae). *Fruits* 47, 81–84.

Caveness, F.E. (1967) Shadehouse host ranges of some Nigerian nematodes. *Plant Disease Reporter* 51, 33–37.

Carrillo, D., Pena, J.E., Crane, J.H. and Duncan, R.E. (2017) Pollinator and insect/mite management in *Annona* spp. ENY-834. IFAS Extension, University of Florida, Gainesville, Florida, p. 6.

Castañeda-Vildózola, A., Nava-Díaz, C., Váldez-Carrasco, J., Ruiz-Montiel, C., Vidal-Hernández, L. and Barrios-Matias, S. (2010) Distribution and host range of *Bephratelloides cubensis* Ashmead (Hymenoptera: Eurytomidae) in Mexico. *Neotropical Entomology* 39, 1053–1055. DOI: 10.1590/s1519-566x2010000600034

de Moura, R.M., Guimaraes, L.M.P. and Pedrosa, E.M.R. (1999) Studies in sudden death of the soursop. *Nematologia Brasileira* 23, 62–68.

de Moura, R.M., Sandra, R.V.L.M. and Guimarães, L.M.P. (2005) Soursop, a new host of *Rotylenchulus reniformis*. *Fitopatologia Brasileira* 30, 437–437. DOI: 10.1590/S0100-41582005000400022

do Nascimento, A.B., Pereira, M.C.T., Oliveira, P.C. do C., Giustolin, T.A. and Alvarenga, C.D. (2023) Dynamics of borer infestation in atemoya fruit in two production periods. *Revista Brasileira de Fruticultura* 45: e021. DOI: 10.1590/0100-29452023021

Farré, J.M., Hermoso, J.M., Guirado, E. and García-Tapia, J. (1999) Techniques of cherimoya cultivation in Spain. *Acta Horticulturae* 497, 91–118. DOI: 10.17660/ActaHortic.1999.497.5

Flores-Canales, R.J., Acuña-Soto, J.A., Santillán-Ortega, C., Isiordia-Aquino, N., Sotelo-Montoya, A.M. and Hernández-Zaragoza, R.D. (2019) Population fluctuation of *Aceria annonae* (Keifer, 1973) (Prostigmata: Eriophyidae), in three municipalities of Nayarit, Mexico. *Revista Mexicana de Ciencias Agricolas* 10, 177–186. DOI: 10.29312/remexca.v10i1.1734

Freitas, V.M., Silva, J.G.P., Gomes, C.B., Castro, J.M.C., Correa, V.R. and Carneiro, R.M.D.G. (2017) Host status of selected cultivated fruit crops to *Meloidogyne enterolobii*. *European Journal of Plant Pathology* 148, 307–319. DOI: 10.1007/s10658-016-1090-8

Fuentes, L.J. (1999) Production of cherimoya (*Annona cherimola* Mill.) in Ecuador. *Acta Horticulturae* 497, 59–70. DOI: 10.17660/ActaHortic.1999.497.3

George, A.P. and Nissen, R.J. (1987) Propagation of *Annona* species: a review. *Scientia Horticulturae* 33, 75–85. DOI: 10.1016/0304-4238(87)90034-3

George A.P. and Nissen R.J. (1992) Edible fruits and nuts: *Annona cherimola* Mill. and *Annona squamosa* L. In: Verherij, E.W.M. and Cornell, R.E. (eds) *Plant Resources of South East Asia, Vol.* 2. Bogor, Indonesia, pp. 71–75.

George, A., Broadley, R., Hutton, D., Redpath, S., Bignell, G. *et al.* (2015) *Integrated Pest and Disease Management Manual for Custard Apple*. Queensland Department of Agriculture and Fisheries, Nambur, Queensland. Available at: www.horticulture.com.au/globalassets/hort-innovation/resource-assets/cu13001-custard-apple-ipdm-manual.pdf (accessed 19 August 2025).

Grissel, E.E. and Schauff, M.E. (1990) A synopsis of the seed-feeding genus *Bephratelloides* (Chalcidoidea: Eurytomidae). *Proceedings of the Entomological Society of America* 92, 177–187.

Hajihassani, A. (2018) *Chemical Nematicides for Control of Plant-parasitic Nematodes in Georgia Vegetable Crops*. Bulletin 1502. College of Agriculture and Environment Sciences, University of Georgia, Athens, Georgia. Available at: https://extension.uga.edu/publications/detail.html?number=B1502 (accessed 19 August 2025).

Hernández-Fuentes, L.M., Urias-López, M.A. and Bautista-Martínez, N. (2010) [Biology and behavior of the seed borer wasp *Bephratelloides cubensis* Ashmead (Hymenoptera: Eurytomidae)]. *Neotropical Entomology* 39, 527–534. DOI: 10.1590/s1519-566x2010000400010

Hernández-Fuentes, L.M., Carrasco, J.M.V., Urías-López, M.A. and Jaimes, R.G.G. (2015) Identification, damage and population fluctuation of *Gonodonta pyrgo* (Lepidoptera: Noctuidae) in *Annona muricata*. *Revista Mexicana de Ciencias Agrícolas* 6, 1001–1012.

Hernández-Fuentes, L.M., Velázquez Monreal, J.J., Hernández, H.G., Illescas Riquelme, C.P. and Manzanilla Ramírez, M.Á. (2022) Insectos y acaros fitófagos asociados con la Guanábana *Annona muricata* L.: una revisión. *Southwestern Entomologist* 47, 747–762. DOI: 10.3958/059.047.0325

Hsu, Y.T. (2014) Investigation of spider mites on *Annona* spp. in Taitung and the toxicity of acaricides to Tetranychus Kan. Zawai (Tetranychidae: Acari). *Formosan Entomology* 34, 11–20.

Kavati, R., Bueno, S.C., Sales Tokunaga, T., Nogueira, E.M.C., Takassaki, J. *et al.* (1997) *Graviola Para Exportação: Aspectos Fitossanitários*. EMBRAPA-SPI, Brasilia.

King, J.R. and Thompson, W.L. (1958) Fruit piercing moth, *Gonodonta nutrix* (Cramer), attacks oranges in Florida. *Florida Entomologist* 41: 61. DOI: 10.2307/3492361

Kondo, T., Campos-Patiño, Y., Tarazona-Velásquez, R., Sarmiento, C.E. and Vergara-Navarro, E.V. (2024) First report of fruit-piercing moth *Gonodonta nutrix* Stoll, 1780 (Lepidoptera: Erebidae) in Colombia, with notes on its morphology and biology. *Applied Entomology* 50, 683–699.

Leite, G.L.D., Souza, M.F., Souza, P.N.S., Fonseca, M.M. and Zanuncio, J.C. (2012) The bagging of *Annona crassiflora* fruits to control fruit borers. *Acta Scientiarum* 34, 253–257. DOI: 10.4025/actasciagron.v34i3.14713

Li, L.Y., Barry, T., Mongar, K. and Wofford, P. (2006) Modeling methyl isothiocyanate soil flux and emission ratio from a field following a chemigation of metam-sodium. *Journal of Environmental Quality* 35, 707–713. DOI: 10.2134/jeq2005.0216

McSorley, R., Campbell, C.W. and Prasad, J.L. (1982) Nematodes associated with tropical and subtropical fruit trees in south Florida. *Proceeding of the Florida State Horticultural Society* 95, 132–135.

Nakasone, H.Y. and Paull R., E. (1998) Annonas. In: Nakasone, H.Y. and Paull, R.E. (eds) *Tropical Fruits*. CAB International, Wallingford, UK, pp. 45–75.

Navale, S.M., Walunj, A.R., Kadam, U.K. and More, P.E. (2024) Assessment of hybrid custard apple (*Annona* spp.) genotypes for fruit fly (*B. dorsalis*) infestation and control strategies. *International Journal of Advanced Biochemistry Research* 8, 759–761. DOI: 10.33545/26174693.2024.v8.i12j.3276

Negrini, M., Morais, E.G.F., Batista, J.S.R. and Chagas, E.A. (2018) Population fluctuations in the pink hibiscus mealybug and its natural enemies in *Annona squamosa* (Annonaceae) in Roraima, Brazil. *Acta Amazonica* 48, 28–31. DOI: 10.1590/1809-4392201701131

Nengi-Benwari, A.O. and Marcus, M.U. (2025) Diversity of nematodes and earthworms associated with tropical trees in an African arboretum. *Asian Journal of Research in Zoology* 8, 17–24. DOI: 10.9734/ajriz/2025/v8i2189

Ochoa, R., Aguilar, H. and Vargas, C. (1994) *Phytophagous Mites of Central America: An Illustrated Guide*. Technical Manual. Centro Agronomico Tropical de Investigation y Ensenanza, Turrialba, Costa Rica.

Oliveira, A. da S., Castellani, M.A., Moreira, A.A., do Nascimento, A.S., Azevedo, M.S. and Oliveira, V.G. (2017) Efficacy of insecticides in fruit borer control and residues on sugar apple fruit. *Revista Ceres* 64, 132–137. DOI: 10.1590/0034-737x201764020004

Oliveira, M.A.S, Genu, P.J. de C., Junqueira, N.T.V. and Pinto, A.C. de Q. (1992) *Pragas da Gravioleira no Cerrado*. Embrapa Cerrado, Brasília.

Oliveira, M.A.S., Icuma I., M. and Alves R., T. (2001) *Insectos-praga e seu Controle. Graviola, Produção*. Embrapa Informação Tecnológica, Brasilia, DF, pp. 34–38.

Oliveira, J.E.M., Lopes, F.S.C., Oliveira, M.D., Pereira, V.S., Freitas, M.T.S. *et al.* (2014) Registro de ocorrencia da cochonilha-rosada *Maconellicoccus hirsutus* no Semiardo, Brasileiro. *Proceedings of XXV Congresso Brasileiro de Entomologia*. In: Goias, Brasil.

Peña, J.E. and Crane, J.H. (2006) Insect/mite management in *Annona* spp. ENY-836 (IG 166). Florida, USA: Florida Cooperative Extension Service, University of Florida.

Peña, J.E., Glenn, H. and Baranowski, R.M. (1984) Important insect pests of *Annona* spp. in Florida. *Proceedings of the Florida State Horticulture Society* 97, 337–340.

Peña, J.E., Nadel, H.M., Pereira, B. and Smith, D. (2002) Pollinators and pests of *Annona* species. In: Pena, D.E., Sharp, J.L. and Wysoki, M. (eds) *Tropical Fruit Pests and Pollinators: Biology, Economics, Natural Enemies and Control*. CABI, Wallingford. UK, pp. 197–221.

Phillips, P., Bekey, R. and Goodall, G. (1987) Argentine ant management in cherimoyas. *California Agriculture* 41, 8–9. DOI: 10.3733/ca.v041n03p8

Ponce, H.L. and Vidal, H. (1981) Asistência Eécnica Sobre o Cultivo de la Guanábana. In: *Memória de III Congreso Nacional de Fruticultura. Guadalajara, Mexico*, p. 171.

Pulavarty, A., Egan, A., Karpinska, A., Horgan, K. and Kakouli-Duarte, T. (2021) Plant parasitic nematodes: a review on their behaviour, host interaction, management approaches and their occurrence in two sites in the Republic of Ireland. *Plants* 10: 2352. DOI: 10.3390/plants10112352

Rahman, M.U., Zhong, X., Uzair, M. and Fan, B. (2024) Application of fungi as biological control strategies for nematode management in horticultural crops. *Phytopathology Research* 6: 38. DOI: 10.1186/s42483-024-00257-6

Rao, M.S., Umamaheshwari, R., Priti, K., Rajinikanth, R., Vidyashree, P.P. *et al.* (2015a) *Nematode Management in Vegetable Crops*. IIHR Technical Bulletin No. 47. ICAR-IIHR, Bengaluru, India.

Rao, M.S., Umamaheshwari, R., Chaya, M.K., Manojkumar, R., Priti, K. *et al.* (2015b) *Management of Nematodes in Fruit Crops*. IIHR Technical Bulletin No. 45. ICAR-IIHR, Bengaluru, India.

Rao, M.S., Umamaheswari, R., Priti, K., Rajinikanth, R., Grace, G.N. *et al.* (2016) Role of biopesticides in the management of nematodes and associated diseases in horticultural crops. In: Hakeem, K., Akhtar, M., Abdullah, S. (eds) *Plant, Soil and Microbes*. Springer, Cham, Switzerland, pp. 117–148.

Rebollar-Alviter, A., Domínguez Alvarez, J.L., Nieto-Angel, D., Nava-Diaz, C. and Delgadillo Sánchez, F. (1997) Principales plagas y enfermedades de las anonaceas en México. In: São José, A.R., Souza, I.V.B., Morais, O.M. and Rebouças, T.N.H. (eds) *Proceedings of the I Brazilian Symposium on Annonaceous*, Universidade Estadual, Vitória da Conquista, Brazil, pp. 222–228.

Reddy, P.P. (2012) *Bio-intensive Nematode Management in Horticultural Crops*. Scientific Publishers, Jodhpur, India.

Romero, H., Aquilino, M., Planelló, R. and de la Peña, E. (2025) Plant–arthropod associations in custard apples, genus *Annona*: a global perspective. *Entomologia Experimentalis et Applicata* 173, 651–660. DOI: 10.1111/eea.13587

Sahu, T., Netam, P.K., Yadu, Y.K., Shanker, D., Dubey, V.K. and Saxena, R.R. (2024) Studies on biology of mealy bug (*Maconellicoccus hirsutus* Green) on custard apple (*Annona squamosa* L). *International Journal of Research in Agronomy* 7, 913–922. DOI: 10.33545/2618060X.2024.v7.i9SI.1627

Sasanelli, N., Konrat, A., Migunova, V., Toderas, I., Iurcu-Straistaru, E. *et al.* (2021) Review on control methods against plant parasitic nematodes applied in Southern Member States (C zone) of the European Union. *Agriculture* 11: 602. DOI: 10.3390/agriculture11070602

Shukla, R.P. and Tandon, P.L. (1984) Insect pests on custard apple. *Plant Protection Bulletin* 32: 31.

Silva, J.F., Alvarenga, C.D., Giustolin, T.A. and Sá, L.P. (2017) *Gonodonta* species (Lepidoptera: Erebidae) defoliating sugar apple and atemoya fruits in the semiarid region of Minas Gerais. *Revista Brasileira de Fruticultura* 39: e370. DOI: 10.1590/0100-29452017370

Silveira, A., Juca, F.F., Melo, P. and Sacramento, C.D. (2008) Phytonematodes associated to soursop tree (*Annona muricata*) in the south of Bahia. *Revista Brasileria Fruticultura* 30, 838–840.

de Sousa, J.M., Gondim, M.G.C. Jr, Lofego, A.C. and de Moraes, G.J. (2015) Mites on Annonaceae species in northeast Brazil and in the state of Para. *Acarologia* 55, 5–18. DOI: 10.1051/acarologia/20152147

Tanimola, A.A. and Ezeunara, E.A. (2021) Response of selected tree seedlings to root-knot nematode (*Meloidogyne incognita*). *FUDMA Journal of Sciences* 5, 151–156. DOI: 10.33003/fjs-2021-0501-548

Torres, W.E. and Sánchez, L.A. (1992) *Fruticultura Colombiana, Guanábano*. Manual de Asistencia Técnica. Instituto Colombiano Agropecuario, Bogotá, Columbia.

Tsai, S.-J. and Hsu, Y.T. (2017) The influence of the population dynamics of spider mites on *Annona* spp. under the pesticides stress. *Acta Horticulturae* 1166, 93–99. DOI: 10.17660/ActaHortic.2017.1166.13

Vidal Hernández, L.V. (1982) *El Cultivo de la Guanábana en México*. Dirección General de Educación Popular, Veracruz, Mexico.

Vitullo, J., Zhang, A., Mannion, C. and Bergh, J.C. (2009) Expression of feedings symptoms from Pink Hibiscus mealybug (Hemiptera: Pseudococcidae) by commercially important cultivars of hibiscus. *Florida Entomologist* 92, 248–254. DOI: 10.1653/024.092.0208

Westerdahl, B.B. (2020) Evaluation of trap cropping for management of root-knot nematode on annual crops. *Acta Horticulturae* 1270, 141–146. DOI: 10.17660/ActaHortic.2020.1270.15

Zhang, Y., Li, S., Li, H., Wang, R., Zhang, K.Q. *et al.* (2020) Fungi–nematode interactions: diversity, ecology and biocontrol prospects in agriculture. *Journal of Fungi* 6: 206. DOI: 10.3390/jof6040206

# 11 Diseases

**Dzarifah Zulperi[1,2*], Norsazilawati Saad[1], Siti Izera Ismail[1] and Muhammad Iqbal Hakim Johari[1]**

*[1]Department of Plant Protection, Faculty of Agriculture, Universiti Putra Malaysia, Serdang, Selangor, Malaysia; [2]Laboratory of Sustainable Resources Management, Institute of Tropical Forestry and Forest Products, Universiti Putra Malaysia, Serdang, Selangor, Malaysia*

## 11.1 Introduction

*Annona* spp. such as *Annona muricata* (soursop), *A. squamosa* (sugar apple) and *A. cherimola* (cherimoya), are tropical and subtropical fruit trees celebrated for their unique flavour profiles and nutritional benefits. These fruits are rich in vitamins, minerals and antioxidants, making them highly valued in various cuisines and traditional medicines. The global market for annona fruits has been growing due to their increasing popularity and demand, particularly in regions where they are not native. However, cultivation of these fruits faces significant challenges due to diseases that can drastically reduce yield and quality.

Among the most common and destructive diseases affecting *Annona* spp. are anthracnose, *Phytophthora* root rot, black spot and leaf blight. Anthracnose disease caused by *Colletotrichum* spp. is prevalent in warm and humid climates and affects the leaves, stems, flowers and fruit, leading to necrotic lesions and fruit rot (Sharma and Lakpale, 2020). *Phytophthora* root rot, caused by *Phytophthora* spp., results in root necrosis, wilting and plant death, particularly under waterlogged conditions (Peralta-Ruiz *et al.*, 2023). Black spot disease, caused by *Asperisporium annonae*, and leaf blight disease, caused by *Alternaria* spp., significantly deteriorate the fruit's quality and marketability (Zakaria, 2022).

Several factors, including environmental conditions and cultural practices, contribute to the high susceptibility of *Annona* spp. to these diseases. In tropical and subtropical regions, high humidity and temperature create ideal conditions for the proliferation of fungal and bacterial pathogens. Moreover, traditional farming practices, such as monocropping and inadequate sanitation, exacerbate the spread of diseases. The complex interactions between these factors necessitate a comprehensive disease management approach incorporating traditional and modern agricultural practices.

Sustainable disease management in *Annona* spp. involves the integration of various strategies, including cultural practices, biological control and the use of resistant varieties. Integrated pest management (IPM) is a holistic approach that combines these methods to achieve effective and sustainable disease control (Deguine *et al.*, 2021). Cultural practices such as crop rotation, intercropping and proper irrigation can reduce disease pressure. Biological control, using natural enemies and beneficial

* Email: dzarifah@upm.edu.my


DOI: 10.1079/9781800627628.0011

microorganisms, can help manage pathogen populations. For example, *Trichoderma* spp. are effective against soil-borne diseases such as *Phytophthora* root rot (Zakaria, 2021).

Breeding and selecting disease-resistant *Annona* spp. cultivars is another crucial aspect of sustainable disease management. Advances in plant breeding and biotechnology have facilitated the development of cultivars with enhanced resistance to crucial pathogens. These resistant cultivars can reduce the need for chemical inputs, lower production costs and provide more stable yields under disease pressure (Zakaria, 2022). Ongoing research into the genetic basis of disease resistance in *Annona* spp. holds the potential for developing even more robust cultivars, combining traditional breeding techniques with modern molecular tools.

In conclusion, the sustainable management of diseases in *Annona* spp. requires a multi-faceted approach that integrates cultural, biological and chemical strategies along with the development of resistant cultivars. By understanding the complex interplay of factors that influence disease dynamics and employing comprehensive management practices, it is possible to mitigate the impact of diseases on *Annona* spp. crops. This chapter explores the diseases affecting *Annona* spp., their symptoms and sustainable management strategies, providing insights and practical approaches to manage these diseases effectively.

## 11.2 Diseases of *Annona* Spp.

### 11.2.1 Anthracnose

Anthracnose is caused by *Colletotrichum* spp. Anthracnose caused by *Colletotrichum gloeosporioides* is characterized by dark, sunken lesions on the leaves of sugar apple (Fig. 11.1A), as well as on the stems, flowers and fruit (Fig. 11.1B) (Sharma and Lakpale, 2020). Infected fruit exhibit black, necrotic spots that can merge, leading to fruit rot and significantly affecting fruit quality and marketability (González-Ruíz *et al.*, 2021; Hernández Fuentes *et al.*, 2021). Small individual lesions may merge to produce large lesions on the fruit surface, which are unattractive to consumers. In atemoya, the symptoms include sunken lesions on leaves, flower abortion, bleaching of branch tips and necrotic lesions on the fruit (dos Santos *et al.*, 2024). The prevalence and impact of anthracnose disease in *Annona* spp. crops have been well documented, highlighting the need for effective management strategies (Borges *et al.*, 2021).

Anthracnose infection commonly occurs in the field during the flowering and fruiting stages (Zakaria, 2021). Climatic factors such as humidity, temperature and rainfall may affect *Colletotrichum* spp. infection. Wet conditions increase the severity of the disease. When the orchards are old, there is likely to have been a

**Fig. 11.1.** (A) Leaf symptoms of anthracnose in sugar apple. (B) Anthracnose symptoms in sugar apple fruit. Photographs courtesy of Dr Y. Balgude.

buildup of anthracnose inoculums. Older trees, particularly those that are lightly pruned, retain moisture on the leaves and fruit for longer, and hence favour infection and fungus development (George *et al.*, 2015).

Although *C. gloeosporioides* was considered the main causal agent of anthracnose in many fruit trees, more than one species causing anthracnose in same host has frequently been reported (Costa *et al.*, 2019; Chung *et al.*, 2020). In custard apple and soursop, *Colletotrichum brevisporum*, *C. cliviae*, *C. fructicola*, *C. gigasporum*, *C. karstii*, *C. siamense*, *C. theobromicola* and *C. tropicale* have also been reported (Costa *et al.*, 2019). dos Santos *et al.* (2024) obtained 19 *Colletotrichum* sp. isolates from leaves and branches of atemoya trees with typical anthracnose symptoms. Seventeen isolates were grouped in the *gloeosporioides* complex and two in the *boninense* complex. It is now accepted that *C. gloeosporioides* is not the most common anthracnose pathogen in tropical fruit crops including annonas, as previously thought.

*Colletotrichum* spp. survive from one favourable period to another on dry branches, old lesions, fruit and affected parts remaining on the ground, on which it sporulates when there are high night temperatures (20–24°C) and high humidity. Dispersal is mainly by wind and rain splashes (Junqueira and Junqueira, 2014).

Effective management of anthracnose involves several strategies. Cultural practices, such as proper sanitation, which includes removing and destroying infected plant debris and pruning infected parts, can significantly reduce the spread of the pathogen (van Bruggen *et al.*, 2016). These practices help minimize inoculum sources and prevent disease recurrence in subsequent growing seasons. In addition, maintaining good field hygiene and avoiding overhead irrigation can further reduce the incidence of anthracnose.

Chemical control methods utilize the application of fungicides, such as copper-based compounds and systemic fungicides, although their use should be minimized to prevent the development of resistance (Garbelotto *et al.*, 2009; Dutta *et al.*, 2023). Application of 0.24% mancozeb and 0.14% thiophanate methyl was found to be effective in controlling the disease (Junqueira and Junqueira, 2014). Prochloraz can be applied during flowering and early fruit set before disease pressure builds, but should not be used at the later stages of fruit development. Spraying of every 2–3 weeks with fungicides such as mancozeb, Cabiro or Biopest is suggested. It is best to rotate each of these fungicides in turn to prevent buildup of resistance (Bignell and Bruun, 2014). Biological control agents such as *Trichoderma* spp. can also be employed to manage anthracnose disease, as they inhibit the growth of *Colletotrichum* spp. through competition and antagonism (Hernandez-Montiel *et al.*, 2021; Rizwana *et al.*, 2021). Combining these strategies provides a comprehensive approach to managing anthracnose effectively (Ramos *et al.*, 2022).

### 11.2.2 Root rot

This disease is caused by multiple fungi including *Rhizoctonia solani*, *Phytophthora* spp., *Pythium* spp. and *Armillaria mellea*. In Chile, root-rot disease in cherimoya ('Concha Lisa' and 'Bronceada') was reported to be caused by *Dactylonectria macrodidyma* (Auger *et al.*, 2015). The symptoms of this disease are initially characterized by a reduction in leaf size and loss of colour. The plants stop growing and produce few shoots. The fruit become smaller and fall off before they are ready to be harvested. In the region of the plant collar or stem, cracks or dark spots can be observed, which reach the cambium. Over time, the lesions increase in size and progress towards the root, becoming dark and rotten (Junqueira and Junqueira, 2014).

*Phytophthora* root rot, caused by *Phytophthora* spp., produces severe symptoms, including root necrosis, wilting and eventual death of the plant. Infected roots exhibit a brownish discoloration and reduced functionality, significantly impacting the plant's ability to absorb water and nutrients, leading to an overall decline (Junqueira and Junqueira, 2014). A recent observation by Wang *et al.* (2025) showed that the fungal and bacterial community in the inter-root soil of root-rot-infected plants had a greater diversity. Illumina MiSeq sequencing results indicated that *Fusarium*, *Trichoderma*, *Bacteroidetes and Proteobacteria* spp. were significantly enriched in the plants with root rot.

Improving soil drainage and avoiding waterlogged conditions are crucial for preventing *Phytophthora* spp. infections. Raised beds and proper irrigation management are essential to reduce the risk of infection. These practices enhance aeration around the root zone, thereby limiting favourable conditions for pathogen proliferation. Maintaining proper field hygiene by removing infected plant debris also helps reduce the inoculum load (van Bruggen *et al.*, 2016).

Developing and planting resistant *Annona* spp. cultivars provide long-term solutions against root-rot disease. Plant breeding programmes focus on identifying and promoting genetic traits that confer resistance to the pathogen, reduce dependency on chemical treatments and ensure sustainable crop production. Resistant cultivars can significantly mitigate the impact of *Phytophthora* spp., offering a reliable method for managing the disease (Singh *et al.*, 2022).

Fungicides containing metalaxyl or phosphonates are effective against the disease when used as soil drenches. In Australia, phosphorus acid spraying every 3 weeks was suggested against the fungi *Phytophthora cinnamomi*, *Pythium* spp. and *Cylindrocladium* spp. (George *et al.*, 2015). These fungicides inhibit the pathogen's growth and prevent further infection. However, scrutinized management is required to avoid resistance to development and potential environmental impacts. Using these chemical controls as part of an IPM strategy is recommended to ensure their effectiveness and sustainability (White *et al.*, 2020; Dutta *et al.*, 2023).

Combining these strategies into an IPM approach is essential for effectively controlling *Phytophthora* root-rot disease. Regular monitoring of soil moisture levels, application of fungicides and incorporation of resistant cultivars are critical components of this approach. Implementing these practices could diminish disease occurrences, leading to better long-term sustainability and productivity of *Annona* spp. (Hernandez-Montiel *et al.*, 2021; Rizwana *et al.*, 2021).

### 11.2.3 *Botryosphaeriaceae* rot and dieback

This infection is favoured by conditions that can reduce plant health such as high temperatures, poor nutrition and poorly conducted pruning (Hanifeh *et al.*, 2017). It can occur at many parts of the annona plant such as stems, branches and fruit. The first symptoms are apparent at the tip of the stems, causing wilting and subsequent leaf fall, and often spread to the trunk causing rot. Furthermore, necrosis of the vascular system similar to black rot can be seen (Fig. 11.2) (Machado *et al.*, 2019). Early symptoms on the fruit are small brownish to black lesions that eventually cover the fruit, causing the fruit to shrivel and dry out (mummify) (Thite *et al.*, 2024). Infected fruit may drop prematurely, reducing yield and quality. In some cases, the disease can also affect the branches and stems, causing lesions that may lead to dieback (Wang *et al.*, 2022). This disease also has been reported

**Fig. 11.2.** Botryosphaeriaceae rot in *Annona muricata*. (A) Trunk rot and cracks in the bark. (B) Leaf fall and death of the plant. (C) Dieback. Photographs courtesy of Dr A.R. Machado.

at the postharvest stage where the lesion begins to develop at the peduncle and the infection progresses to the flesh, causing discoloration (Cardoso *et al.*, 2017). In the past, dieback of Annonaceae was associated with *Lasiodiplodia theobromae* only (Junqueira and Junqueira, 2014). However, based on phylogenetic analysis, Machado *et al.* (2019) identified eight different species of Botryosphaeriaceae: *Lasiodiplodia brasiliense* (on soursop), *L. crassispora* (on soursop, sugar apple and *Annona leptopetala*), *L. hormozganensis* (on sugar apple), *L. iraniensis* (on soursop, sugar apple and atemoya), *L. pseudotheobromae* (on soursop, sugar apple, atemoya and *A. crassiflora*), *L. subglobose* (on soursop, sugar apple and *A. leptopetela*), *L. theobromae* (on soursop and sugar apple) and *Pseudofusicoccum stromaticum* (on soursop, atemoya and *A. crassiflora*).

### 11.2.4 *Armillaria* root rot

This disease occurs sporadically in older orchards. The fungus *Armillaria luteobubalina* can survive on stumps and roots of various trees for many years, and causes a slow decline of trees with dieback of twigs and leaves. The disease can be diagnosed by the white plaques of fungal growth under the bark of the trunk and the black 'shoestrings' of the fungus on the major roots (George *et al.*, 2015). In California, this disease is caused by the mushroom fungus *Armillaria mellea*, which is native to Californian oak trees. The infected trees show poor growth, shoot dieback, small yellowing leaves and excessive leaf drop, followed by canopy collapse and death (Bender, 1998).

### 11.2.5 Wilt

Wilt caused by *Phytophthora* spp. was reported from the states of São Paulo, Goiás, Minas Gerais and the Federal District in Brazil. In São Paulo, the disease is caused by *Phytophthora nicotianae* var. *parasitica*. It has mainly been observed in sugar apple, but is also seen in atemoya and soursop (Junqueira and Junqueira, 2014).

The initial symptoms of the disease are a slight discoloration of the leaves, which gradually become chlorotic. Incipient wilting appears before the leaves began to dry. The dead leaves remain attached to the plant. The roots become necrotic and darken. The lesions can reach the collar of the plant, reaching the soil line (Junqueira *et al.*, 2003; Junqueira and Junqueira, 2014).

The disease can affect any age of tree and is observed more often in clayey soil with poor drainage. In soils with a pH between 6.0 and 7.0 and those rich in organic matter, the incidence of the disease is low compared with trees growing in acid soil. Planting of annonas should not be done in areas subject to water logging or flooding.

The first report of *Ceratocystis fimbriata* causing wilt disease of soursop in South Sumatra, Indonesia, and worldwide was documented by Pratama *et al.* (2023). Trees that are infected with *C. fimbriata* typically show gradual dropping of their leaves, branches or even death of the entire plant (Fig. 11.3). This withering can occur rapidly and is often a prominent sign of disease. The genus *Ceratocystis* comprises a group of pathogenic fungi that target the xylem of plants, initiating hyphal growth and spore production within the vessels. These vessels, in turn, obstruct tracheal elements and induce water stress to plants. When infected by this pathogen, plants first lose their dark green leaf colour, followed by wilting, drying and eventual death, with dry leaves persisting on the plant. Morphological and phylogenic analysis revealed that all pathogen isolates were identified as *C. fimbriata* (Pratama *et al.*, 2023).

### 11.2.6 *Alternaria* brown spot

*Alternaria* spp. have been reported as a plant pathogens of many important crops worldwide. *Alternaria alternata* causes brown spot disease in *Annona* spp., as reported from India (Kumar and Singh 2015), Ecuador (Guevara *et al.*, 2019) and Brazil (Veloso *et al.*, 2025).

The disease affects both leaves and fruit. The leaves show black to brown necrotic lesions bordered by chlorotic tissue around the central vein and apex, while the damage to the fruit is limited to the epidermis (Guevara *et al.*, 2019). Veloso *et al.* (2025) recorded dark brown necrotic lesions in an inverted V-shape and with irregular edges, starting at the apex and progressing to the centre of the leaves. As the disease progresses, the lesions coalesce and affect a large part of the leaf blade, leading to deformation and defoliation (Fig. 11.4).

**Fig. 11.3.** (A) Wilting and yellowing leaves in a diseased tree infected with *Ceratocystis fimbriata*. (B) Soursop tree affected by wilt disease. Photographs courtesy of Dr A. Muslim.

**Fig. 11.4.** *Alternaria* brown spot in annona twigs (A) and leaves (B). Photographs courtesy of Dr J.S. Velosco.

No specific measures for control of *Alternaria* spp. are available. However, a number of fungicides such as captan, mancozeb, iprodione, captafol and carbendazime are used to control different *Alternaria* spp. in fruit crops. Kumar and Singh (2020) reported maximum disease control of *Alternaria* leaf spot caused by *A. alternata* in *Ziziphus mauritiana* by three sprays of the fungitoxicants tebuconazole 25% and trifloxystrobin 50% in water-dispersible granules (0.05%).

### 11.2.7 Black spot

Black spot disease caused by *Asperisporium annonae* manifests as small, black lesions on the leaves, stems and fruit of annona crops. These lesions can merge, causing significant damage to the leaves and fruit, which reduces the plant's photosynthetic capability and market value (Borges *et al.*, 2021). The disease is particularly problematic in regions with high humidity, as these conditions favour proliferation of the pathogen (González-Ruíz *et al.*, 2021).

Management of black spot disease involves several strategies. Regular monitoring and removal of infected plant parts are essential to reduce the inoculum load in the field (Leal and Paull, 2023). Proper spacing and pruning practices improve air circulation, which helps reduce the humidity levels around the plants, thereby limiting the conditions that favour the pathogen (van Bruggen *et al.*, 2016). This cultural practice is crucial for preventing the disease's spread and maintaining the health of the crops.

Chemical control is another effective strategy for managing black spot disease. The application of protective fungicides, such as mancozeb or chlorothalonil, can help control spread of the pathogen (White *et al.*, 2020; Dutta *et al.*, 2023). However, employing IPM strategies to minimize chemical usage and reduce the risk of developing fungicide resistance is essential. IPM strategies incorporate various control methods to manage pests and diseases sustainably.

Biological control offers a sustainable approach to managing black spot disease. Utilizing antagonistic microorganisms such as *Bacillus subtilis* can help manage the disease through competitive exclusion and the production of antimicrobial compounds (Hernandez-Montiel *et al.*, 2021). These biocontrol agents can suppress the pathogen population and reduce the incidence of black spots, providing an environmentally friendly alternative to chemical fungicides (Ramos *et al.*, 2022).

### 11.2.8 Leaf spot and fruit spot

Leaf spots on soursop caused by *Colletotrichum siamense* was recently reported in Mexico (Beltrán-Peña *et al.*, 2023). Symptoms appeared as round, black, necrotic spots that grew up to 6 mm in diameter with a brown or grey colour at the centre.

*Cercospora* leaf spot caused by *Cercospora annonae* affects soursop, resulting in defoliation (Junqueira and Junqueira, 2014). It also affects atemoya, sugar apple and *Rollinia* spp. *Pseudocercospora* rot caused by *Pseudocercospora annomicola* was first reported from Taiwan in atemoya. *Pseudocercospora* rot is a major problem in Australia, especially in wet seasons. Early symptoms consist of diffuse spots, which appear as indentations on the fruit surface. The spots are small (ranging in size up to 15 mm) and dark purple to grey. The spots often coalesce to form large disfigured areas. Large necrotic spots are commonly seen as the disease progress. In advanced cases, superficial cracks appear in the skin and the hardened skin prevents normal fruit development (George *et al.*, 2015). A regular protective spray programme using mancozeb and Cabrio is suggested as a control measure with the frequency to be determined by the prevailing weather condition.

### 11.2.9 Rust (*Phakopsora cherimoliae* Cumm.)

The leaves of the atemoya, cherimoya, custard apple, sugar apple and ilama (*Annona diversifolia*) are commonly attacked by rust fungus in the summer and autumn. When infestation becomes severe, the leaves turn yellow and fall from the tree. The disease is recognized by numerous small, raised, yellow spots covered with a golden-yellow mass of spores, which are on the undersurface of the leaves. On the upper surface of the leaves, the small spots are yellow and more or less smooth (Cook, 1975; McMillan, 1986). Spaying of chlorothalonil, triadimenol and azoxystrobin may be effective to control rust disease.

### 11.2.10 Leaf blight

Leaf blight disease caused by *Alternaria* spp. poses a significant challenge for *Annona* spp. cultivation. This disease is characterized by irregular brown to black leaf lesions, which can lead to premature defoliation and reduced fruit yield (Smith *et al.*, 2019; Leal and Paull, 2023). The lesions often start as small spots that expand, causing large necrotic areas that weaken the leaves and reduce the plant's overall photosynthetic capacity (Borges *et al.*, 2021; González-Ruíz *et al.*, 2021).

Implementing crop rotation and avoiding monoculture are essential cultural practices for reducing the incidence of leaf blight disease. Crop rotation helps break the disease cycle by removing the host plant for a period, thereby reducing the pathogen load in the soil. Ensuring balanced fertilization, in particular avoiding excessive nitrogen, can help prevent the disease. This is because excess nitrogen can make plants more susceptible to infection by promoting tender growth that is more easily invaded by pathogens (White *et al.*, 2020).

Chemical control is another effective strategy for managing leaf blight. Foliar sprays with fungicides such as azoxystrobin and difenoconazole can control the disease (Gur *et al.*, 2020), but their use should be part of an integrated disease management plan to avoid the development of resistance (Rizwana *et al.*, 2021). These fungicides work by inhibiting the fungal pathogen's ability to infect and spread, thus protecting the leaves and improving plant health.

Biological control offers a sustainable approach to managing leaf blight. Applying compost teas and other organic amendments can enhance soil microbial diversity, suppressing natural disease (Hernandez-Montiel *et al.*, 2021; Ramos *et al.*, 2022). Organic treatments increase the beneficial microbial population in the soil, which competes with and suppresses the pathogen, reducing its ability to cause disease.

### 11.2.11 Purple blotch

Purple blotch is caused by *Phytophthora capsicii* (it was previously thought to be caused by *Phytophthora palmivora*). The symptoms first appear on small fruit. Small purple spots quickly grow to cover the entire fruit surface (Fig. 11.5). On larger fruit, the spots or blotches usually reach 2 cm in diameter and gradually expand to 5 cm wide. They have a distinct margin and sometimes a halo around the initial spot. The fruit may be covered by white fungal growth, which develops under moist conditions. Affected fruit drop readily (Bignell and Brunn, 2014).

Once *P. capsici* is in the soil, it remains there indefinitely. The disease is usually more of a problem in heavy soils and is more prevalent after periods of continuous wet weather. Mancozeb and copper sprays should be used for control. Fungicides sprays of mancozeb should be started at fruit set. Ridomil Gold is believed to be more effective because it provides both the protective properties of copper hydroxide and the curative and systemic properties of Metalaxyl-M (Bignell and Brunn, 2014).

### 11.2.12 *Pseudocercospora* spot

This disease is caused by *Pseudocercospora annonicola*. It was first reported in custard apple in Taiwan. Early symptoms consist of a diffuse spot, which appears as indentations on the fruit surface. The spots are small (ranging in size up to 15 mm) and dark purple to grey. The spots often coalesce to form large disfigured areas. Large necrotic spots are commonly seen as the disease progresses. In advanced cases, superficial cracks appear in the skin, and the hardened skin prevents normal fruit development. (Fig. 11.6). A regular protective spray programme using mancozeb and Cabrio with the frequency determined by prevailing weather conditions has been recommended (Bignell and Brunn, 2014).

### 11.2.13 Root rot

In Chile, cherimoya trees were found to be affected by root-rot disease The pathogen was identified as *Dactylonectria macrodidyma*. The infected trees showed reduced vigour, shortened internodes, and dying of shoots, along with a necrotic root crown and the development of secondary roots. Internal necrosis extended from the bark to the pith in diseased parts of the plant. Other symptoms included black discoloration of wood gum inclusions in the xylem vessels, black streaks in the vascular tissue, and a reduction in root biomass, with sunken, necrotic root lesions (Auger *et al.*, 2015).

## 11.3 Postharvest Diseases

Annona fruits are climacteric in nature, with a high respiration rate and ethylene production

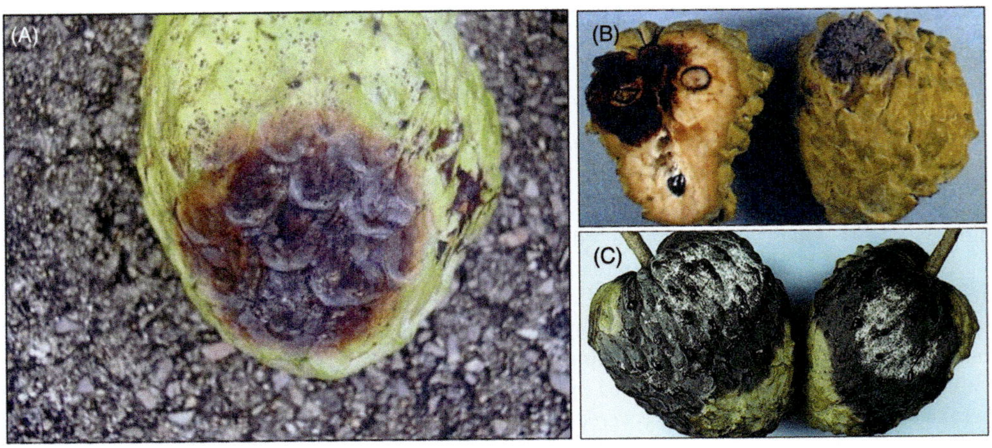


**Fig. 11.5.** Early (A), moderate (B) and advanced (C) symptoms of purple blotch. Photographs courtesy of Dr Grant Bignell.

**Fig. 11.6.** Mild (A) and advanced symptoms (B) of *Pseudocercospora* spot. Photographs courtesy of Dr Grant Bignell.

during ripening, and a short postharvest shelf life. The fruit pulp consists mainly of water, non-reducing sugar and carbohydrates, which makes it susceptible to attack by microorganisms. In soursop, members of the genera *Lasiodiplodia*, *Colletotrichum*, *Rhizopus*, *Fusarium* and *Gliocladium* have been identified as potential pathogens. These fungi are known primarily for invading, penetrating and colonizing the fruit tissue, causing damage and secreting secondary metabolites known as mycotoxins, and can cause alterations in physical appearance, nutritional values and organoleptic characteristics. In addition, the mycotoxins can cause allergies and poisoning when consumed (González-Ruíz *et al.*, 2021). Fungi reported to attack cherimoya fruit include *Botrytis cinerea*, *Rhizopus stolonifer*, *Penicillium expansum*, *Phomopsis vexans*, *Alternaria* and *Fusarium*. *Botrytis*, *Rhizopus*, *Penicillium* and *Phomopsis* spp., which penetrate the fruit through wounds in the epidermis and produce more damage at higher temperatures. *Botrytis* spp. are probably the most aggressive, because of their ability to penetrate undamaged epidermis and their resistance to low temperatures (Morales and Cooper, 1977). *Penicillium*, *Botrytis* and *Rhizopus* rot cause soft and wet rots, while *Phomopsis* rot causes spots on the epidermis that give the skin a leather-like and eventually scaly appearance (Merodio and de la Plaza, 1997). Postharvest diseases are described in detail in Chapter 12 (this volume).

## 11.4. Sustainable Management Strategies for *Annona* Spp.

### 11.4.1 Integrated pest management

IPM is crucial for maintaining the health and productivity of *Annona* spp. IPM utilizes various strategies such as biological control, cultural practices, mechanical methods and chemical treatments in a coordinated manner to manage pest populations. This approach not only minimizes the use of harmful chemicals but also promotes the use of natural predators and beneficial insects. Studies have shown that these methods can effectively reduce pest infestations, leading to healthier crops and a reduced environmental footprint (Foster *et al.*, 2020).

One of the key components of IPM is the use of biological control agents. These include natural predators, parasitoids and pathogens that target specific pests without harming the plants or other beneficial organisms. For example, introducing predatory insects or using microbial pesticides can significantly lower pest populations, reducing the need for chemical interventions (Gavamukulya *et al.*, 2020). Microbial pathogens such as the fungus *Beauveria bassiana* and the bacterium *Bacillus thuringiensis* have been identified as key agents in targeting specific pests while minimizing the impact on the ecosystem. These agents infect and eventually kill the pests, offering a precise pest control mechanism (Anggrain *et al.*, 2022).

Cultural practices such as crop rotation, intercropping and maintaining field hygiene also play a vital role in IPM. These practices disrupt the life cycle of pests and reduce their habitat, lowering their population numbers. Additionally, mechanical methods such as hand-picking pests, using traps and employing barriers can provide immediate control measures. Chemical treatments are used as a last resort, ensuring they are targeted and minimally invasive to the environment (El Zawawy, 2015).

### 11.4.2 Soil health and fertility

Soil health is fundamental to the sustainable cultivation of *Annona* spp. Healthy soil provides essential nutrients, supports root development and enhances water retention, all critical for robust plant growth. Organic farming techniques, such as compost and green manure, significantly enhance soil fertility and structure. These organic materials improve the physical properties of the soil, increase microbial activity and provide a slow release of nutrients, which is beneficial for long-term soil health (El Zawawy, 2015).

Compost derived from decomposed organic matter is rich in nutrients and beneficial microorganisms. Compost improves texture, aeration and water-holding capacity when added to the soil. This enhances soil fertility and promotes a healthy root environment, leading to more substantial and resilient *Annona* spp. crops. Green manure, which involves growing specific crops to be ploughed back into the soil, adds organic matter and nutrients. This practice improves soil structure, prevents erosion and increases the availability of essential nutrients such as nitrogen. By incorporating these organic farming techniques, farmers can reduce their reliance on chemical fertilizers and promote a more sustainable and eco-friendly approach to cultivation of annonas (Kumar *et al.*, 2021).

Additionally, crop rotation is a crucial practice for maintaining soil health. Farmers can break the life cycles of pests and diseases that affect *Annona* spp. by rotating different crops in the same field over successive seasons. Crop rotation also helps balance the nutrient demands of different crops, preventing the depletion of specific nutrients from the soil. For instance, leguminous crops, which fix atmospheric nitrogen, can be rotated with annona cultivation to enhance soil nitrogen levels naturally. This practice improves soil fertility, promotes biodiversity, and reduces the risk of soil degradation (El Zawawy, 2015).

The integration of cover crops is another effective strategy for improving soil health and fertility. Cover crops, such as legumes, grasses and clover, are planted primarily to cover the soil rather than for harvest. These crops prevent erosion, suppress weeds and enhance soil organic matter. When cover crops are ploughed back into the soil, they decompose and release nutrients, enriching the soil. Additionally, cover crops can improve soil structure and porosity, leading to better root penetration and water infiltration. By adopting these practices, farmers can create a more sustainable and productive environment for the growth of *Annona* spp. (Kumar *et al.*, 2021).

### 11.4.3 Water management practices

Efficient water management is essential for optimizing the growth of *Annona* spp., particularly in areas with inconsistent rainfall. Advanced irrigation techniques such as drip irrigation directly target the plant root zones, significantly reducing water wastage through minimized evaporation and runoff. This method ensures a consistent water supply during vital growth phases to reduce stress and promote healthy plant development. Drip irrigation has been recognized for its efficiency in water usage and its positive impact on plant health across various agricultural contexts (United Nations, 2024).

Rainwater harvesting is another effective strategy for sustainable water management in the growth of *Annona* spp. By collecting and storing rainwater, farmers can ensure a reliable water source during dry periods, help maintain appropriate soil moisture levels and reduce reliance on local water supplies. This practice supports agricultural water needs and contributes to overall water conservation efforts (Tarranum *et al.*, 2023).

Recent advancements recommend the integration of soil moisture sensors and automated irrigation systems. These technologies optimize irrigation based on real-time soil moisture data, ensuring water is applied efficiently and effectively. This approach minimizes water use while maximizing productivity and supports the sustainable management of water resources (Schwamback *et al.*, 2023).

### 11.4.4 Postharvest management

Effective postharvest management of annona fruits is crucial for minimizing losses, maintaining quality and maximizing market value. This involves a comprehensive approach that incorporates several essential practices, described below.

#### *11.4.4.1 Improved handling and storage techniques*

Proper handling of annona fruit from harvest to the market is vital. Techniques that minimize physical damage and bruising can significantly reduce the incidence of spoilage. Controlled-atmosphere storage and refrigeration are critical in extending the shelf life of the fruit. Controlled-atmosphere storage involves modifying the composition of the air surrounding the fruit by reducing oxygen levels while increasing carbon dioxide levels. This method has been proven to delay ripening and extend freshness (Indiarto *et al.*, 2020).

#### *11.4.4.2 Innovative packaging technologies*

Recent advancements in packaging technologies include the development of smart packaging solutions that can respond to changes in the environment or the state of the fruit. These packages can include indicators for temperature and freshness, helping to ensure optimal conditions throughout the supply chain.

#### *11.4.4.3 Quality management and value addition*

Recent research has also focused on value-addition techniques, such as applying edible coatings to enhance the marketability of annona fruit. Edible coatings help retain moisture, reduce the fruit's exposure to pathogens and slow down the metabolic processes that lead to spoilage (Naik *et al.*, 2022).

#### *11.4.4.4 Postharvest treatments*

Various treatments such as natural waxes and edible coatings have been explored to enhance appearance and shelf life of the fruit. These coatings can hinder moisture and gas exchange, further slowing the ripening process and reducing microbial growth (Adhikari and Aarati, 2021).

## 11.5 Conclusion

Sustainable management of diseases of *Annona* spp. is critical for the global cultivation and marketability of these nutritionally rich fruit. By integrating traditional agricultural practices with modern scientific approaches and advanced breeding techniques, growers can effectively manage and mitigate the impact of diseases such as anthracnose, *Phytophthora* root rot, black spot and leaf blight. Adopting comprehensive disease management strategies ensures the preservation of crop yields and quality, promotes environmental sustainability and supports the agricultural economy.

As research progresses, particularly in genetic resistance and biocontrol methods, the agricultural community will be better equipped to face the challenges posed by pathogens and environmental factors. This evolving landscape of agricultural science promises to enhance the resilience of *Annona* spp. crops, securing a reliable supply of these valuable fruits for future generations.

## References

Adhikari, B. and Aarati, G.C. (2021) Post-harvest practices of horticultural crops in Nepal: Issues and management. *Archives of Agriculture and Environmental Science* 6, 227–233. DOI: 10.26832/24566632.2021.0602015

Anggraini, S., Chatterjee, S., Murali, E and Aravindhan, S. (2022) Baculoviruses for insect pest management: application, challenges and opportunities. *Pest Management Science* 78, 1473–1483. DOI: 10.1002/ps.6713

Auger, J., Pérez, I. and Esterio, M. (2015) Occurrence of root rot disease of cherimoya (*Annona cherimola* Mill.) caused by *Dactylonectria macrodidyma* in Chile. *Plant Disease* 99, 1282–1282. DOI: 10.1094/PDIS-03-15-0257-PDN

Beltrán-Peña, H., Marquez-Licona, G., Garcia-Leon, E., Aguilar-Perez, V.H. and Solano-Baez, A.R. (2023) Occurrence of *Colletotrichum siamense* causing leaf spots on soursop (*Annona muricata*) in Mexico. *Plant Disease* 107: 2866. DOI: 10.1094/PDIS-03-23-0592-PDN

Bender, G. (1998) Insect pests and diseases. In: *The CCA Cherimoya Handbook*. California Cherimoya Association, California.

Bignell, G. and Bruun, D. (2014) Accelerating the development of the Australian custard apple industry. Final Report for HAL Project CU 11000. Department of Agriculture, Fisheries and Forestry, Queensland, Australia.

Borges, R.C.F., Rossato, M., Santos, M.D.M., Macedo, M.A., Fonseca, M.E.N. *et al.* (2021) *Colletotrichum siamense* as a causal agent of leaf anthracnose in seedlings of *Annona muricata* in nurseries from the Federal District, Brazil. *Journal of Plant Diseases and Protection* 128, 583–588. DOI: 10.1007/s41348-020-00406-x

Cardoso, J.E., Lima, J.S., Viana, F.M.P., Ootani, M.A., Araújo, F.S.A. *et al.* (2017) First report of *Lasiodiplodia brasiliense* causing postharvest fruit rot of custard apple (*Annona squamosa*) in Brazil. *Plant Disease* 101, 1542–1542. DOI: 10.1094/PDIS-03-17-0369-PDN

Chung, P.-C., Wu, H.-Y., Wang, Y.-W., Ariyawansa, H.A., Hu, H.-P. *et al.* (2020) Diversity and pathogenicity of *Colletotrichum* species causing strawberry anthracnose in Taiwan and description of a new species, *Colletotrichum miaoliense sp. Scientific Reports* 10: 16. DOI: 10.1038/s41598-020-70878-2

Cook, A.A. (1975) *Diseases of Tropical and Subtropical Fruits and Nuts*. Hafner Press, New York.

Costa, J.F.O., Kamei, S.H., Silva, J.R.A., Miranda, A.R.G.S., Netto, M.B. *et al.* (2019) Species diversity of *Colletotrichum* infecting *Annona* spp. in Brazil. *European Journal of Plant Pathology* 153, 1119–1130. DOI: 10.1007/s10658-018-01630-w

Deguine, J.-P., Aubertot, J.-N., Flor, R.J., Lescourret, F., Wyckhuys, K.A.G. *et al.* (2021) Integrated pest management: good intentions, hard realities. A review. *Agronomy for Sustainable Development* 41: 38. DOI: 10.1007/s13593-021-00689-w

dos Santos, J.M.C., de Oliveira Costa, J.F., da Silva, J.L., Lima, T.S., dos Santos, T.F. *et al.* (2024) *Colletotrichum* species associated with atemoya anthracnose in Northeast Brazil. *Comunicata Scientiae* 15: e4111. DOI: 10.14295/cs.v15.4111

Dutta, P., Kumari, A., Mahanta, M., Upamanya, G.K., Heisnam, P. *et al.* (2023) Nanotechnological approaches for management of soil-borne plant pathogens. *Frontiers in Plant Science* 14: 1136233. DOI: 10.3389/fpls.2023.1136233

El Zawawy, N.A. (2015) Antioxidant, antitumor, antimicrobial studies and quantitative phytochemical estimation of ethanolic extracts of selected fruit peels. *International Journal of Current Microbiology and Applied Sciences* 4, 98–309.

Foster, K., Oyenihi, O., Rademan, S., Erhabor, J., Matsabisa, M. *et al.* (2020) Selective cytotoxic and anti-metastatic activity in DU-145 prostate cancer cells induced by *Annona muricata* L. bark extract and phytochemical, annonacin. *BMC Complementary Medicine and Therapies* 20: 375. DOI: 10.1186/s12906-020-03130-z

Garbelotto, M., Harnik, T.Y. and Schmidt, D.J. (2009) Efficacy of phosphoric acid, metalaxyl-M and copper hydroxide against *Phytophthora ramorun in vitro* and *in planta*. *Plant Pathology* 58, 111–119. DOI: 10.1111/j.1365-3059.2008.01894.x

Gavamukulya, Y., Maina, E.N., El-Shemy, H.A., Wamunyokoli, F. and Magoma, G. (2020) Synthesis and characterization of nanoparticles from extracts of fruits of *Annona muricata*: a green nanobiotechnology approach. *African Journal of Science, Technology and Engineering* 1, 153–171.

George, A., Broadleg, R., Hutton, D., Redpath, S., Bignell, G. *et al.* (2015) *Integrated Pest and Disease Management Manual for Custard Apple*. Queensland Department of Agriculture and Fisheries, Nambur, Queensland, Australia.

González-Ruíz, A.V., Palomino-Hermosillo, Y.A., Balois-Morales, R., Ochoa-Jiménez, V.A., Casas-Junco, P.P. *et al.* (2021) Pathogenic fungi associated with soursop fruits (*Annona muricata* L.) during postharvest in Nayarit, Mexico. *Horticulturae* 7: 471. DOI: 10.3390/horticulturae7110471

Guevara, F.E., Oviedo, M.A., Corral, M., Vicra, W.F., Garrido, P. A. *et al.* (2019) First report of *Alternaria* brown spot in cherimoya (*Annona cherimola* Mill.) caused by *Alternaria alternate* in Ecuador. *Plant Disease* 103: 2949. DOI: 10.1094/PDIS-02-19-0416-PDN

Gur, L., Reuveni, M. and Cohen, Y. (2020) Control of *Alternaria* fruit rot in "Pink Lady" apples by fungicidal mixtures. *Crop Protection* 127: 104947. DOI: 10.1016/j.cropro.2019.104947

Hanifeh, S., Zafari, D. and Soleimani, M.J. (2017) Reaction of some apple cultivars to *Diplodia bulgarica* in Iran. *Mycosphere* 8, 1253–1260. DOI: 10.5943/mycosphere/8/2/9

Hernández Fuentes, L.M., Montalvo González, E., García Magaña, M. de L., Anaya Esparza, L.M., Nolasco González, Y. *et al.* (2021) Current situation and perspectives of fruit Annonaceae in Mexico: biological and agronomic importance and bioactive properties. *Plants* 11: 7. DOI: 10.3390/plants11010007

Hernandez-Montiel, L.G., Droby, S., Preciado-Rangel, P., Rivas-García, T., González-Estrada, R.R. *et al.* (2021) A sustainable alternative for postharvest disease management and phytopathogens biocontrol in fruit: antagonistic yeasts. *Plants* 10: 2641. DOI: 10.3390/plants10122641

Indiarto, R., Izzati, A.N. and Djali, M. (2020) Post-harvest handling technologies of tropical fruits: a review. *International Journal of Emerging Trends in Engineering Research* 8, 3951–3957. DOI: 10.30534/ijeter/2020/165872020

Junqueira, N.T.V. and Junqueira, K.P. (2014) Major diseases of Annonaceae in Brazil: description and control. *Revista Brasileira de Fruticultura* 36, 55–64. DOI: 10.1590/S0100-29452014000500006

Junqueira, N.T.V., Cunha, M.M., Junqueira, K.P. *et al.* (2003) Diseases and pests of Annonaceae. In: Manica, I. (ed.) *Annonaceous Fruits: Ata or Pinha, Atemolia, Cherimoya, and Soursop: Production Technology, Post-Harvest and Market, Vol.* 1. Cinco Continentes Editora, Porto Alegre, Brazil, pp. 387–440.

Kumar, A. and Singh, H.K. (2020) Management of *Alternaria* leaf spot [*Alternaria alternata* (Fr.) Keissler] of ber (*Zizyphus mauritiana* Lamk.). *Journal of Pharmacognosy and Phytochemistry* 9, 155–1555.

Kumar, M., Changan, S., Tomar, M., Prajapati, U., Saurabh, V. *et al.* (2021) Custard apple (*Annona squamosa* L.) leaves: nutritional composition, phytochemical profile, and health-promoting biological activities. *Biomolecules* 11: 614. DOI: 10.3390/biom11050614

Kumar, S. and Singh, R. (2015) *Annona muricata*, a new host record for *Alternaria alternata* causing leaf spots in Uttar Pradesh, India. *Plant Pathology & Quarantine* 5, 44–47. DOI: 10.5943/ppq/5/2/2

Leal, F. and Paull, R.E. (2023) The soursop (*Annona muricata*): botany, horticulture, and utilization. *Crop Science* 63, 362–389. DOI: 10.1002/csc2.20894

Machado, A.R., Custódio, F.A., Cabral, P.G.C., Capucho, A.S. and Pereira, O.L. (2019) Botryosphaeriaceae species causing dieback on Annonaceae in Brazil. *Plant Pathology* 68, 1394–1406. DOI: 10.1111/ppa.13060

McMillan, R.T. Jr (1986) Serious diseases of tropical fruits in Florida. *Proceedings of the Florida State Horticulture Society* 99, 224–227.

Merodio, C. and de la Plaza, J.L. (1997) Cherimoya. In: Mitra, S. (ed.) *Postharvest Physiology and Storage of Tropical and Subtropical Fruits*. CAB International, Wallingford, UK, pp. 269–293.

Morales, A. and Cooper, T. (1977) Identification de hongos causantes de pudriciones en cherimoyas (*Annona cherimola* Mill.) almacenadas a baja temperature. *Investigation Agricola* 3, 75–76.

Naik, H.R., Amin, T. and Sheraz Mahdi, S. (2022) Post-harvest management and value addition of food crops. In: Bahar, F.A., Bhat, M.A., and Mahdi, S.S. (eds) *Secondary Agriculture: Sustainability and Livelihood in India*. Springer, Cham, pp. 131–146. DOI: 10.1007/978-3-031-09218-3_11

Peralta-Ruiz, Y., Rossi, C., Grande-Tovar, C.D. and Chaves-López, C. (2023) Green management of postharvest anthracnose caused by *Colletotrichum gloeosporioides*. *Journal of Fungi* 9: 623. DOI: 10.3390/jof9060623

Pratama, R., Muslim, A., Suwandi, S. and Shk, S. (2023) First report of *Ceratocystis fimbriata* causing wilt disease of soursop in South Sumatra, Indonesia. *Biodiversitas: Journal of Biological Diversity* 24, 6711–6721. DOI: 10.13057/biodiv/d241233

Ramos, A.L.C.C., Mazzinghy, A.C.D.C., Correia, V.T.D.V., Nunes, B.V., Ribeiro, L.V. *et al.* (2022) An integrative approach to the flavonoid profile in some plants' parts of the *Annona* genus. *Plants* 11: 2855. DOI: 10.3390/plants11212855

Rizwana, H., Bokahri, N.A., Alsahli, S.A., Al Showiman, A.S., Alzahrani, R.M. *et al.* (2021) Postharvest disease management of *Alternaria* spots on tomato fruit by *Annona muricata* fruit extracts. *Saudi Journal of Biological Sciences* 28, 2236–2244. DOI: 10.1016/j.sjbs.2021.01.014

Schwamback, D., Persson, M., Berndtsson, R., Bertotto, L.E., Kobayashi, A.N.A, and Wendland, E.C. (2023) Automated low-cost soil moisture sensors: trade-off between cost and accuracy. *Sensors* 23: 2451. DOI: 10.3390/s23052451

Sharma, G.L. and Lakpale, N. (2020) Diseases of custard apple and their management. In: Chand, G., Akhtar, M.N. and Kumar, S. (eds) *Diseases of Fruits and Vegetable Crops*. Apple Academic Press, Oakville, Canada, pp. 75–86.

Singh, Y., Bhatnagar, P., Singh, J., Sharma, Y.K. and Arya, C.K. (2022) Characterizing yield, nutrient elimination and fruit quality analysis of custard apple (*Annona squamosa*) under consortium of *Azospirillum brasilense* and vermicompost. *Current Horticulture* 10, 30–33. DOI: 10.5958/2455-7560.2022.00024.3

Tarranum, R., Babu, M., Kulkarni, P.S., Reddy, G.V.S., Palled, V. and Ramesh, G. (2023) Performance evaluation of inverted umbrella type rainwater harvesting system at Raichur campus. *International Journal of Environment and Climate Change* 13, 3436–3442. DOI: 10.9734/ijecc/2023/v13i113518

Thite, S., Patil, K., Jadhav, R., Suryawanshi, Y. and Chumchu, P. (2024) Empowering agricultural research: a comprehensive custard apple (*Annona squamosa*) disease dataset for precise detection. *Data in Brief* 53: 110078. DOI: 10.1016/j.dib.2024.110078

United Nations (2024) United Nations World Water Development Report 2024: Water for Prosperity and Peace. United Nations, New York.

van Bruggen, A.H., Gamliel, A. and Finckh, M.R. (2016) Plant disease management in organic farming systems. *Pest Management Science* 72, 30–44. DOI: 10.1002/ps.4145

Veloso, J.S., do Nascimento, E.R.C., Duarte, I.G., dos Santos Vieira, W.A., Silva, A.M.F. *et al.* (2025) *Alternaria* brown spot: an emerging problem on annonaceous orchards in Northeast Brazil. *Plant Pathology* 74, 730–745. DOI: 10.1111/ppa.14046

Wang, C., Wang, M., Xu, L. and Yang, Y. (2022) First report of *Lasiodiplodia theobromae* causing dieback in custard apple (*Annona squamosa*) tree in China. *Plant Disease* 106: 327. DOI: 10.1094/PDIS-05-21-1034-PDN

Wang, Y., Wang, W., Liu, X., Chen, M., Jiang, J. and An, F. (2025) Analysis of inter-root soil habitat and physiological traits in healthy and root rot-infested custard apple (*Annona squamosa* L.) plants. *Journal of Soil Science and Plant Nutrition* 25, 3484–3497. DOI: 10.1007/s42729-025-02347-6

Zakaria, L. (2021) Diversity of *Colletotrichum* species associated with anthracnose disease in tropical fruit crops – a review. *Agriculture* 11: 297. DOI: 10.3390/agriculture11040297

Zakaria, L. (2022) Fungal and oomycete diseases of minor tropical fruit crops. *Horticulturae* 8: 323. DOI: 10.3390/horticulturae8040323

# 12 Postharvest Management

**Vittorio Farina, Roberta Passafiume and Ilenia Tinebra***
*Università degli Studi di Palermo, Department of Agricultural, Food and Forest Sciences, Viale delle Scienze, Palermo, Italy*

## 12.1 Introduction

The family Annonaceae Juss. encompasses over 119 genera and more than 2000 species, which are distributed across subtropical and tropical regions (Pino, 2010). In recent years, members of this family have also been cultivated in a Mediterranean climate, producing promising quantities of high-quality fruit (Puccio *et al.*, 2019). The genus *Annona*, described by Linnaeus in his *Systema Naturae* (1735), includes approximately 120 species. This chapter discusses the postharvest behaviour of the most significant species within the genus: *Annona cherimola* Mill. (cherimoya), *A. muricata* L. (soursop), *A. squamosa* L. (sugar apple or sweetsop), *A. reticulata* L. (custard apple) and the interspecific hybrid atemoya (*A. cherimola* × *A. squamosa*).

## 12.2 Maturity Indices

All *Annona* spp. are climacteric fruit, meaning that they continue to ripen after harvest. This characteristic impacts their harvesting and postharvest handling. The fruit are typically harvested when fully mature but still firm, as they are known for their delicate, sweet flavour, a trait that often leads to their interchangeable naming as custard apples or sugar apples, and sometimes results in confusion between the cultivars (Lizana and Reginato, 1990). Maturity indices for these fruit include changes in skin colour, texture and internal characteristics. For cherimoya, atemoya, sweetsop and custard apple, the primary indicators of maturity are colour changes from dark green to light green or greenish-yellow. Some cultivars may even turn brown (Manica *et al.*, 1994; Pal and Kumar, 1995; Pino, 2010). Other visual cues include the appearance of a cream colour between the segments on the skin and increased surface smoothness of the carpels (Salunkhe and Desai, 1984; Kader and Arpaia, 1999). The optimal eating stage for these fruit occurs around 5–6 days postharvest, coinciding with peak ethylene production (Paull, 1982). Soursop ripens rapidly once harvested. Fruit picked too early exhibit poor quality, such as losing their sheen and becoming yellowish-green, and may not ripen properly, leading to a bitter pulp (Paull and Duarte, 2011). In India, sugar apple is typically harvested when it reaches a creamish-green skin colour with yellowish valleys between the areoles (Vishnu Prasanna *et al.*, 2000). This indicates that the fruit are mature enough to undergo ripening without significant quality loss. Research into

* Email: ilenia.tinebra@unipa.it


DOI: 10.1079/9781800627628.0012

the optimal harvest time for *Annona* spp. has provided valuable insights into maturity indicators. For soursop, Worrell *et al.* (1994) identified key maturity indicators, including changes in fruit size, firmness and colour. Soursop fruit are best harvested when they have reached full size but before significant ripening symptoms such as softening occur. A slight colour change and a firm yet slightly yielding texture are ideal for ensuring quality and shelf life. Further studies by Bulathkandage *et al.* (2019) developed specific harvesting indices based on fruit growth patterns for soursop. Monitoring fruit diameter and weight are effective methods for determining optimal harvest time, ensuring the fruit is at its full growth potential but has not begun to deteriorate. For sugar apples, Moura *et al.* (2020) used carpel interspace development as a maturity index. Harvesting when the gaps between carpels are sufficiently developed helps ensure that the fruit will ripen properly and maintain its quality. Márquez Cardozo *et al.* (2012) provided a detailed analysis of soursop's physiological and physico-chemical characteristics, identifying firmness, colour changes and chemical composition as critical parameters for assessing ripeness. They emphasized that harvesting just before full ripeness ensures the best texture and flavour retention. Jiménez-Zurita *et al.* (2023) extended our understanding of soursop's growth stages using the BBCH scale, which provides a standardized method for tracking fruit development. Accurate timing based on these stages is crucial for harvesting at peak quality. Determining the optimal harvest time for *Annona* spp. involves careful monitoring of fruit size, firmness, colour and other physiological changes. By adhering to these maturity indices, producers can optimize harvest timing, enhance fruit quality and extend shelf life. Accurate harvest timing ensures that the fruit are at their peak, ready for market, and suitable for extended storage and transport.

## 12.3 Postharvest Physiology

The study of postharvest physiology is vital for *Annona* spp. as it directly impacts their marketability, economic value and consumer acceptance. By understanding the physiological and biochemical changes that occur after harvest, it is possible to develop strategies to minimize spoilage, extend shelf life and maintain nutritional quality.

The high respiration rates observed in *Annona* spp., including cherimoya, soursop and custard apple, accelerate the depletion of stored carbohydrates and can hasten senescence. Studies by Bruinsma and Paull (1984) and Brown *et al.* (1988) revealed that soursop exhibits a decline in respiration rate as it ripens, which correlates with increased ethylene production. Pareek *et al.* (2011) discussed the role of ethylene in postharvest physiology, emphasizing its impact on texture, colour and flavour changes. Managing ethylene exposure through optimized storage conditions or the use of ethylene inhibitors can significantly extend the shelf life of these fruit.

Optimal storage conditions are crucial for maintaining the quality of annona fruit postharvest. Pareek *et al.* (2011) highlighted that temperature and humidity control are fundamental in extending shelf life. Lower temperatures slow down respiration and ethylene production, thereby delaying ripening and prolonging the fruit's shelf life. Refrigeration is commonly used for soursop and cherimoya to extend storage life. However, excessive cold can lead to chilling injuries, which must be balanced with the benefits of extended shelf life. The challenge lies in finding the optimal storage conditions that minimize damage while maximizing longevity. The optimum temperature of storage is 8–12°C depending on cultivar, ripeness stage and duration (Table 12.1), and the optimum relative humidity is 90–95% (Kader, 2022).

Different *Annona* spp. respond variably to postharvest treatments. For example, Márquez Cardozo *et al.* (2012) observed that specific storage conditions impact the physiological and physico-chemical characteristics of soursop. Postharvest treatments, such as modified-atmosphere packaging (MAP) and a controlled atmosphere (CA), play a critical role in enhancing fruit preservation. These treatments are tailored to address the unique physiological responses of each *Annona* spp., improving overall fruit quality and marketability.

### 12.3.1 Respiration and ethylene production

Respiration and ethylene production are fundamental physiological processes that significantly

**Table 12.1.** Respiration rates, ethylene production, storage temperature, relative humidity and shelf life of *Annona* spp. (Brown *et al.*, 1988; Paull, 1982; Espinosa *et al.*, 2013; Pareek *et al.*, 2011; Jomngam *et al.*, 2015).

| Species | Respiration rate | Ethylene production | Storage temperature (°C) | Relative humidity (%) | Shelf life |
|---|---|---|---|---|---|
| *A. cherimola* | 25–40 ml $CO_2$ $kg^{-1}$ $h^{-1}$ during ripening | Climacteric peak of 100 μl $kg^{-1}$ $h^{-1}$ | 8–12 | 90–95 | Approximately 2–3 weeks |
| *A. muricata* | 20–65 mg $CO_2$ $kg^{-1}$ $h^{-1}$ during ripening | 110–130 μl $kg^{-1}$ $h^{-1}$, peaks as fruit ripens | 23 | 65 | Approximately 1 week |
| *A. reticulata* | Approximately 63–180 ml $CO_2$ $kg^{-1}$ $h^{-1}$. Moderate, declines with hot water and calcium treatments | Climacteric peak of 110–135 μl $kg^{-1}$ $h^{-1}$ | 20 | 85–90 | Approximately 1–2 weeks |
| *A. squamosa* | 55–180 mg $CO_2$ $kg^{-1}$ $h^{-1}$ during ripening | Climacteric peak of 100–180 μl $kg^{-1}$ $h^{-1}$ | 20 | 85–90 | Approximately 1–2 weeks |
| Atemoya | 60–300 mg $CO_2$ $kg^{-1}$ $h^{-1}$ | Climacteric peak of 240–250 μl $kg^{-1}$ $h^{-1}$ | 20 | 90–95 | Approximately 1–2 weeks |

influence the ripening, quality and shelf life of fruits. Ethylene is a plant hormone that plays a crucial role in the ripening of fruits, influencing their texture, colour and flavour. The production of ethylene in *Annona* spp. varies, which is critical for understanding their ripening and senescence processes. Cherimoya, soursop, custard apple, sugar apple and atemoya exhibit diverse patterns in these processes. Respiration involves the consumption of oxygen ($O_2$) and the release of carbon dioxide ($CO_2$), water and energy, which varies among *Annona* spp. and significantly impacts their postharvest physiology. *Annona* spp. exhibit different ethylene production rates, with cherimoya and atemoya producing moderate amounts, whereas soursop and sugar apple exhibit higher ethylene production, leading to faster ripening (Wongs-Aree and Noichinda, 2011). Cherimoya, for example, has a climacteric respiration pattern with a significant increase during ripening, a trait shared with other climacteric fruits (González *et al.*, 2014). Soursop and sugar apple also exhibit high respiration rates, indicative of their rapid ripening and shorter shelf life (Chunprasert *et al.*, 2006). Generally, high levels of ethylene (up to 100–300 μl $kg^{-1}$ $h^{-1}$, depending on cultivar) are produced during ripening at 20°C (Kader, 2022).

#### 12.3.1.1 *A. cherimola* Mill.

Studies have shown that the respiration rate of cherimoya increases from 25 to 40 ml $CO_2$ $kg^{-1}$ $h^{-1}$ during the ripening phase (Alique, 1995). Similarly, ethylene production increases significantly during the climacteric peak, which is typical of climacteric fruits (Silva *et al.*, 2016). Cherimoya respiration rate peaks at about 200–250 mg $CO_2$ $kg^{-1}$ $h^{-1}$ when ripening at room temperature after harvest, while ethylene production peaks at approximately 100–300 μl $kg^{-1}$ $h^{-1}$ (Brown *et al.*, 1988). Recent studies have highlighted the application of ethylene inhibitors such as 1-methylcyclopropene (1-MCP) to delay ripening and extend shelf life (Silva *et al.*, 2016). Temperature management also significantly reduces these processes (Alique, 1995).

#### 12.3.1.2 *A. muricata* L.

Soursop has been observed to have high respiration and ethylene production rates, which result in rapid ripening and a shorter shelf life. The respiration rate of soursop ranges from 50 to 70 ml $CO_2$ $kg^{-1}$ $h^{-1}$ (Silva *et al.*, 2016). Ethylene production also increases significantly as the fruit ripens, contributing to the accelerated ripening process (Yahia, 2006). Research has shown that controlled-atmosphere (CA) storage and ethylene blockers such as 1-MCP can effectively manage these processes, extending shelf life (Silva *et al.*, 2016).

#### 12.3.1.3 *A. reticulata* L.

Custard apple shows moderate to high respiration and ethylene production rates. The respiration rate ranges from 30 to 45 ml $CO_2$ $kg^{-1}$ $h^{-1}$ during the ripening process (Torres *et al.*, 2010). Ethylene production in custard apple also increases during ripening, although not as dramatically as in soursop or sugar apple (Vyas *et al.*, 2015). Recent investigations have emphasized the importance of hot-water dips and calcium treatments in reducing respiration and delaying ethylene production, thereby extending shelf life (Torres *et al.*, 2010).

#### 12.3.1.4 *A. squamosa* L.

Sugar apple exhibits high respiration and ethylene production rates, contributing to its rapid ripening and highly perishable nature. The respiration rate for sugar apple is between 35 and 55 ml $CO_2$ $kg^{-1}$ $h^{-1}$ (Chunprasert *et al.*, 2006). Ethylene production is also high, which further accelerates the ripening process (Wongs-Aree and Noichinda, 2011). Studies have found that Modified Atmosphere Packaging (MAP) and ethylene inhibitors, alongside temperature management, are effective strategies for extending the marketable life of the fruit (Wongs-Aree and Noichinda, 2011).

#### 12.3.1.5 *Atemoya*

Atemoya, a hybrid of cherimoya and sugar apple, exhibits intermediate characteristics in terms of respiration and ethylene production. The respiration rate ranges from 28 to 50 ml $CO_2$ $kg^{-1}$ $h^{-1}$ (Jomngam *et al.*, 2017). Ethylene production in atemoya increases during ripening, although it shows a more balanced pattern compared with its parent species (Jomngam *et al.*, 2017). Research has shown that integrated postharvest approaches such as the use of MAP and ethylene inhibitors, and optimal storage temperatures are essential for managing these processes and maintaining fruit quality (Jomngam *et al.*, 2017).

### 12.3.2 Physico-chemical changes during ripening

The ripening process in *Annona* spp. involves complex physico-chemical transformations. These changes affect colour, texture, starch content, soluble solids, acidity, polyphenols and volatile compounds. Understanding these changes is crucial for better utilization and enhancement of the quality of the fruit.

#### 12.3.2.1 *Changes in colour*

Colour transformation is a significant ripening indicator in *Annona* spp. During ripening, the chlorophyll content decreases, leading to a reduction in green colour and the emergence of yellow, orange or red hues due to carotenoid accumulation. For instance, in cherimoya, the green skin colour fades as the fruit ripens, revealing a yellowish tint indicative of chlorophyll degradation and carotenoid synthesis (González *et al.*, 2014). In soursop, the green skin changes to a yellowish-green as ripening progresses, a result of chlorophyll breakdown and the accumulation of carotenoids and flavonoids (Silva *et al.*, 2016). This colour change is not only a visual indicator of ripeness but also plays a role in consumer preference and marketability (Table 12.2) (Opara *et al.*, 2007).

#### 12.3.2.2 *Changes in texture*

Textural changes during ripening are critical for consumer acceptance of annona fruit and its storage (Table 12.3). Ripening involves cell-wall degradation, leading to softening of the fruit. This process is driven by the activity of cell-wall-modifying enzymes such as pectinases, cellulases and polygalacturonases (Sanchez *et al.* 1998). In sugar apple, the firm texture transitions to a softer consistency as these enzymes break down pectin and cellulose, facilitating conversion of the cell-wall structure (Wongs-Aree and Noichinda, 2011). Similarly, in custard apple, the firm flesh becomes softer due to enzymatic hydrolysis of cell-wall components (Vyas *et al.*, 2015). Studies have shown that these enzymatic activities are crucial for the desirable mouthfeel and palatability of the ripe fruit (Balois-Morales *et al.*, 2019). Márquez Cardozo *et al.* (2012) detailed the physiological and physico-chemical changes in

**Table 12.2.** Colour transformation of *Annona* spp. during ripening.

| *Annona* spp. | Ripening colour change | Reference |
|---|---|---|
| *A. cherimola* | Skin colour changes from green to a more yellowish hue as it ripens | Gregorio *et al.* (2019) |
| *A. muricata* | Skin colour transitions from dark green to light green or yellowish-green | Leal and Paull (2023) |
| *A. reticulata* | Skin colour changes from green to reddish-brown | Pareek *et al.* (2011) |
| *A. squamosa* | Skin colour changes from green to yellowish-green or brown | Wongs-Aree and Noichinda (2011) |
| Atemoya | Blends various colours as it ripens, often from green to yellow | Wongs-Aree and Noichinda (2011) |

**Table 12.3.** Textural changes of *Annona* spp. during ripening.

| *Annona* spp. | Softening characteristics | Reference |
|---|---|---|
| *A. cherimola* | Becomes significantly softer and more palatable upon ripening | Silva *et al.* (2016) |
| *A. muricata* | Rapidly softens, making it highly perishable | Yahia (2006) |
| *A. reticulata* | Softens moderately, retaining some firmness even as it ripens | Pareek *et al.* (2011) |
| *A. squamosa* | Softens quickly to a creamy texture | Pareek *et al.* (2011) |
| Atemoya | Exhibits intermediate softening characteristics, balanced texture | Pareek *et al.* (2011) |

soursop during ripening, noting that texture softening is a critical aspect. This softening results from the degradation of cell-wall components such as pectin and hemicelluloses, which can be accelerated by high temperatures and humidity, leading to quicker spoilage if not managed properly. Similarly, cherimoya undergoes significant texture changes as it ripens, primarily due to the breakdown of structural polysaccharides. Palma *et al.* (1993) emphasized that proper handling and storage conditions are essential to mitigate undesirable texture changes and extend the fruit's marketable life.

#### 12.3.2.3 Changes in starch

Starch degradation is a key metabolic change during the ripening of *Annona* spp. Starch is hydrolysed into simpler sugars, contributing to the sweetness of the ripe fruit. In cherimoya, starch content decreases significantly as ripening progresses, with amylase enzymes breaking down starch into glucose and fructose (Alique, 1995). In soursop, the starch content diminishes as it is converted to simpler sugars, enhancing the fruit's sweetness and flavour profile (Siang *et al.*, 2019). This transformation is essential for the fruit's energy supply and is a critical aspect of its postharvest physiology (Table 12.4) (González *et al.*, 2014).

#### 12.3.2.4 Changes in soluble solids content and total sugars

The increase in soluble solids content (SSC) and total sugars is a hallmark of ripening in annona fruits (Table 12.5). As a typical climacteric fruit, it has high respiration rates and a significant amount of soluble solids, which causes it to ripen quickly (Moreno and de la Plaza, 1983; Paull, 1996; Torres *et al.*, 2010; Pareek *et al.*, 2011; Cordeiro *et al.*, 2013). SSC, which includes sugars, organic acids and other soluble compounds, increases as starch is converted to sugars. In custard apple, SSC rises significantly during ripening, with glucose and fructose being the predominant sugars contributing to the sweet taste (Torres *et al.*, 2010). Atemoya also exhibits an increase in SSC and total sugars, reflecting the enzymatic breakdown of starch into soluble sugars (Jomngam *et al.*, 2017). These changes are critical for the flavour and nutritional quality of the fruit (Pott *et al.*, 2020).

#### 12.3.2.5 Changes in acidity

Acidity changes during ripening affect the taste and flavour of annona fruits (Table 12.6). The titratable acidity generally decreases as organic acids are metabolized. In sugar apple, titratable

**Table 12.4.** Starch content change of *Annona* spp. during ripening.

| *Annona* spp. | Starch content change | Reference |
|---|---|---|
| *A. cherimola* | Starch content decreases significantly, leading to increased sweetness | Alique (1995) |
| *A. muricata* | Starch is rapidly converted to sugars, enhancing sweetness | Siang *et al.* (2019) |
| *A. reticulata* | Exhibits a moderate decrease in starch content during ripening | Vyas *et al.* (2015) |
| *A. squamosa* | Rapid breakdown of starch enhances sweetness | Nwokocha and Williams (2009) |
| Atemoya | Significant decrease in starch content, similar to cherimoya | Chen *et al.* (2019) |

**Table 12.5.** Increase in soluble solids content (SSC) of *Annona* spp. during ripening.

| *Annona* spp. | SSC change | References |
|---|---|---|
| *A. cherimola* | SSC increases significantly, with higher levels of glucose and fructose | Gutiérrez *et al.* (1994) |
| *A. muricata* | Marked increase in SSC, primarily due to sugar accumulation | Silva *et al.* (2016) |
| *A. reticulata* | Exhibits a moderate increase in SSC during ripening | Vyas *et al.* (2015); Torres *et al.* (2010) |
| *A. squamosa* | SSC rises significantly during ripening | Wongs-Aree and Noichinda (2011) |
| Atemoya | Inherits high SSC from its parent species | Jomngam *et al.* (2017); Chen *et al.* (2019) |

**Table 12.6.** Acidity changes in *Annona* spp. during ripening.

| *Annona* spp. | Acidity change | Reference(s) |
|---|---|---|
| *A. cherimola* | Decreases in acidity, becoming less tart | Gutiérrez *et al.* (1994) |
| *A. muricata* | Acidity reduces significantly, enhancing sweetness | Badrie and Schauss (2010) |
| *A. reticulata* | Shows a moderate decrease in acidity | Gu *et al.* (2024); Vyas *et al.* (2015) |
| *A. squamosa* | Acidity decreases, making the fruit taste sweeter | Zhang *et al.* (2023) |
| Atemoya | Exhibits a decrease in acidity similar to cherimoya | Zhang *et al.* (2023) |

acidity decreases as citric and malic acids are degraded, resulting in a sweeter taste (Chunprasert *et al.*, 2006). Similarly, in cherimoya, the reduction in acidity is associated with the metabolism of organic acids, which balances the fruit's sweetness and enhances its palatability (Silva *et al.*, 2016). This decrease in acidity is a common trait in many climacteric fruits and is crucial for the overall taste profile (Pareek *et al.*, 2011).

#### *12.3.2.6 Changes in polyphenols*

Polyphenolic compounds, which contribute to the antioxidant capacity and astringency of the fruit, undergo significant changes during ripening (Table 12.7). In soursop, the concentration of polyphenols decreases as ripening progresses, leading to a reduction in astringency and an increase in sweetness (Silva *et al.*, 2016). In cherimoya, the decline in polyphenol content is correlated with the fruit's maturation and the development of its characteristic flavour (Alique, 1995). These compounds are also linked to the health benefits associated with fruit, including their antioxidant properties (Gupta-Elera *et al.*, 2011).

#### *12.3.2.7 Changes in volatile compounds*

Volatile compounds play a crucial role in the aroma profile of ripening annona fruits (Table 12.8). During ripening, the synthesis of esters, alcohols

**Table 12.7.** Changes in polyphenols during ripening of *Annona* spp.

| *Annona* spp. | Polyphenol content change | Reference(s) |
|---|---|---|
| *A. cherimola* | Polyphenol content decreases slightly during ripening | Loizzo *et al.* (2012) |
| *A. muricata* | Exhibits a reduction in polyphenol content | Do *et al.* (2024) |
| *A. reticulata* | Shows a moderate decrease in polyphenols | Jayaprakash (2017) |
| *A. squamosa* | Polyphenol levels decline during ripening | Nguyen *et al.* (2020) |
| Atemoya | Demonstrates a decrease in polyphenol content during ripening | Santos *et al.* (2016) |

**Table 12.8.** Changes in volatile compounds during ripening of *Annona* spp.

| *Annona* spp. | Volatile compounds change | Reference |
|---|---|---|
| *A. cherimola* | Increases in esters and alcohols, enhancing aroma | Manríquez *et al.* (2014) |
| *A. muricata* | Significant increase in volatile esters | Márquez Cardozo *et al.* (2012) |
| *A. reticulata* | Volatile compounds increase, improving aroma | Thang *et al.* (2013) |
| *A. squamosa* | Marked increase in aromatic volatiles | Cuevas-Glory *et al.* (2013) |
| Atemoya | Inherits aromatic profiles from both parents, with an increase in volatiles | Wongs-Aree and Noichinda (2011) |

and aldehydes increases, contributing to the fruit's fragrance. In custard apple, the production of volatile esters such as ethyl acetate and ethyl butyrate intensifies as the fruit ripens, enhancing its aromatic appeal (Vyas *et al.*, 2015). Similarly, in atemoya, the concentration of volatile compounds such as ethyl hexanoate and hexyl acetate increases, contributing to the complex aroma profile of the ripe fruit (Jomngam *et al.*, 2017). These aromatic compounds are essential for the sensory quality and consumer acceptance of the fruit.

## 12.4 Storage

### 12.4.1 Controlled-atmosphere storage

Controlled-atmosphere storage (CA) storage is a technique used to prolong the shelf life and maintain the quality of various horticultural products by regulating the concentrations of gases such as $O_2$ and $CO_2$ around the produce. This method has been studied extensively in *Annona* spp., particularly cherimoya, due to its perishable nature and high commercial value. Palma *et al.* (1993) investigated the respiratory behaviour of cherimoya under different CA conditions. They found that reducing $O_2$ levels and increasing $CO_2$ concentrations effectively slowed down the respiration rate of the fruit. Lowering the respiration rate is crucial as it decelerates the metabolic processes that lead to ripening and senescence, thereby extending the fruit's shelf life. This study showed that the optimal CA conditions for cherimoya involve $O_2$ levels of around 3–5% and $CO_2$ levels of 5–10%: cherimoyas can be kept for up to 6 weeks at 10°C (50°F) in 5% $O_2$ and then ripened with good flavour at 20°C (68°F); exposure to lower than 1% $O_2$ and/or more than 15% $CO_2$ can result in the development of off-flavours and uneven ripening (Kader, 2022). Alique and Oliveira (1994) examined the impact of CA storage on the sugar and organic acid content of cherimoya. Their findings indicated that CA storage significantly influences the metabolic pathways associated with sugar and acid metabolism. In particular, controlled reduction of $O_2$ and elevation of $CO_2$ levels helped in maintaining higher levels of sugars such as glucose and fructose while also preserving organic acids such as citric and malic acid. This preservation is vital for maintaining the fruit's flavour and overall sensory quality.

Research by Berger *et al.* (1993) explored the combined effects of CA and waxing on the postharvest life of cherimoya cultivar 'Bronceada'. The results demonstrated that the application of CA storage in conjunction with waxing significantly prolonged the shelf life of the fruit by reducing water loss and delaying ripening processes. Waxing alone provided a barrier to moisture loss, but when combined with CA, the benefits were synergistically enhanced, leading to better retention of firmness and visual appeal. Celis and Marcela (1993) studied the behaviour of cherimoya 'Concha Lisa' under CA storage. Similar to findings in other studies, CA storage was effective in maintaining the quality of this cultivar. The research underscored that different cultivars might have varying responses to CA conditions, necessitating tailored CA protocols to optimize storage outcomes for each specific cultivar.

Alique (1995) investigated the residual effects of short-term high $CO_2$ treatments on the ripening of cherimoya. The study found that brief exposures to elevated $CO_2$ levels (up to 20%) prior to CA storage could have lasting benefits in delaying ripening and senescence. This suggests that integrating high-$CO_2$ pre-treatments could be a valuable strategy in enhancing the effectiveness of CA storage for cherimoya. Yahia (2006) provided a comprehensive review of MAP and CA storage for tropical fruits, including *Annona* spp. The review highlighted the broader applicability of CA storage techniques across various tropical fruits, emphasizing the need for species-specific and sometimes cultivar-specific adjustments in CA conditions to achieve optimal storage results. The overarching theme is the potential of CA storage to significantly reduce postharvest losses and maintain the quality of tropical fruits during extended storage and transportation periods.

CA storage is also beneficial for other *Annona* spp., such as soursop, custard apple, sugar apple and atemoya. For soursop, CA storage involves slightly higher $O_2$ levels (3–5%) and moderate $CO_2$ levels (5–7%). This atmosphere helps in slowing down the ripening process and reduces the occurrence of postharvest diseases, thereby extending the fruit's marketable life significantly. Custard apples benefit from CA storage by maintaining a delicate balance of reduced $O_2$ (2–3%) and elevated $CO_2$ (3–5%). This setup helps in maintaining firmness and reducing decay, thus prolonging the shelf life while preserving sensory qualities. Sugar apple storage under CA typically involves $O_2$ levels of around 3% and $CO_2$ levels of 5% (Yahia, 2006). This method reduces respiration rates and delays senescence, ensuring that the fruit remains fresh and marketable for extended periods. Atemoya benefits from CA storage conditions, similar to its parent species. Optimal storage conditions include low $O_2$ levels (2–3%) and moderate $CO_2$ levels (5–7%), which help in maintaining fruit quality and extending shelf life (Yahia, 1998).

Overall, CA storage is a crucial postharvest technique for *Annona* spp., offering significant benefits in terms of shelf-life extension and quality preservation. By carefully managing the atmospheric conditions, postharvest losses can be minimized, and the fruit can reach distant markets in better condition.

### 12.4.2 Modified-atmosphere packaging

MAP is a postharvest technology designed to extend the shelf life and preserve the quality of the fruit by altering the atmospheric composition surrounding the produce. This method has shown promising results in maintaining the physico-chemical properties and extending the marketability of various *Annona* spp., including custard apple and cherimoya. Patil *et al.* (2015) conducted a study on sugar apples, revealing that MAP effectively maintains the physico-chemical properties of the fruit. The research demonstrated that sugar apples stored in MAP exhibited significantly lower weight loss, and showed delayed ripening and preserved firmness compared with those stored under regular atmospheric conditions. The controlled environment created by MAP helped in reducing respiration rates and ethylene production, which are critical factors in slowing down the ripening process and extending shelf life. Additionally, the MAP-treated fruit maintained higher levels of ascorbic acid and overall sensory quality, making them more appealing to consumers over a longer period. Tinebra *et al.* (2022) explored the impact of MAP combined with low-temperature storage on cherimoya. Their findings indicated that this combined approach significantly extended the marketability of cherimoya. The modified-atmosphere conditions, characterized by reduced $O_2$ and elevated $CO_2$ levels, along with low temperatures, were effective in slowing down the metabolic activities that lead to fruit deterioration. The study highlighted that cherimoya stored in MAP had better texture, flavour and visual quality, making them more suitable for long-distance transportation and extended storage (Fig. 12.1). This combination of

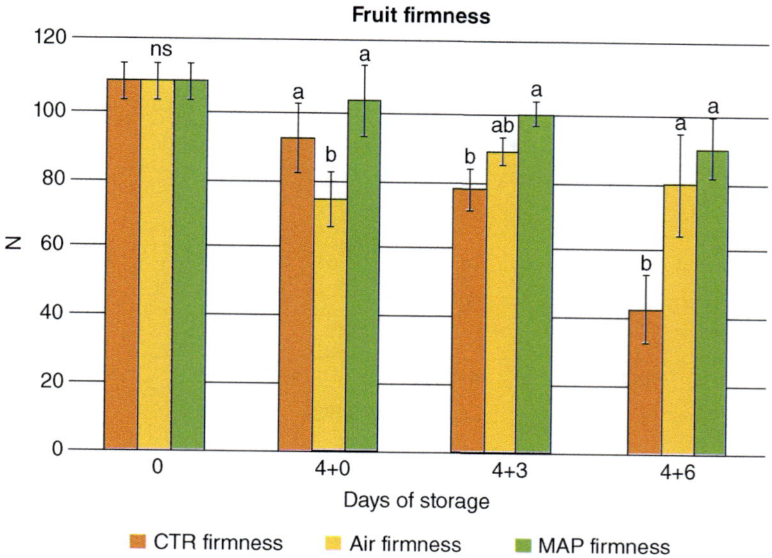


**Fig. 12.1.** Evolution of cherimoya fruit firmness over storage time and during simulated shipping periods (4 + 0, 4 + 3 and 4 + 6 days). All values are shown as mean ± sd ($N$ = 6). Different letters at each evaluation time indicate a significant difference between the treatments for Tukey's HSD test ($P < 0.05$), while ns indicates there is no significant difference between any two treatments for Tukey's HSD test ($P < 0.05$). The figure shows the differences between the control (CTR: fruit stored in unsealed bags); treatment in air: 21% $O_2$ + 1% $CO_2$ + 78% $N_2$ (passive-MAP); and modified-atmosphere packaging (MAP) treatment: 30% $CO_2$ + 10% $O_2$ + 60% $N_2$. From Tinebra *et al.* (2022); used with permission.

MAP and refrigeration was particularly effective in minimizing decay and maintaining the fruit's overall quality (Fig. 12.2). Venkatram *et al.* (2015) investigated the effects of antioxidants and MAP on the physico-chemical characteristics of sugar apple 'Balanagar'. The study found that the use of MAP, in conjunction with antioxidant treatments, provided substantial benefits in preserving the quality of sugar apples. The MAP environment helped in maintaining the firmness and reducing enzymatic browning, which is often a significant issue in postharvest handling. The antioxidants further enhanced the fruit's resistance to oxidative stress, thereby complementing the benefits of MAP in extending shelf life and maintaining quality. Reis *et al.* (2014) examined the effect of MAP on the texture properties of atemoya during refrigerated storage. Their research demonstrated that MAP significantly preserved the textural integrity of atemoya fruits. The modified-atmosphere conditions of reduced $O_2$ and increased $CO_2$ levels helped in maintaining the cell structure and reducing the softening of the fruit. This preservation of texture is crucial for consumer acceptance and marketability, as texture is a key quality attribute for atemoya.

In conclusion, MAP is a highly effective method for extending the shelf life and preserving the quality of *Annona* spp. By regulating the atmospheric composition around the fruit, MAP can significantly reduce respiration rates, delay ripening, and minimize physical and chemical deterioration. The integration of MAP with other treatments, such as low-temperature storage or antioxidants, further enhances its effectiveness. This technology offers a robust solution for the postharvest handling of sugar apple, cherimoya, atemoya, and potentially other *Annona* spp., ensuring that the fruit remain fresh and marketable for longer periods.

### 12.4.3 Edible coatings

Interest in developing biologically based materials to prolong shelf life and improve the quality of fresh, ready-to-eat and formulated food products

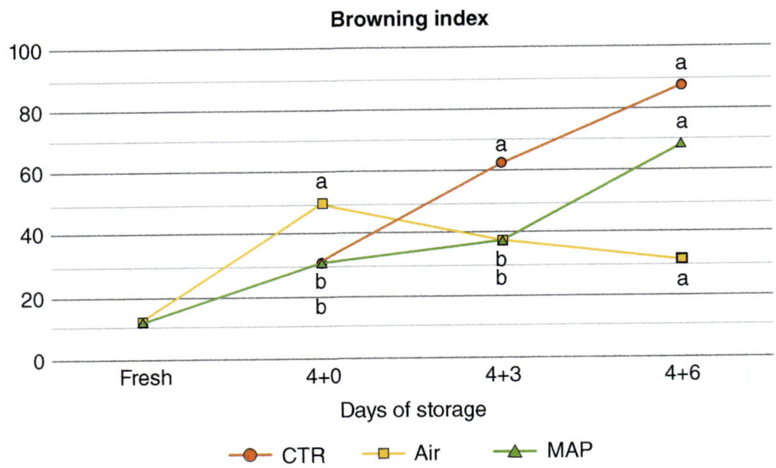


**Fig. 12.2.** Evolution of the browning index (Wang *et al.*, 2005) of cherimoya fruit over storage time and during simulated shipping periods (4 + 0, 4 + 3 and 4 + 6 days). The figure shows the differences between the control (CTR: fruit stored in unsealed bags); treatment in air: 21% $O_2$ + 1% $CO_2$ + 78% $N_2$ (passive-MAP); and modified-atmosphere packaging (MAP) treatment: 30% $CO_2$ + 10% $O_2$ + 60% $N_2$. From Tinebra *et al.* (2022); used with permission.

has increased rapidly in recent decades (Diab *et al.*, 2001). Among these materials, edible coatings have proven effective in food protection, and the recent use of multi-layered coatings has further extended the shelf life of fresh fruits (Armghan Khalid *et al.*, 2022; Passafiume *et al.*, 2022). The edible coating can be applied by dipping, spraying or brushing, and effectively extends fresh fruit shelf life without causing anaerobiosis, reducing decay or compromising quality. They must meet specific requirements such as gas and moisture barriers, colour, mechanical properties and non-toxicity. Applying edible coatings slows cellular respiration and senescence and inhibits the development of fungi and bacteria, thus enhancing the fruit's appearance (Dhall, 2013; Mahajan *et al.*, 2014). With sustainability in mind, it is essential to find alternatives to chemicals while maintaining product quality. European Directives (95/2/EC and 98/72/EC) (Agencia Estatal Boletín Oficial del Estado, 1995) and the US Food and Drug Administration regulations (FDA, 2006) define edible coatings as those formulated with food additives, and their use for fresh produce is continually evolving.

Physico-chemical analyses of fruits treated with edible coatings have shown positive results, such as keeping firmness by maintaining cell-wall pectin and increasing titratable acidity and vitamin content due to antioxidant and antimicrobial substances. Few studies have focused on improving the shelf life of cherimoya. Application of 0.75% chitosan (a non-toxic high-molecular-weight polymer derived from the outer shell of crustaceans) as an edible coating could be considered a useful approach to maintain product quality during preservation. Minh *et al.* (2019) showed the best results in terms of weight loss, pH, total soluble solids (TSS), titratable acidity and ascorbic acid, total plate count and sensory characteristics could be achieved by applying a coating to soursop fruit.

Rossi-Márquez *et al.* (2023) evaluated the application of multi-layer protein and chitosan coating on fresh cherimoya to determine the effect on the shelf life on the fruit, total phenolic content, pH, °Brix, weight loss and hardness. According to this study, the multi-layered coating maintained the physico-chemical and nutraceutical characteristics over time while reducing water loss. An increase in soluble solids was correlated with the ripening process.

In Mexico, Salomon-Castaño *et al.* (2022) analysed the postharvest behaviour of various quality parameters of ilama (*Annona diversifolia* Saff.) fruit covered with an edible hawthorn pectin coating. The flesh of this fruit is exposed to biotic and abiotic environmental factors when the peel starts cracking. Under storage at 21 ± 2°C, the uncoated fruit had a shelf life of 3 days, while the coated fruit lasted for 7 days. Fruit stored at 4°C and −20°C remained fresh for 11 days; however, the peels became hard, particularly in uncoated fruit. Across all three storage temperatures, applying coatings reduced the accumulation of TSS, lowered respiration rates and minimized peel discoloration, showing a decrease in fruit metabolism. A lower storage temperature also reduced the accumulation of TSS.

Liu *et al.* (2016) investigated the effects of citric acid and chitosan coating on the fruit-ripening attributes of cherimoya. Application of citric acid and chitosan coating delayed the browning, splitting and decay of cherimoya fruit, preserved firmness and prevented an increase in respiration rate. This effect is due to the reduction of pectin methyl esterase activities and polygalacturonase, limiting pectin degradation. Additionally, the coating regulated carboxymethylcellulase activity and starch–sugar conversion, thus delaying softening and splitting. Throughout storage, coated fruit showed lower polyphenol oxidase activity and higher activities of superoxide dismutase, peroxidase and catalase, along with reduced superoxide radical production and malondialdehyde content compared with controls.

Another study evaluated the effects of calcium chloride ($CaCl_2$), benzyl adenine and chitosan on the shelf life of sugar apple 'Arka Sahan'. The combination of 6% $CaCl_2$, 150 p.p.m. benzyl adenine and 1.5% chitosan significantly improved the shelf life by inhibiting ripening and maintaining the quality of the fruit up to 12 days at ambient temperature. The treated fruit exhibited better colour, taste, flavour and appearance compared with untreated controls (Patidar *et al.*, 2021).

Torres *et al.* (2010) explored the impact of postharvest warm dipping with $CaCl_2$ on atemoya fruit. The study found that treatment at 40°C maintained the fruit's edible condition for up to 6 days, although the peel's appearance was affected by calcium as early as 4 days. Untreated fruit exhibited flesh browning after 8 days, which was correlated with increased polyphenol oxidase (PPO) and peroxidase activities, higher total phenolics and reduced ascorbic acid content. Weight loss occurred continuously throughout storage, with no significant differences between the treated and untreated fruit. Total titratable acidity and TSS increased, while pH decreased during the experiment. The results indicated that $CaCl_2$ dipping effectively reduced flesh browning, and the warm treatment had a synergistic effect, probably linked to decreased PPO activity. Additionally, variations in ascorbic acid content suggested that warm dipping with $CaCl_2$ contributed to the fruit's antioxidant capacity.

These studies suggest that the use of edible coatings on annona fruit can be an effective strategy for improving postharvest quality and extending the shelf life of the fruit. However, application of edible coatings on cherimoya is largely unknown today. By addressing the challenges of rapid ripening and physical damage, these coatings could significantly enhance the fruit's commercial value and marketability. Sustainable and natural coatings offer an eco-friendly alternative to synthetic preservatives, aligning with global trends towards reducing chemical use in fruit preservation.

### 12.4.4 Hot-water treatment

Hot-water treatment (HWT) is a common postharvest technique used to control diseases and pests in various fruits, including *Annona* spp. This treatment involves immersing the fruit in hot water at a specific temperature for a certain duration.

HWT is effective in controlling surface pests such as fruit flies, which are common in tropical and subtropical fruits. It helps in reducing the incidence of fungal and bacterial diseases that can affect the fruit during storage and transportation. By reducing the microbial load, HWT can extend the shelf life of cherimoya. González-Aguilar *et al.* (2000) examined the impact of HWT on the postharvest quality of cherimoya fruit. The results of their study showed that HWT at 47°C for 15 min was effective in controlling fungal infections while maintaining fruit quality.

Research conducted by Balakoti (2012) examined the effectiveness of HWT in managing fruit fly larvae in cherimoya and demonstrated that treating cherimoya at 46°C for 20 min effectively controlled the fruit flies without causing significant damage to the fruit.

Torres *et al.* (2010) investigated the effects of different HWT conditions on the shelf life and quality of cherimoya fruit. Their findings suggested that treating the fruit at 50°C for 10 min significantly reduced decay and extended shelf life with minimal impact on quality. Similarly, Jomngam *et al.* (2017) reported that while HWT can induce stress responses, these do not necessarily lead to significant quality degradation if properly managed. Ramos-Guerrero *et al.* (2018) identified optimal conditions that balance effective disease control with minimal fruit quality loss.

### 12.4.5 γ-Irradiation and growth regulators

Only one study has investigated the combined effect of γ-irradiation and gibberellic acid ($GA_3$) on the quality and shelf life of sugar apple. The experiment demonstrated that treating fruit with 0.4 kGy of γ-irradiation combined with 50 p.p.m. of $GA_3$ significantly reduced physiological weight loss and spoilage while maintaining higher levels of TSS and sugars. This treatment also delayed ripening and reduced the incidence of spoilage, effectively extending the shelf life of the fruit (Ambika *et al.*, 2019).

### 12.4.6 Treatment with 1-methylcyclopropene

A promising postharvest treatment to mitigate these losses is the application of 1-methylcyclopropene (1-MCP) a synthetic compound that inhibits ethylene action, thereby delaying ripening and senescence. Given the perishable nature of *Annona* spp., 1-MCP presents a valuable tool in postharvest management. Several studies have demonstrated that 1-MCP treatment significantly extends the shelf life of annona fruits. For instance, research by Wongs-Aree and Noichinda (2011) showed that 1-MCP-treated sugar apples exhibited delayed ripening and reduced weight loss over a storage period of 21 days compared with untreated fruits. Similarly, a study on soursop by Benassi *et al.* (2003) reported a notable extension of shelf life by up to 50% with 1-MCP application, attributed to slowed metabolic activities. Application of 1-MCP has been found to maintain firmness in annona fruits by inhibiting the ethylene-induced softening process. According to a study by González-Aguilar *et al.* (2000), 1-MCP-treated *A. muricata* retained higher firmness levels throughout the storage period compared with untreated fruit, which softened rapidly. Physiological disorders such as chilling injury and browning are common in annona fruit during storage. Research indicates that 1-MCP can mitigate these issues. For example, a study by Espinosa *et al.* (2013) found that 1-MCP treatment reduced the incidence of browning and chilling injury in cherimoya stored at low temperatures, thereby enhancing marketability. Studies have shown that 1-MCP treatment helps preserve the nutritional quality of annona fruit. A study by Siang *et al.* (2019) demonstrated that 1-MCP-treated sugar apple maintained higher levels of ascorbic acid and phenolic compounds during storage compared with untreated fruit, which experienced significant degradation of these nutrients. The efficacy of 1-MCP treatment depends on the concentration and duration of exposure. Most studies recommend a concentration range of 500–1000 p.p.b. of 1-MCP for effective results. For instance, a study by Moreno-Hernández *et al.* (2014) suggested that a 12 h exposure to 1000 p.p.b. of 1-MCP significantly delayed ripening in sugar apple. The application of 1-MCP is also influenced by the storage conditions. Optimal results are observed when treated fruit are stored at low temperatures. For example, a study by Espinosa *et al.* (2013) found that 1-MCP-treated soursop fruit stored at 10°C exhibited better quality retention compared with those stored at higher temperatures.

## 12.5 Postharvest Pathology

Annona fruit are highly perishable and susceptible to various postharvest diseases that significantly reduce their marketability and consumer acceptance. *Colletotrichum gloeosporioides* is one of the most prevalent pathogens causing anthracnose in annona fruit. This fungal disease manifests as dark, sunken lesions on the fruit surface, which can expand and coalesce, leading to extensive rot. Research by Narayanasamy (2010) highlighted that anthracnose not only affects the visual quality of the fruit but also compromises its nutritional value by degrading nutrients. For instance, anthracnose was shown to decrease the levels of ascorbic acid and other antioxidants in the fruit.

*Botrytis cinerea*, commonly known as grey mould, is another significant pathogen affecting *Annona* spp. during postharvest storage. This fungus thrives in high humidity and causes soft, watery decay covered with grey mould. According to Elad (2016), *B. cinerea* infection can lead to rapid deterioration of fruit quality, resulting in substantial economic losses. *Rhizopus stolonifer*, the causative agent of *Rhizopus* rot, is frequently reported in stored annona fruit. This pathogen causes soft, watery rot with a characteristic fermented odour. According to Mari *et al.* (2016), *Rhizopus* rot can spread quickly under warm, humid conditions, making it a significant concern during the postharvest phase.

Various *Penicillium* spp. are known to infect annona fruit, leading to blue mould decay. This disease is characterized by blue-green spore masses on the fruit surface and soft rot underneath. A study by Tian *et al.* (2015) indicated that *Penicillium* spp. infections not only spoil the fruit but can also produce mycotoxins, posing health risks to consumers.

Black canker is caused by *Phomopsis anonacearum* and appears as purple spots on the fruit, which become hard and cracked, followed by the development of small black bodies containing spores.

*Botryodiplodia* rot is caused by *Botryodiplodia theobromae*. It first appears purple and later becomes pimpled with black pycnidia. The fruit flesh becomes brown and corky (Kader, 2022).

Proper orchard management, such as pruning, sanitation and adequate spacing, can reduce the incidence of diseases by improving air circulation and reducing humidity around the fruit (Mari *et al.*, 2016). The use of disease-resistant annona cultivars can significantly reduce the impact of postharvest pathogens. Breeding programmes focused on developing resistant cultivars are essential for sustainable disease management.

The appearance of lesions, mould and rot detracts from the fruit's visual appeal, making it less attractive to consumers. Pathogen-induced decay leads to softening and breakdown of fruit tissues, resulting in a mushy texture. Infections can alter the fruit's flavour and aroma, often leading to off-flavours and unpleasant odours. Pathogens can degrade essential nutrients, reducing the fruit's health benefits.

Fungicides such as thiophanate-methyl and fludioxonil are commonly used to control postharvest pathogens in annona fruit. However, their use is limited by concerns over chemical residues and resistance development (Elad, 2016). The use of biocontrol agents, such as antagonistic bacteria and fungi, has shown promise in managing postharvest diseases. For example, *Bacillus subtilis* has been effective against *Botrytis cinerea* and *Penicillium* spp. in annona fruit (Tian *et al.*, 2015). Natural antifungal compounds, such as essential oils and plant extracts, offer an eco-friendly alternative to synthetic fungicides. Studies have shown that extracts from neem, thyme and clove can inhibit the growth of common postharvest pathogens (Narayanasamy, 2006).

Keeping annona fruit at optimal low temperatures can slow down the growth of pathogens. For example, storing soursop at 10°C significantly reduced the incidence of postharvest diseases (Mari *et al.*, 2016). Maintaining low humidity levels during storage can help prevent the proliferation of moisture-loving pathogens such as *Botrytis cinerea* and *Rhizopus stolonifer* (Elad, 2016).

George and Nissen (1987) suggested dipping annona fruit in a heated benomyl suspension (0.5 g active ingredient $l^{-1}$) or an equivalent concentration of guazatine solution at 50–52°C for 5 min to manage postharvest rots. Alternatively, immersing the fruit in unheated prochloraz (0.125 g active ingredient $l^{-1}$) for 1 min at 25°C effectively controlled postharvest rotting. However, high concentrations of some treatments, especially prochloraz, can cause skin injury and should be avoided.

## 12.6 Physiological Disorders

Annona fruit occasionally present abnormalities due to physiological disorders (George and Nissen, 1987):

- Fruit splitting is probably due to sudden changes in moisture content or temperature, with some species and cultivars being less susceptible than others (Paull, 1996).

- Chilling injury can be caused by exposure of annona fruit to temperatures below 8–12°C, depending on cultivar and ripeness stage. Symptoms include darkening and hardening of the skin, pitting, failure to develop full flavour and mealy flesh.
- Russeting refers to the superficial russet-coloured patches on the fruit skin. This condition is probably caused by low night temperatures (<13°C) combined with low humidity. The almost-mature fruit are more prone to russeting, with cherimoya being less susceptible than sugar apple and atemoya.
- Hard seed casings and brown lumps can be caused by boron deficiency or sudden changes in fruit water content. In north-eastern Brazil, sugar apple pulp frequently exhibits hard seed casing disorder when trees are grown in dry areas without irrigation.

## 12.7 Conclusions

Understanding the postharvest physiology and pathology of *Annona* spp. is pivotal for optimizing fruit quality and extending shelf life. Table 12.9 summarizes these features. Precise maturity indices are crucial to determining the optimal harvest time, ensuring the best possible flavour, texture and nutritional value upon consumption. Respiration and ethylene production are central to the postharvest behaviour of these fruit. The rate of respiration impacts shelf life and quality, while ethylene plays a significant role in ripening and senescence. Effective management of these physiological processes can lead to improved storage outcomes and prolonged marketability. Physico-chemical changes during ripening are intricate and species specific, involving modifications in texture, sweetness, acidity and volatile compounds. These changes are critical for consumer acceptance and market value. By understanding these changes, postharvest handling protocols can be better tailored to maintain fruit quality. Storage techniques are essential for mitigating postharvest losses. Different *Annona* spp. may require specific storage conditions to maximize shelf life and minimize quality degradation.

Postharvest pathology, including fungal, bacterial and physiological disorders, poses significant challenges to the maintenance of fruit quality. Understanding the aetiology and progression of these diseases enables the development of effective control measures, reducing losses and preserving fruit integrity.

**Table 12.9.** Summary of the important features of postharvest management.

| Characteristic | Key information | References |
|---|---|---|
| Role of respiration and ethylene | Respiration and ethylene production are essential in ripening, quality and shelf life of *Annona* spp. fruits, influencing texture, colour and flavour during ripening | Wongs-Aree and Noichinda (2011); Anaya-Esparza *et al.* (2017); Kader (2022) |
| Variation among species | Ethylene and respiration rates vary: cherimoya and atemoya show moderate ethylene levels, while soursop and sugar apple produce high levels, leading to quicker ripening and reduced shelf life | Chunprasert *et al.* (2006); Anaya-Esparza *et al.* (2017) |
| Cherimoya specifics | Ethylene production in cherimoya is preceded by a rise in respiration and a decrease in flesh firmness, which impacts its texture as it ripens | Anaya-Esparza *et al.* (2017) |
| Soursop and sugar apple | Both exhibit high respiration rates and ethylene levels, promoting faster ripening and shorter shelf life | Chunprasert *et al.* (2006) |
| General ethylene range | Ethylene levels for *Annona* spp. during ripening can reach up to 100–300 μl $kg^{-1}$ $h^{-1}$ at 20°C, affecting ripening speed and quality | Kader (2022) |

Overall, integrating knowledge of maturity indices, postharvest physiology, respiration and ethylene production, physico-chemical changes during ripening, storage techniques and postharvest pathology is fundamental for enhancing the postharvest life of *Annona* spp. This comprehensive approach ensures that these fruit can reach consumers with optimal quality, thereby increasing their market potential and profitability. Future research should focus on refining these strategies and developing innovative solutions to address emerging postharvest challenges in *Annona* spp.

## References


Agencia Estatal Boletín Oficial del Estado (1995) Directiva 95/2/CE del Parlamento Europeo y del Consejo de 20 de febrero de 1995 relativa a aditivos alimentarios distintos de los colorantes y edulcorantes. Available at: www.boe.es/doue/1995/061/L00001-00040.pdf (accessed 20 August 2025).

Alique, R. (1995) Residual effects of short-term treatments with high $CO_2$ on the ripening of cherimoya (*Annona cherimola* Mill.) fruit. *Journal of Horticultural Science* 70, 609–615. DOI: 10.1080/14620316.1995.11515333

Alique, R. and Oliveira, G.S. (1994) Changes in sugars and organic acids in cherimoya (*Annona cherimola* Mill.) fruit under controlled atmosphere storage. *Journal of Agricultural and Food Chemistry* 42, 799–803. DOI: 10.1021/jf00039a040

Ambika, C.P., Joshi, V., Kumar, T.S. and Goud, C.R. (2019) Combined effect of gamma irradiation and post-harvest treatments on quality and shelf life of custard apple (*Annona squamosa* L.) cv. Balanagar. *International Journal of Current Microbiology and Applied Sciences* 8, 2731–2743. DOI: 10.20546/ijcmas.2019.807.335

Anaya-Esparza, L.M., Ramírez-Marez, M.V., Montalvo-González, E. and Sánchez-Burgos, J.A. (2017) Cherimoya (*Annona cherimola* Mill.). In: Yahia, E.M. (ed.) *Fruit and Vegetable Phytochemicals: Chemistry and Human Health*, 2nd ed. Wiley, Hoboken, New Jersey, pp. 993–1002.

Armghan Khalid, M., Niaz, B., Saeed, F., Afzaal, M., Islam, F. *et al.* (2022) Edible coatings for enhancing safety and quality attributes of fresh produce: A comprehensive review. *International Journal of Food Properties* 25, 1817–1847. DOI: 10.1080/10942912.2022.2107005

Badrie, N. and Schauss, A.G. (2010) Soursop (*Annona muricata* L.): composition, nutritional value, medicinal uses, and toxicology. In: Watson, R.R. and Preedy, V.R. (eds) *Bioactive Foods in Promoting Health*. Academic Press, Cambridge, UK, pp. 621–643.

Balakoti, G. (2012) Effect of modified atmosphere packaging and hot water treatments on shelf life of custard apple (*Annona squamosa* L.) stored at ambient temperature. PhD thesis, Dr YSR Horticultural University, Andhra Pradesh, India.

Balois-Morales, R., Jiménez-Zurita, J.O., Alia-Tejacal, I., López-Guzmán, G.G., Palomino-Hermosillo, Y.A. and Sánchez-Herrera, L.M. (2019) Antioxidant enzymes and antioxidant activity in two soursop selections (*Annona muricata* L.) from Nayarit, Mexico stored at 15°C. *Revista Brasileira de Fruticultura* 41: e083. DOI: 10.1590/0100-29452019083

Benassi, G., Correa, G.A.S.F., Kluge, R.A. and Jacomino, A.P. (2003) Shelf life of custard apple treated with 1-methylciclopropene: an antagonist to the ethylene action. *Brazilian Archives of Biology and Technology* 46, 115–120. DOI: 10.1590/S1516-89132003000100016

Berger, H., Galletti, L., Marin, J., Fichet, T. and Lizana, L.A. (1993) Effect of controlled atmosphere and waxing on the postharvest life of cherimoya (*Annona cherimola* Mill.) cv. Bronceada. *Proceedings of the Interamerican Society for Tropical Horticulture* 37, 121–130.

Brown, B.I., Wong, L.S., George, A.P. and Nissen, R.J. (1988) Comparative studies on the postharvest physiology of fruit from different species of *Annona* (custard apple). *Journal of Horticultural Science* 63, 521–528. DOI: 10.1080/14620316.1988.11515887

Bruinsma, J. and Paull, R.E. (1984) Respiration during postharvest development of soursop fruit, *Annona muricata* L. *Plant Physiology* 76, 131–138. DOI: 10.1104/pp.76.1.131

Bulathkandage, M., Eeswara, J.P., Madujith, T., Gunawardena, R., Dinesh, C. and Jayapala, R. (2019) Determination of harvesting indices and harvesting stage of *Annona muricata* L. (soursop) based on fruit growth. *Tropical Agriculturist* 167, 18–33. DOI: 10.4038/ta.v167i4.99

Celis, T. and Marcela, C. (1993) Behavior of cherimoya fruits (*Annona cherimola* Mill.), cultivar Concha Lisa under controlled atmosphere storage. *AGRIS – International System for Agricultural Science and Technology* 46: 81.

Chen, J., Duan, Y., Hu, Y., Li, W., Sun, D. *et al.* (2019) Transcriptome analysis of atemoya pericarp elucidates the role of polysaccharide metabolism in fruit ripening and cracking after harvest. *BMC Plant Biology* 19: 219. DOI: 10.1186/s12870-019-1756-4

Chunprasert, A., Uthairatanakij, A. and Wongs-Aree, C. (2006) Storage quality of 'Neang' sugar apple treated with chitosan coating and map. *Acta Horticulturae* 712, 857–864. DOI: 10.17660/ActaHortic.2006.712.112

Cordeiro, N., Sousa, L., Freitas, N. and Gouveia, M. (2013) Changes in the mesocarp of *Annona cherimola* Mill. 'Madeira' during postharvest ripening. *Postharvest Biology and Technology* 85, 179–184. DOI: 10.1016/j.postharvbio.2013.05.014

Cuevas-Glory, L., Ortiz-Vazquez, E., Sauri-Duch, E. and Pino, J. (2013) Characterization of aroma-active compounds in sugar apple (*Annona squamosa* L.). *Acta Alimentaria* 42, 102–108. DOI: 10.1556/aalim.42.2013.1.10

Dhall, R.K. (2013) Advances in edible coatings for fresh fruits and vegetables: a review. *Critical Reviews in Food Science and Nutrition* 53, 435–450. DOI: 10.1080/10408398.2010.541568

Diab, T., Biliaderis, C.G., Gerasopoulos, D. and Sfakiotakis, E. (2001) Physicochemical properties and application of pullulan edible films and coatings in fruit preservation. *Journal of the Science of Food and Agriculture* 81, 988–1000. DOI: 10.1002/jsfa.883

Do, Y.V., Le, Q.N.T., Nghia, N.H., Vu, N.D., Tran, N.T.Y. *et al.* (2024) Assessment of the changes in product characteristics, total ascorbic acid, total flavonoid content, total polyphenol content and antioxidant activity of dried soursop fruit tea (*Annona muricata* L.) during product storage. *Food Science & Nutrition* 12, 2679–2691. DOI: 10.1002/fsn3.3949

Elad, Y. (2016) Cultural and integrated control of *Botrytis* spp. In: Fillinger, S. and Elad, Y. (eds) *Botrytis – The Fungus, the Pathogen and its Management in Agricultural Systems*. Springer, Cham, Switzerland, pp. 149–164. DOI: 10.1007/978-3-319-23371-0_8

Espinosa, I., Ortiz, R.I., Tovar, B., Mata, M. and Montalvo, E. (2013) Physiological and physicochemical behavior of soursop fruits refrigerated with 1-methylcyclopropene. *Journal of Food Quality* 36, 10–20. DOI: 10.1111/jfq.12011

FDA (US Food and Drug Administration) (2006) 21 CFR Part 172: Food additives permitted for direct addition to food for human consumption. Subpart C: coatings, films and related substances. Available at: www.ecfr.gov/current/title-21/chapter-I/subchapter-B/part-172 (accessed 20 August 2025).

George, A.P. and Nissen, R.J. (1987) Propagation of *Annona* species: a review. *Scientia Horticulturae* 33, 75–85. DOI: 10.1016/0304-4238(87)90034-3

González, E.M., Fernández, A.E.L., Paez, H.R., Oca, M.M.M.D. and Gómez, B.T. (2014) Uso combinado de 1-meticiclopropeno y emulsiones de cera en la conservación de guanábana (*Annona muricata*). *Revista Brasileira de Fruticultura* 36, 296–304. DOI: 10.1590/S0100-29452014000500035

González-Aguilar, G.A., Gayosso, L., Cruz, R., Fortiz, J., Báez, R. and Wang, C. Y. (2000) Polyamines induced by hot water treatments reduce chilling injury and decay in pepper fruit. *Postharvest Biology and Technology* 18, 19–26. DOI: 10.1016/S0925-5214(99)00054-X

Gregorio, G., Antonio, D. and Rocco, Z. (2019) Comparative study of some fruit quality characteristics of two of *Annona cherimola* Mill. grown in southern Italy. *AIMS Agriculture and Food* 4, 658–671. DOI: 10.3934/agrfood.2019.3.658

Gupta-Elera, G., Garrett, A.R., Martinez, A., Robison, R.A. and O'Neill, K.L. (2011) The antioxidant properties of the cherimoya (*Annona cherimola*) fruit. *Food Research International* 44, 2205–2209. DOI: 10.1016/j.foodres.20DOI: 10.DOI: 10.038

Gutiérrez, M., del Mar Sola, M., Pascual, L. and Vargas, A.M. (1994) Postharvest changes of sugar concentrations in chilled-injured cherimoya (*Annona cherimola* Mill.). *Journal of Plant Physiology* 143, 27–32. DOI: 10.1016/S0176-1617(11)82093-6

Jayaprakash, A. (2017) Phytochemicals, antimicrobial and antioxidant properties of *Annona reticulata* Linn. *Journal of Academia and Industrial Research* 6, 90–95.

Jiménez-Zurita, J.O., Alia-Tejacal, I., Balois-Morales, R., Villarreal-Fuentes, J.M., Núñez-Colín, C.A. *et al.* (2022) Phenological growth stages of soursop trees (*Annona muricata* L.) based on the extended BBCH-scale. *Revista Chapingo Serie Horticultura* 29, 5–18. DOI: 10.5154/r.rchsh.2022.03.006

Jomngam, P., Chen, K.Y., Chumpookam, J. and Shiesh, C.C. (2017) Effect of ripening and storage temperatures on quality of atemoya (*Annona cherimola* Mill. × *A. squamosa* L.) fruit. *Acta Horticulturae* 1166, 161–166. DOI: 10.17660/ActaHortic.2017.1166.23

Kader, A.A. (2022) *Cherimoya: Recommendations for Maintaining Postharvest Quality*. Available at: https://postharvest.ucdavis.edu/produce-facts-sheets/cherimoya-Atemoya-sweetsop

Kader, A.A. and Arpaia, M.L. (1999) Cherimoya, atemoya and sweet sweetsop. Produce facts. *UC Perishables Handling Quarterly* 97, 17–18.

Leal, F. and Paull, R.E. (2023) The soursop (*Annona muricata*): botany, horticulture, and utilization. *Crop Science* 63, 362–389. DOI: 10.1002/csc2.20894

Liu, K., Liu, J., Li, H., Yuan, C., Zhong, J. and Chen, Y. (2016) Influence of postharvest citric acid and chitosan coating treatment on ripening attributes and expression of cell wall related genes in cherimoya (*Annona cherimola* Mill.) fruit. *Scientia Horticulturae* 198, 1–11. DOI: 10.1016/j.scienta.2015.11.008

Lizana, L.A. and Reginato, G. (1990) Cherimoya. In: Nagy, S., Shaw, P.E. and Wardowski, W.F. (eds) *Fruit of Tropical and Subtropical Origin: Composition, Properties and Uses*. Florida Science Source, Lake Alfred, Florida, pp. 131–138.

Loizzo, M.R., Tundis, R., Bonesi, M., Menichini, F., Mastellone, V. *et al.* (2012) Radical scavenging, antioxidant and metal chelating activities of *Annona cherimola* Mill. (cherimoya) peel and pulp in relation to their total phenolic and total flavonoid contents. *Journal of Food Composition and Analysis* 25, 179–184. DOI: 10.1016/j.jfca.2011.09.002

Mahajan, P.V., Caleb, O.J., Singh, Z., Watkins, C.B. and Geyer, M. (2014) Postharvest treatments of fruits. *Philosophical Transactions A: Mathematical, Physical and Engineering Sciences* 372: 20130309. DOI: 10.1098/rsta.2013.0309

Manica, I., Accorsi, M.R., Delloto, F.A., Fioravanço, J.C.F., Paiva, M.C. *et al.* (1994) *Fruticultura: Cultivo das Anonáceas – Ata – Cherimólia – Graviola*. Cinco Continentes, Porto Alegre, Brazil.

Manríquez, D.A., Muñoz-Robredo, P., Gudenschwager, O., Robledo, P. and Defilippi, B.G. (2014) Development of flavor-related metabolites in cherimoya (*Annona cherimola* Mill.) fruit and their relationship with ripening physiology. *Postharvest Biology and Technology* 94, 58–65. DOI: 10.1016/j.postharvbio.2014.03.004

Mari, M., Bautista-Baños, S. and Sivakumar, D. (2016) Decay control in the postharvest system: role of microbial and plant volatile organic compounds. *Postharvest Biology and Technology* 122, 70–81. DOI: 10.1016/j.postharvbio.2016.04.014

Márquez Cardozo, C.J., Villacorta Lozano, V., Yepes Betancur, D.P., Ciro Velásquez, H.J. and Cartagena Valenzuela, J.R. (2012) Physiological and physico-chemical characterization of the soursop fruit (*Annona muricata* L. cv. Elita). *Revista Facultad Nacional de Agronomía Medellín* 65, 6477–6486. DOI: 10.13031/2013.41985

Minh, N.P., Vo, T.S., Tram, D.N., Nga, N.H., Bang, N.T. and Diem, T.M.T.. (2019) Application of chitosan edible coating for soursop (*Annona muricata*) storage. *Journal of Pharmaceutical Sciences and Research* 11, 284–288.

Moreno, J. and de la Plaza, J.L. (1983) The respiratory intensity of cherimoya during refrigerated storage: a special case of climacteric fruit? *Acta Horticulturae* 138, 179–186. DOI: 10.17660/ActaHortic.1983.138.20

Moreno-Hernández, C.L., Sáyago-Ayerdi, S.G., García-Galindo, H.S., Mata-Montes de Oca, M. and Montalvo-González, E. (2014) Effect of the application of 1-methylcyclopropene and wax emulsions on proximate analysis and some antioxidants of soursop (*Annona muricata* L.). *Scientific World Journal* 2014: 896853. DOI: 10.1155/2014/896853

Moura, E.A., Chagas, P.C., Oliveira, R.R., Taveira, D.L.L., Grigio, M.L. and Araújo, W.F. (2020) Determination of the harvest time of sugar apples (*Annona squamosa* L.) in function of carpel interspace. *Acta Scientiarum. Agronomy* 43: e48732. DOI: 10.4025/actasciagron.v43i1.48732

Narayanasamy, P. (2006) Traditional wisdom of plants in pest control in agriculture. In: Trivedi, P.C. (ed.) *Medicinal Plants: Traditional Knowledge*. IK International Publishing House, New Delhi, pp. 159–184.

Narayanasamy, P. (2010) *Microbial Plant Pathogens – Detection and Disease Diagnosis: Viral and Viroid Pathogens, Vol.* 3. Springer, Dordrecht, Netherlands.

Nguyen, M.T., Nguyen, V.T., Le, V.M., Trieu, L.H., Lam, T.D. *et al.* (2020) Assessment of preliminary phytochemical screening, polyphenol content, flavonoid content, and antioxidant activity of custard apple leaves (*Annona squamosa* Linn.). In: *IOP Conference Series: Materials Science and Engineering*, Vol. 736. IOP Publishing, Bristol, UK, p. 062012. DOI: 10.1088/1757-899X/736/6/062012

Nwokocha, L.M. and Williams, P.A. (2009) New starches: physicochemical properties of sweetsop (*Annona squamosa*) and soursop (*Anonna muricata*) starches. *Carbohydrate Polymers* 78, 462–468. DOI: 10.1016/j.carbpol.2009.05.003

Opara, L.U., Al-Said, F.A. and Al-Abri, A. (2007) Assessment of what the consumer values in fresh fruit quality: case study of Oman. *New Zealand Journal of Crop and Horticultural Science* 35, 235–243. DOI: 10.1080/01140670709510190

Pal, D.K. and Sampath Kumar, P. (1995) Changes in the physico-chemical and biochemical compositions of custard apple (*Annona squamosa* L.) fruits during growth, development and ripening. *Journal of Horticultural Science* 70, 569–572. DOI: 10.1080/14620316.1995.11515328

Palma, T., Aguilera, J.M. and Stanley, D.W. (1993) A review of postharvest events in cherimoya. *Postharvest Biology and Technology* 2, 187–208. DOI: 10.1016/0925-5214(93)90047-7

Pareek, S., Yahia, E.M., Pareek, O.P. and Kaushik, R.A. (2011) Postharvest physiology and technology of *Annona* fruits. *Food Research International* 44, 1741–1751. DOI: 10.1016/j.foodres.2011.02.016

Passafiume, R., Tinebra, I., Gaglio, R., Settanni, L., Sortino, G. *et al.* (2022) Fresh-cut mangoes: how to increase shelf life by using neem oil edible coating. *Coatings* 12: 664. DOI: 10.3390/coatings12050664

Patidar, M., Kanwar, J., Kanpure, R.N., Singh, O.P. and Rathore, G.P.S. (2021) Effect of different post-harvest parameters on physiological and biochemical parameter of custard apple var. Arka Sahan. *Journal of Pharmacognosy and Phytochemistry* 10, 1465–1468.

Patil, R.A., Rao, D.V.S. and Manasa, B. (2015) Modified atmosphere packaging maintains physico-chemical properties of custard apple (*Annona squamosa* L.) fruits. *Indian Journal of Agricultural Research* 49, 489–495. DOI: 10.18805/ijare.v49i6.6674

Paull, R.E. (1982) Postharvest variation in composition of soursop (*Annona muricata* L.) fruit in relation to respiration and ethylene production. *Journal of the American Society for Horticultural Science* 107, 582–585. DOI: 10.21273/JASHS.107.4.582

Paull, R.E. (1996) Postharvest atemoya fruit splitting during ripening. *Postharvest Biology and Technology* 8, 329–334. DOI: 10.1016/0925-5214(96)00012-9

Paull, R.E. and Duarte, O. (2011) *Tropical Fruits*, 2nd edn. Crop Production Science in Horticulture, Vol. 20. CABI, Wallingford, UK.

Pino, J.A. (2010) Annona fruits. In: Hui, Y.H. (ed.) *Handbook of Fruit and Vegetable Flavors*. Wiley, Hoboken, New Jersey, pp. 229–246.

Pott, D.M., Vallarino, J.G. and Osorio, S. (2020) Metabolite changes during postharvest storage: effects on fruit quality traits. *Metabolites* 10: 187. DOI: 10.3390/metabo10050187

Puccio, S., Perrone, A., Sortino, G., Gianguzzi, G. and Gentile, C. (2019) Yield, pomological characteristics, bioactive compounds and antioxidant activity of *Annona cherimola* Mill. grown in mediterranean climate. *AIMS Agriculture and Food* 4, 592–603. DOI: 10.3934/agrfood.2019.3.592

Ramos-Guerrero, A., González-Estrada, R., Montalvo-González, E., Miranda-Castro, S.P. and Gutiérrez-Martínez, P. (2018) Effect of the application of inducers on soursop fruit (*Annona muricata* L.): postharvest disease control, physiological behaviour and activation of defense systems. *Emirates Journal of Food and Agriculture* 30, 1019–1026. DOI: 10.9755/ejfa.2018.v30.i12.1883

Reis, M.F.T., Corrêa, P.C., Diniz, M.D.M.S., Baptestini, F.M. and Fernandes, L.S. (2014) Efeito do uso de atmosferas modificadas sobre as propriedades texturais de atemoia durante o armazenamento refrigerado. *Revista Magistra* 26, 1478–1482.

Rossi-Márquez, G., Dávalos-Saucedo, C.A., Mayek-Pérez, N. and di Pierro, P. (2023) Multilayered edible coatings to enhance some quality attributes of ready-to-eat cherimoya (*Annona cherimola*). *Coatings* 13: 41. DOI: 10.3390/coatings13010041

Salomon-Castaño, J., Martínez-Mendoza, A.A., Franco-Mora, O., Castañeda-Vildózola, Á. and Sánchez-Pale, J.R. (2022) Application of edible coating for keeping postharvest quality of ilama (*Annona diversifolia* Saff.) fruits. *Acta Horticulturae* 1340, 55–60. DOI: 10.17660/ActaHortic.2022.1340.8

Salunkhe, D.K. and Desai, B.B. (1984) Custard apple and jujube. In: Salunkhe, D.K. and Desai, B.B. (eds) *Postharvest Biotechnology of Fruits, Vol.* 2. CRC Press, Boca Raton, Florida, pp. 133–135.

Sanchez, J.A., Zamorano, J.P. and Alique, R. (1998) Polygalacturonase, cellulase and invertase activities during cherimoya fruit ripening. *Journal of Horticultural Science and Biotechnology* 73, 87–92. DOI: 10.1080/14620316.1998.11510948

Santos, W.N.D., Sauthier, M.C.S., Cavalcante, D.D., Benevides, C.M., Dias, F.S. and Santos, D.C. (2016) Mineral composition, nutritional properties, total phenolics and flavonoids compounds of the atemoya fruit (*Annona squamosa* L. × *Annona cherimola* Mill.) and evaluation using multivariate analysis techniques. *Anais da Academia Brasileira de Ciências* 88, 1243–1252. DOI: 10.1590/0001-3765201620150537

Siang, L.M., Ding, P. and Mohamed, M.T.M. (2019) Response of 1-methycyclopropene on postharvest quality of local soursop (*Annona muricata* L.). *Sains Malaysiana* 48, 571–579. DOI: 10.17576/jsm-2019-4803-09

Silva, G.M., Cordeiro, M.H.M., Mizobusti, G.P., Sobral, R.R., Mizobutsi, E.H. *et al.* (2016) 1-MCP extends atemoya shelf life. *Journal of Food, Agriculture and Environment* 14, 35–39.

Thang, T.D., Dai, D.N., Hoi, T.M. and Ogunwande, I.A. (2013) Study on the volatile oil contents of *Annona glabra* L., *Annona squamosa* L., *Annona muricata* L. and *Annona reticulata* L., from Vietnam. *Natural Product Research* 27, 1232–1236. DOI: 10.1080/14786419.2012.724413

Tian, J., Wang, Y., Zeng, H., Li, Z., Zhang, P. *et al.* (2015) Efficacy and possible mechanisms of perillaldehyde in control of aspergillus niger causing grape decay. *International Journal of Food Microbiology* 202, 27–34. DOI: 10.1016/j.ijfoodmicro.2015.02.022

Tinebra, I., Scuderi, D., Busetta, G., Palazzolo, E., Settanni, L. *et al.* (2022) Modified atmosphere packaging and low temperature storage extend marketability of cherimoya (*Annona cherimola* Mill.). *Italus Hortus* 29: 115. DOI: 10.26353/j.itahort/2022.1.115137

Torres, L.M.A.R., da Silva, M.D., Guaglianoni, D.G. and Neves, V.A. (2010) Effects of heat treatment and calcium on postharvest storage of atemoya fruits. *Alimentos e Nutrição* 20, 359–368.

Venkatram, A., Bhagwan, A., Thirupathi, JR and Kumar, J.S. (2015) Effect of antioxidants and modified atmospheric packaging on physico-chemical characteristics of Balanagar custard apple (*Annona squamosa* L.) fruits. *Environment and Ecology* 33, 1513–1518.

Vishnu Prasanna, K.N., Sudhakar Rao, D.V. and Krishnamurthy, S. (2000) Effect of storage temperature on ripening and quality of custard apple (*Annona squamosa* L.) fruits. *Journal of Horticultural Science and Biotechnology* 75, 546–550. DOI: 10.1080/14620316.2000.11511283

Vyas, P.B., Rao, T.V.R. and Thakkar, V.R. (2015) Combined effects of chemical and physical elicitors on postharvest quality of custard apple (*Annona squamosa* L., cv. Balanagar). *Scientia Horticulturae* 187, 50–57. DOI: 10.1016/j.scienta.2015.02.040

Wang, Y.S., Tian, S.P. and Xu, Y. (2005) Effects of high oxygen concentration on pro- and anti-oxidant enzymes in peach fruits during postharvest periods. *Food Chemistry* 91, 99–104. DOI: 10.1016/j.foodchem.2004.05.053

Wongs-Aree, C. and Noichinda, S. (2011) Sugar apple (*Annona squamosa* L.) and atemoya (*A. cherimola* Mill. × *A. squamosa* L.). In: Yahia, E.M. (ed.). *Postharvest Biology and Technology of Tropical and Subtropical Fruits*. Woodhead Publishing, Sawston, UK, pp. 399–427e. DOI: 10.1533/9780857092618.399

Worrell, D.B., Sean, C.M. and Huber, D.J. (1994) Growth, maturation and ripening of soursop (*Aannona muricata* L.) fruit. *Scientia Horticulturae* 57, 7–15. DOI: 10.1016/0304-4238(94)90030-2

Yahia, E.M. (1998) Modified and controlled atmospheres for tropical fruits. *Horticultural Reviews* 22, 123–183.

Yahia, E.M. (2006) Modified and controlled atmospheres for tropical fruits. *Stewart Postharvest Review* 5: 6. DOI: 10.2212/spr.2006.5.6

Zhang, L., Huang, C., Zhao, Y., Zheng, C. and Hu, C. (2023) Post-ripening and senescence behavior of atemoya (*Annona cherimola* × *A. squamosa*) under two typical storage temperatures. *Postharvest Biology and Technology* 200: 112336. DOI: 10.1016/j.postharvbio.2023.112336

# Index

Note: Page numbers in **bold** type refer to **figures**, page numbers in *italic* type refer to *tables*.

Printed and bound by CPI Group (UK) Ltd, Croydon, CR0 4YY

13/09/2026

14961117-0001